Disaster Risk Reduction

Methods, Approaches and Practices

For further volumes:
http://www.springer.com/series/11575

About the Series

Scope of The Series

Disaster risk reduction is a process, which leads to the safety of community and nations. After the 2005 World Conference on Disaster Reduction, held in Kobe, Japan, the Hyogo Framework for Action [HFA] was adopted as a framework of risk reduction. The academic research and higher education in disaster risk reduction has made / is making gradual shift from pure basic research to applied, implementation oriented research. More emphasis is given on the multi-stakeholder collaboration and multi-disciplinary research. Emerging university networks in Asia, Europe, Africa and Americas have urged for the process-oriented research in disaster risk reduction field. Keeping this in mind, this new series will promote the outputs of action research on disaster risk reduction, which will be useful for a wider range of stakeholders including academicians, professionals, practitioners, and students and researchers in the related field. The series will focus on some of emerging needs in the risk reduction field, starting from climate change adaptation, urban ecosystem, coastal risk reduction, education for sustainable development, community based practices, risk communication, human security etc. Through academic review, this series will encourage young researchers and practitioners to analyze field practices, and link it to theory and policies with logic, data and evidences. Thus, the series emphasizes evidence based risk reduction methods, approaches and practices.

Series Editor

Editorial Advisory Group

Rajib Shaw
Editor

Disaster Recovery

Used or Misused Development Opportunity

Springer

Editor
Rajib Shaw
International Environment and Disaster Management
Graduate School of Global Environmental Studies
Kyoto University
Kyoto, Japan

ISSN 2196-4106 ISSN 2196-4114 (electronic)
ISBN 978-4-431-54254-4 ISBN 978-4-431-54255-1 (eBook)
DOI 10.1007/978-4-431-54255-1
Springer Tokyo Heidelberg New York Dordrecht London

Library of Congress Control Number: 2013949862

Printed on acid-free paper

Springer is part of Springer Science+Business Media (www.springer.com)

Preface

Disaster recovery has been discussed for many years. Quite often, post-disaster recovery leads to rebuilding risk; recovery efforts are not informed by lessons learned and experience from previous disasters; recovery needs assessment has not been demand-driven; stakeholder consultative processes are weak; and institutions set up to manage recovery have not led to sustained national and local capacity for disaster reduction.

There is increasing evidence from recent disasters that well-informed and well-prepared local governments and local communities can minimize the impacts of disasters. It is a well-accepted fact that communities vary from place to place, and their perception and ways of responding to disaster also vary. Therefore, it is important to decentralize policy and to customize it based on local needs and priorities. For large countries in Asia such as China, India, and Indonesia, different provinces have different cultural, socio-economic, and ethnic contexts. Thus, risk-reduction activities also need to be customized based on local contexts.

Disaster recovery can be a development opportunity, provided the recovery process is used in combination with an appropriate governance system. A strong local government is often found to be effective for a successful recovery process. However, governance alone is not enough; it needs to be linked to appropriate technology, innovation, and education systems. In this book, a framework of GET (governance–education–technology) is proposed for a successful recovery program. Examples from different post-disaster situations are examined and a common framework for GET and cross-cutting issues is described here.

This book is written for students and young researchers aspiring to a career in disaster risk reduction and environmental studies. I hope that they will find the book useful and relevant to their work.

Kyoto, Japan Rajib Shaw

Contents

Contributors

Amjad Ali Center for Disaster Preparedness and Management, Peshawar University, Peshawar, Pakistan

K. Chandrasekar National Remote Sensing Centre, Indian Space Research Organization, Hyderabad, India

Sahba Chauhan SEEDS India, Delhi, India

Wen-Ni Chen National Yunlin University of Science and Technology, Yunlin, Taiwan

Kapil Gupta Department of Civil Engineering, Indian Institute of Technology Bombay, Mumbai, India

Mahima Gupta Faculty of Planning and Public Policy, CEPT University, Ahmedabad, India

Manu Gupta SEEDS, Delhi, India

Junichi Hibino Radio FM YY, Kobe, Japan

Chia-Chen Hsu National Yunlin University of Science and Technology, Yunlin, Taiwan

Shao-Yang Huang National Yunlin University of Science and Technology, Yunlin, Taiwan

Kenji Isayama Department of Medical Science and Technology, Hiroshima International University, Hiroshima, Japan

Mikio Ishiwatari Japan International Cooperation Agency, Tokyo, Japan

Aminul Islam United Nations Development Prorgamme, Dhaka, Bangladesh

Naim Kapucu School of Public Administration, University of Central Florida, Orlando, FL, USA

Amir Nawaz Khan Center for Disaster Preparedness and Management, Peshawar University, Peshawar, Pakistan

R.R. Krishnamurthy University of Madras Arts and Science College and University Constituent College, Kanchipuram Dist, India

Dyah Kusumastuti Faculty of Civil and Environmental Engineering, Institute of Technology Bandung, Bandung, Indonesia

Chi-Feng Lin National Yunlin University of Science and Technology, Yunlin, Taiwan

Fuad Mallick Postgraduate Programs in Disaster Management, BRAC University, Dhaka, Bangladesh

Shohei Matsuura International Environment and Disaster Management, Graduate School of Global Environmental Studies, Kyoto University, Kyoto, Japan

Vinay Nikam Department of Civil Engineering, Indian Institute of Technology Bombay, Mumbai, India

Yukihiko Oikawa Kesennuma City Education Board, Kesennuma, Japan

R. Parthasarathy Faculty of Planning and Public Policy, CEPT University, Ahmedabad, India

Krishna S. Pribadi Faculty of Civil and Environmental Engineering, Institute of Technology Bandung, Bandung, Indonesia

Juan M. Pulhin College of Forestry and Natural Resources, University of the Philippines Los Baños, Los Baños, Philippines

Saut A.H. Sagala School of Architecture, Planning and Policy Development, Institute of Technology Bandung, Bandung, Indonesia

Rajib Shaw International Environment and Disaster Management, Graduate School of Global Environmental Studies, Kyoto University, Kyoto, Japan

Yukiko Takeuchi Global Survivability Studies (GSS), Kyoto University, Kyoto, Japan

V. Thiruppugazh Gujarat Institute of Disaster Management, Gandhinagar, India

D. Shanmugam Department of Applied Geology, University of Madras, Chennai, India

Anshu Sharma SEEDS India, Delhi, India

Akhilesh Surjan Global Survivability Studies (GSS), Kyoto University, Kyoto, Japan

Andrew Eusebio S. Tan West Visayas State University, Iloilo City, Philippines

Jet-Chau Wen National Yunlin University of Science and Technology, Yunlin, Taiwan

Ramanditya Wimbardana School of Architecture, Planning and Policy Development, Institute of Technology Bandung, Bandung, Indonesia

Chapter 1
Post Disaster Recovery: Issues and Challenges

Rajib Shaw

Abstract While pre-disaster preparedness is important, it is equally important to focus on post disaster recovery. Recovery, if properly conducted can be a development opportunity for the local communities. The recovery plan aims at reducing future risk and effectively enables post-disaster recovery using an integrated disaster risk reduction approach, is an integral part of risk reduction. In fact, the inclusion of recovery planning in disaster risk reduction signals a shift away from rehabilitation and brings into fore the interconnection between risk reduction and sustainable development. There are several issues on recovery, including local community involvement, social capital of communities, strong local governments etc. To understand the post disaster recovery, GET (Governance-Education-Technology) framework has been proposed, and the book is structured as per this framework.

Keywords Balance of speed and quality • Community resilience • Development perspective of recovery • GET framework • Social capital

1.1 Introduction

Disasters are unavoidable in many instances. However, through pre-disaster risk reduction and preparedness activities, it is possible to reduce the impacts of disasters. The disaster specialists have lobbied for risk reduction over past two decades. However, very unfortunately, funding commitments at all levels (from international, national to local) becomes higher after the disasters. Thus, these opportunities need to be utilized properly and effectively to make the post disaster recovery as a development opportunity.

R. Shaw (✉)
International Environment and Disaster Management, Graduate School of Global Environmental Studies, Kyoto University, Kyoto, Japan
e-mail: shaw.rajib.5u@kyoto-u.ac.jp

R. Shaw (ed.), *Disaster Recovery: Used or Misused Development Opportunity*, Disaster Risk Reduction, DOI 10.1007/978-4-431-54255-1_1, © Springer Japan 2014

While there has been significant focus on the pre-disaster preparedness and mitigation aspects, post-disaster reconstruction issues should not be discarded. Rehabilitation and reconstruction programs are development opportunities, and therefore their sustainability is an important issue. There are as many rehabilitation programs as there are numbers of natural disasters. Each disaster has different characteristics and disasters like earthquakes can be particularly destructive especially for lives and properties. Earthquakes affect all, including rich, middle-class and poor. When they destroy an urban area, massive re-planning of the city is required. Thus, the recovery process is a learning exercise on what is safe and sustainable for the community. Governments (national, provincial, city or local) and NGOs (both international and local) put tremendous efforts into reducing vulnerability and to enhancing sustainability in the reconstruction and rehabilitation programs (Shaw et al. 2003). However, the key question is: even though lots of effort is put into disaster recovery programs, why have some communities carried out faster (in terms of time frame) and more satisfying (in terms of holistic and participatory) recovery programs while others have not? Where do such differences come from? There is possibly no straightforward answer, since it is a complex mixture of social, economic, religious, political and other issues.

Quite often, post-disaster recovery leads rebuilding risk; recovery efforts are not informed by lessons learnt and experiences from previous disasters; recovery needs assessment has not been demand driven; stakeholder consultative processes is weak; institutions set up to manage recovery have not led to sustained national and local capacities for disaster reduction (Sinha and Srivastava 2009). Further, the event of disasters attracts huge financial, logistic, other physical and technological supports, which often go beyond the ability of high-risk developing countries to manage and channel them effectively in risk reduction. It is therefore felt necessary to have recovery plan in place providing insights and guidelines to support post-disaster recovery interventions/operations—their structural and non-structural dimensions, which could lead to risk reduction as envisaged in Hyogo Framework of Action (HFA). Structural dimensions like roads, bridges, embankments, drainage, and buildings etc., and nonstructural dimensions like existing vulnerabilities and capacities, rules, regulations, planning, skills, knowledge, awareness, community support systems, social transactions in terms of reciprocity, trust and exchange of labor and skills etc. are essentially to contextualize and package in form of recovery planning. Recovery planning could be both a pre- and post-disaster task. While pre-disaster could be based on the lessons and best practices, the post-disaster makes adjustments depending on the situations on the ground and may not be holistic.

The recovery plan aims at reducing future risk and effectively enables post-disaster recovery using an integrated disaster risk reduction approach, is an integral part of risk reduction. In fact, the inclusion of recovery planning in disaster risk reduction signals a shift away from rehabilitation and brings into fore the interconnection between risk reduction and sustainable development. This shift enables looking beyond realizing the pre-disaster situation in the spirit of "build back and better" than before. This is what brings in the concept of Pre-disaster Recovery Planning (Pre DRP) for risk reduction in recovery and reconstruction efforts. The

Pre DRP is also to take into account the provision of assistance or intervention during or immediately after a disaster to address short-term crisis management as well as long-term risk reduction strategies in integrated and holistic manner (Sinha and Srivastava 2009).

This chapter provides an overview of some of the emerging issues of post disaster recovery. The chapter also provides an overview of the GET (governance-education-technology) matrix, which is used as the base of the recovery issues in the book. Finally, the chapter gives an overview of the chapters presented in the book.

1.2 Disaster Recovery and Local Resilience

As stated in Robert B. Olshansky's article *Planning for the Rebuilding of New Orleans*, "…Katrina disrupted urban systems, economies, and lives, and pose huge problems for local governments and planners trying to reorganize and finance reconstruction quickly and effectively as possible." This statement really summarizes the pressure of rebuilding the city and how important it is to have a completed comprehensive plan in times like these. The disaster of the city was due to flooding of the levee systems take in this storm was only a category 3 hurricane, which happened to flood 80 % of the city (LRA 2006). However if substantial post-disaster recovery planning such as: quick external funding, strong local leadership, cooperation between city, state and federal officials, local, citizen based processes for making and reviewing reconstruction decisions, previous planning for consensus policies for future development and pre-existing planning institutions (Olshansky 2006), the community could have been involved and funds would have been quickly distributed. In New Orleans case, the promised funds from FEMA of a $7.5 million was not provided to start off the recovery, instead other plans and agencies came together to establish funding to the victims of this disaster (Olshansky 2006). The help from the Louisiana Recovery Authority (LRA) was established to communicate between state to federal governments to provide adequate funding. One of the major challenges that presented pressure was money, "money derives recovery," as well as the quickness of recovery to sustain the social and economic networks that were destroyed (Olshansky 2006).

Regardless of community size or the nature of the disaster, local government leaders are responsible for overseeing all four phases of emergency management—preparedness, response, recovery, and mitigation. Federal and state governments play a supporting role in the immediate aftermath and in providing funding and guidance for long-term recovery and mitigation. Preparation and response—half of the emergency management cycle—generally get the most attention, particularly in high-risk areas. Preparing to respond usually involves significant training and practice to ensure that key local employees and supporting resources are ready to jump into action quickly and that local residents understand their roles and responsibilities in preparing for and responding to disasters. Local government leaders—particularly those who have been through a major community disaster—recognize that preparing for long-term disaster recovery demands as much attention as preparing for short-term response. After a major disaster, the recovery process takes months and even years to bring a community back to a "new normal" and as strong as or better than before the disaster (Becker 2012).

There are increasing evidences from the recent disasters that the well aware and well prepared local governments and local communities can minimize the impacts of disasters. Japan is a classic example, which was had a tradition central government dominated disaster management system. However, in case of Kobe earthquake of 1995, it was the local communities and neighbors, which helped in 98 % of the survivors, and rest 2 % were rescued by the formal rescue operation. After the Kobe earthquake, Japan's focus was not only to enhance community preparedness, but to make networks of local governments in the form of alliance, to help each other in time of disaster. An outcome of the alliance was to help the city of Toyooka after the 2004 typhoon disaster to overcome the disaster debris and waste by the neighboring cities and prefectures (UNEP 2005). After the 2011 East Japan Earthquake and Tsunami, the Union of Kansai local Governments (Kansai Union) provided support to dispatch of experts and personnel, provision of emergency relief goods, and temporary accommodation of affected populations. Kansai Union is composed of seven prefectures (Osaka-fu, Kyoto-fu, Hyogo-ken, Wakayama-ken, Shiga-ken, Tokushima-ken and Tottori-ken) in the Kansai region.

A comparative analysis of Aceh Tsunami (2004) and Yogjakarta earthquake (2005) shows that Yogjakarta had the unique community participation culture (locally called Gotang Royang or mutual help), which was very effective in the recovery process. Thus, the role of housing facilitators was different in case of Yogjakarata (housing facilitators were used in technical problem solving) than that of Aceh (where housing facilitator were used for social problem solving). Thus, it can be argued that the community character defines the level of community involvement and participation (Ochiai and Shaw 2009).

Even within the same city, the analysis of Nakagawa and Shaw (2004) reveals that the social capital changes over place to place. In Kobe, in the district of Mano, the social capital was properly utilized in the recovery process due to a strong community leadership, which helped in the faster decision-making compared to other areas of the same city. Similar observation is also found in case of Gujarat earthquake recovery (Nakagawa and Shaw 2004), which pointed out relatively faster and better quality recovery with a specific community group as compared with others.

It is a well-accepted fact that the community varies from places to places, and its perception and ways to respond to disaster also varies. Therefore, it is important to decentralize the policy and to customize it based on the local needs and priorities. For a large country in Asia like China, India, Indonesia, different provinces have different cultural, socio-economic and ethnic context. Thus, the risk reduction activities also need to be customized based on the local context.

McCreight (2010) stated five dimensions to resilience after disaster: (1) personal and familial socio-psychological well being, (2) organizations and institutional restoration, (3) economic and commercial resumption of services and productivity, (4) restoring infrastructural systems and integrity, and (5) operational regularity of public safety and government. Aldrich (2012) in his analysis emphasized communal resilience at the neighborhood, ward or area to engage in positive, networked adaptation after a crisis. He also argued that it is possible to use public policy programs and local initiatives to build resilience, i.e., to increase a neighborhood's capacity to recovery effectively.

1.3 Post Disaster Recovery Issues

In an interesting analysis, Flynn (2011) has proposed seven cracks in the foundation of the disaster studies, especially in behavior health, which are: (1) lack of understanding of the seriousness of behavioral health factors, (2) lack of understanding of the broad roles the experts can play, (3) leadership challenges, (4) personality dependent challenges, (5) resources, (6) culture, and (7) failure to meaningfully engage the public. Based on the argument, Flynn (2011) also pointed out ten priority areas: (1) preparedness and planning, (2) system integration, (3) expand and apply evidence base, (4) leadership, (5) communication, (6) engaging communities, (7) public-academic linkages, (8) public private linkage, (9) redefining special population, and (10) prevention.

The above issues are quite important and needs to be addressed for a holistic post disaster recovery program. In the overall dynamics of recovery, emergency phase still continues predominantly. Recovery is yet to become an integral component of preparedness and mitigation. Post-disaster recovery is often understood to return to where it was before the disaster, which too often means rebuilding pre-existing conditions of disaster risk, thus preparing the ground for future disaster. The dynamics of recovery highlights the "gaps" as well as "speed" with the backdrop of scale of unmet needs (basically indicating quality of recovery) and timeline (shows the speed of recovery). The challenge lies in realizing recovery as an opportunity for building the resilient society with the notion of build back and better than before with the reasonable speed.

At the aftermath of a disaster, there has always been pressure to quickly restore support systems, livelihood and repair damages. In most of the cases, this undermines the quality of relief, reconstruction and rehabilitation works. The pressure of time and other constrains such as the difficulties in communication and transport in the post-disaster environment make it difficult to restore and lives and livelihoods with enhanced resilience. However, recovery is a balance between speed and quality. The speed is higher when it is done in a centralized way, by single agency. However, when it comes to cooperation and collaboration among different stakeholders, departments and agencies, the process becomes slow. However, the question is do we always need a fast recovery? A community based consultation and decision-making needs time, it needs to be linked to different culture and local situation. Thus, the recovery should not be measured just on the speed. The quality becomes equally important.

Local communities ties and social bonds become another important issue. Aldrich (2012) also argued that while it is easy to blame the poor or unsteady recovery on politicians, bureaucrats, or culture of corruption, empirical data from disasters show that regardless of the quality of governance, different neighborhoods under the same leadership come back at different rates over the medium to longer time. His argument also supports the findings of Nakagawa and Shaw (2004). Two case studies of Kobe and Gujarat Earthquakes show that although the local socio-economic and cultural backgrounds are different in these two areas, the recovery process of urban area is quite similar. At every stage of the disaster cycle (rescue, relief and rehabilitation), the communities played the most important roles among other concerned stakeholders. In both cases, the communities with social capital are found to be efficient in rescue and relief. The most challenging part was during reconstruction, where town planning and rezoning was applied, and collective decision-making was needed. In Hyogo, as municipal governments submitted the town planning without any consultation with the local community, it took from several months to a few years to finalize the reconstruction plans in the "black-zone" areas. In Bhuj, after finalizing the town planning in November 2002, massive protests from the property owner against the plan took place. Reacting to that, people who were living in temporary shelters and wished for the earliest reconstruction of the city became frustrated at further delays and demonstrated strongly against those who were opposed to the town planning.

In the post disaster recovery period, the local government and communities get external aid from different sources. However, if the local government is not able to have a strong regulation, recovery packages, the recovery process often goes in different directions. The recovery becomes non-uniform: too much of the aid reaches one places, where some other remote areas often lacks resources. Also, quality control becomes another issue, which is often compromised if there is no strong influence of local governments.

1.4 GET Framework

The issues described above are some selected issues, which pose significant challenges in the post disaster situation. To make a systematic approach, a framework called GET (Governance-Education-Technology) is proposed (Shaw 2011). This is elaborated as follow:

1.4.1 Governance

Governance is focusing on the policy issue. Simply putting, governance is the art/science of decision-making. The concept of governance refers to the complex set of values, norms, processes, and institutions by which society manages its development and resolves conflict, formally and informally. It involves the state, but also the civil society at the local, national, regional and global levels. Four pillars of governance are: accountability, accessibility, transparency and efficiency. The governance related issues do not necessarily be linked only to the national level, it can go to local governance or even in the community level. Governance needs to emphasize how a healthy partnerships are essential between Governments and NGO's, however in many situations there is a lack of cooperation, and a spirit of competition. Often Governmental and NGO officials accentuate what divides them rather than recognise their shared values. One of the key challenges is how to prioritize the DRM issues in the local government priorities. This can only be possible with a strong legal and institutional framework (as exemplified by several recent Disaster Management Law or Regulations in different countries), which provides specific support to the local governments.

1.4.2 Education

Only the provision of legal framework to institutionalize DRM in local level is not enough, unless it is linked to the education and awareness raising mechanism. Education is a cultural issue. Education related policy-need to focus on the contents and process of education. The contents part needs to be culturally calibrated based on the local context of Asia. The region has different socio-economic, and political diversity, and the content needs to be very much customized to the local context. The process of education is equally important. The involvement of different stakeholders in the education process is an important aspect for successful implementation of policy and practice at the local level. The delivery mechanism of the education should be strengthened, and needs to be customized as per the local issues. Education has two specific components, one in the formal education system (at all levels from primary to higher education) and training and capacity building (pre-service and/or in-service training system). Awareness raising focuses more to enhance the perception and knowledge base of the local communities.

1.4.3 Technology

Technology does not necessarily means high-tech issues including the IT (information technology). It is required to recognize the local knowledge and traditional knowledge as well as local processes (*Process Technology*) to enhance local skills and actions. Technology has different meaning to definition. Kameda (2009) defines technology as "a set of rational means and knowledge pertinent to realizing specific objectives that have solid logical bases and stability." In a conventional recognition, technology meant just engineering products. But when we consider implementation strategies, technologies should involve not only products but processes as well. This requires innovation of research community to reform from "product focused research" to "process oriented research," or "product-process linked research." Kameda (2009) classified technology as:

- *Implementation Oriented Technology (IOT)*: Products from modern research and development that are practiced under clear implementation strategies
- *Process Technology (PT)*: Know-how for implementation and practice, capacity building and social development for knowledge ownership
- *Transferable Indigenous Knowledge (TIK)*: Traditional art of disaster reduction that is indigenous to specific region (s) but having potential to be applied to other regions and having time-tested reliability

Thus, to enhance a sustainable recovery, a balanced approach of GET (Governance-Education-Technology) is very much required. Based on the country and local context, there may be specific challenges related to governance, where more policy formulation and its implementation mechanism need to be focused. In other cases, there are education related challenges, which need to cover awareness raising both for the policy makers as well as communities for successful implementation of recovery plans. In some other countries or communities, it may be the need of the appropriate technology or recognizing the local and indigenous knowledge to be incorporated in policy issues. The chapters of the book are designed based on this framework.

1.5 About the Book

This book has 21 chapters, and is divided into four parts: Governance and institutional issues (five chapters), education and learning issues (four chapters), technology and innovations issues (five chapters) and cross-cutting issues (five chapters). The final chapter provides an analysis of the key issues.

Chapter 2 examines the institutional mechanisms setup for reconstruction following the Gujarat earthquake 2001, one of the largest disasters in terms of the intensity and diversity of impacts. Extra-ordinary mechanisms are usually set up in the aftermath of disasters to coordinate and speedup reconstruction process. There are many examples of such Extra-ordinary mechanisms that played a key role in successful reconstruction. But, only a few of them have survived as successful institutions for long-term disaster risk reduction. Several factors such as political will, availability of resources, requirements of the International Financial Institutions, bureaucratic and political leadership determine the nature and mandate of the institutions set up.

Threat of natural disasters will continue, but their consequences can be minimized if communities and people reduce their vulnerabilities and increase their resilience. In the field of disaster and crisis management, too much emphasis is placed on response to disasters. Research on disaster recovery is limited. Chapter 3 focuses on collaborative governance principles applied to disaster recovery. The study uses the recent National Disaster Recovery Framework (NDRF) in the U.S. as an example.

Chapter 4 focuses on Taiwan. Due to the serious disaster caused by Typhoon Morakot, the disaster prevention and response mechanism in Taiwan were induced to be improved, following are several transformations: (1) Planning establishing Disaster Prevention and Rescue Agency, (2) Special regulations of Typhoon Morokot post-disaster reconstruction was established by Executive Yuan, (3) Amended the Disaster Prevention and Rescue Law, (4) Combined the Disaster Prevention and Rescue Report and National Disaster Prevention and Response Committee, and (5) Office of Disaster Prevention and Rescue was established by Executive Yuan. The government promoted several transformations after Typhoon Morakot in order to effectively response similar disaster in the future, to ensure the safety of people lives and sustainable development of Taiwan. Citing specific examples of changes in the national, county, and city level, this chapter provides the key lessons and suggests future directions of institutional perspectives of DRR in Taiwan.

Much confusion is inevitable on the ground in the immediately aftermath of mega-disasters in developing countries, since local governments are greatly overstretched because of a wide range of recovery activities. The local governments often face various difficulties in post disaster situation, such as shortages of dedicated agencies and skilled and experienced staff, and finance resources. Also, for development agencies it takes time to respond quickly to disasters in developing countries. Chapter 5 aims to propose practical institutional and governance mechanisms of quick recovery for developing countries and development agencies by examining practices in the East Japan Earthquake and Tsunami. The study observes that institutional and legislative features of Japanese governmental system contributed to prompt recovery. For example, roads leading to cities on devastated coasts were opened in less than a week. It was found that quick recovery can be attributed to various measures prepared at normal times, such as advance financial arrangements, prior agreements with private sector, and mobilization of expert teams from other areas through national networks. These have been improved by learning lessons from past disasters.

After the disaster of East Japan Earthquake and Tsunami, education sectors such as schools and board of education (BOE) in tsunami-affected area, faced many hazards and challenges immediately and continuously. Chapter 6 describes how educational institutions are implementing the strategies for recovering schools and education from the disaster of earthquake and tsunami.

Flood 2010 stretched to an area of 100,000 km^2, which comprised of 78 districts of Pakistan. Chapter 7 gives an insight into the implications of the 2010 flood on education sector. Section 1.1 of the paper draws a general sketch of the 2010 flood, which is followed by a brief of the factors enhancing the vulnerability of the country to floods in Sect. 1.2. Section 1.3 discusses the direct and indirect impacts of the flood on education sector whereas the way forward and conclusion are summed up in the final section.

Education plays an enabling role in ensuring that a positive transformation takes place within communities undergoing recovery. Chapter 8 explores possibilities based on initiatives carried out in the past where in civil society organizations have engaged in learning and education with disaster affected communities. Collaboration with professional institutions, provided the necessary knowledge resource and validation adding value to the exercise. Experiences also highlighted challenges that civil societies had to face in term of capacity and impact. The chapter presents a framework for the role of education in recovery with the civil society playing the role of an enabler.

Chapter 9 focuses on following four points: (1) Find out issue of school location, (2) Find out t issue of continuity of education, (3) Role of school and school facility in community, and (4) Adaptation to future affected area in Japan and thought network of Asian universities. This research method is mainly literature review, questionnaire survey, hearing survey, develop the network thought workshop and make publications for outreach. This research output will be helpful to Iwate prefecture, Miyagi prefecture, Fukushima prefecture and etc. for recovery process, and will be help to future affected area in Japan and abroad.

The concept, "School Centered Community Building," is composed of four main pillars, which are: (1) Securing safety of schools and relocating them to safe areas as necessary, (2) Retrofitting schools to improve their functions as evacuation center and hub for disaster management, (3) Making school building eco-friendly and (4) Making schools multi-functional public facilities to become the center of communities. Initial findings of the school recovery process in Toni District, Kamaishi City, Iwate Prefecture will be presented as a study case in Chap. 10 to look into the actual implementation of School Centered Recovery and Community Building. Kamaishi is one of the disaster-affected cities in which in its city recovery plan has pledged to implement this concept.

Chapter 11 provides an ex-post review of the past experiences and challenges in post-disaster housing reconstruction after earthquakes in Aceh (2004), Yogyakarta (2006), West Java (2009) and West Sumatra (2009), and reveals some strategic issues in implementing safer housing reconstruction that have to be addressed in the future for achieving "build back better" post-disaster reconstruction programs. Past experiences showed that training and capacity building of construction personnel and home-owners through the dissemination of guidelines and manuals as well as building codes and standards for anti-seismic design which have been implemented in the housing reconstruction programs by the national and local governments, NGOs and aid agencies have not much changed the attitudes toward construction practices by builders and masons, nor the attitude of local government building administrators in ensuring the earthquake safety of the houses.

Considerable efforts have been made during the last two decades especially in making quantitative assessment on resource potential and the alteration of coastal environment through multispectral remote sensing data and spatial analysis tools such as GIS. Scientific database generated through these technology tools are used as a decision-support system by incorporating stakeholder's perception and field data, which helped considerably in effective coastal zone management in Tamil

Nadu. Also efforts have been made in training and capacity building in different levels, which are mainly focused to implement ICZM in the field. Chapter 12 is aimed to provide the genesis, different stages and present scenario of coastal zone management practices in Tamil Nadu coast.

In the aftermath of Aila there wasn't response from relief agencies as was the case with Sidr. At the same time there was the rethinking about the actual usefulness of cyclone shelters and possible alternatives. The state of habitat in the Aila affected areas, although deplorable provided the opportunity to try out innovations, one such being the concept of the Disaster Resilient Habitat (DRH). DRH called for strengthening an entire village, houses and infrastructure so that they are able to withstand winds and storm surges and even if there is destruction, getting back to life is easier. Chapter 13 describes how this innovative idea came into being and how it was implemented.

The increasing focus on climate change impacts, both in terms of catastrophic climate events and day to day environmental stresses, has brought to the fore the need to look at traditional wisdom as a source for climate coping mechanisms too. The issue gets highlighted in post disaster situations, particularly post hydrometeorological disaster situations, as the sensitivity and realization of the link between disaster risk reduction and climate change adaptation becomes clearer than in any other context. Chapter 14 uses a number of case studies, including that of the unprecedented flash floods in the tropical desert of Rajasthan and the mountain desert of Leh in India. It looks traditional wisdom based local practices such as weather forecasting, water harvesting, snow capture, as well as new innovative practices such as artificial glaciers and local climate knowledge management.

Chapter 15 provides a case study from Mumbai, which has an area of 437 km^2 with a population of 12 million. The city came to a complete halt due to the unprecedented rainfall of 944 mm during the 24 h starting 0830 on 26 July 2005. At least 419 people (and 16,000 cattle) were killed due to the ensuing flash floods and landslides in Mumbai municipal area. At that time, there was no real time warning system in place and the instrumentation with the authorities could give rainfall values only once every 24 h. Following this event, weather stations comprising state-of-the-art automatic weather stations comprising tipping bucket rain gauges were installed at 35 locations in the municipal limits and one ultrasonic flow gauge on the Mithi River. These rain gauges have been programmed to relay rainfall intensity in real time (every 15 min) through LAN to the disaster emergency control room. In addition, two innovative flood control structures have also been designed and implemented on the city's main Mithi River. A flood simulation model based on real time rainfall data has been developed to forecast 15-min ahead rainfall and forecast water levels at critical points 30-min ahead. This has been of immense help in evacuating people during subsequent high rainfall events. Other measures undertaken to make Mumbai more flood resistant will also be described.

Chapter 16 maps the role of mangrove ecosystem in protecting the coast with the increasing threats from extreme events and also identifies the threats to mangroves from anthropogenic activities along the coast of Kutch in Gujarat, which is found to be highly vulnerable to cyclones and storm surges. The analysis is based on an empirical study of 176 households covering nine villages along the Kutch coast and draws cases from other studies too. The empirical study shows that the coastal communities are mostly aware of the protection afforded by mangroves during the cyclones and the multiple benefits of mangrove restoration. However, based on their socio-economic conditions and livelihood structure, the communities have varied perception of the benefits from mangroves leading to differences in the outlook towards conservation of these natural resources.

Disaster recovery from industrial pollution such as major oil spill presents formidable challenge for governance and disaster management considering their adverse environmental and socioeconomic impacts. The community's social dynamics could explain how responses to such perturbation are made and what inherent traits of the households and communities contribute to their adaptive capacity to disasters. Using both qualitative and quantitative approaches in research, Chap. 17 examines the adaptive and transformative capacities of communities affected by oil spill in Guimaras, Philippines—one of the worst cases in the country's history.

Environment management especially in the context of urban areas, where high density of population and infrastructure is concentrated, is not an option but a pre-requirement for reducing risk from major disasters and towards swift recovery. Chapter 18 argues that in an urban area, disaster recovery is anyway a challenging task. However, sound urban planning and environment management practices can help in multifarious ways. For example, it can reduce damage from disasters, enhance response efficiency after the disaster and reduce recovery time and efforts, and even can act as launching pad towards "building back better" towards resilience.

Previous research suggests that primary prevention is the most effective means of reducing disaster damage. However, Japan has the highest proportion of older aging population in the world. The super-aging society is not only an individual issue, but also a salient factor crucial for public policies, such as pensions, health, and long-term care. In addition, although we can expect another mega earthquake anytime soon, it is difficult to maintain awareness of disaster prevention. Strengthening community network through integrated healthcare may play an important role when an actual disaster happens. Chapter 19 aims to propose building disaster resilient communities by utilizing systems and concepts of integrated healthcare and to strengthen community network by building health network with specific case studies. The chapter also proposes to improve local disaster risk reduction capacity by strengthening coping mechanisms of local communities in emergency care.

The use of community radio in several countries shows that the radio can play an important role in every phase of disaster management because information and communication are two of the most important factors within the phases. As time passes after a disaster, the role of Community Radio changes little by little. In post disaster response and recovery period, community Radio is expected to reflect the voice of the various people in the community in recovery and revitalization policy in order that not one person is left out in the recovery from the disaster that hit the community. The participation of various groups of victims in various ways makes community radio of the victims, by the victims and for the victims. Therefore, Community radio can strengthen a community tie to toward recovery. Citing examples from Indonesia and Japan, Chap. 20 provides an outline of future expected role of community radio.

Chapter 21 provides an analysis on how disaster recovery can act as a development vehicle. The chapter also emphasizes the GET framework for future disaster recovery.

Acknowledgements The author acknowledges JSPS (Japan Society for Promotion of Science), CWS (Church World Services) Asia Pacific, and MERCY Malaysia for financial support.

References

Aldrich D (2012) Building resilience: social capital in post disaster recovery. University of Chicago Press, Chicago, 232 pp

Becker C (2012) Disaster recovery: a local government responsibility. http://webapps.icma.org/pm/9102/public/cover.cfm?author=christine%20becker&title=disaster%20recovery%3A%20%20a%20local%20government%20responsibility. Accessed 28 April 2013

Flynn B (2011) Community and organizational response to disasters. In: Kaufman R, Edwards R, Mirsky J, Avgar A (eds) Crisis as an opportunity. University Press of America, Maryland, pp 1–15

Kameda H (2009) Implementation technology for disaster reduction. In: Shaw R, Krishnamurthy R (eds) Disaster management: global challenges and local solutions. University Press, Hyderabad, pp 206–219

LRA (2006) Disaster recovery initiative, Louisiana recovery authority, 24 pp. http://www.doa.la.gov/cdbg/dr/plans/. Amend2_ED-Infra_Approved_06-09-13.pdf. Accessed 29 April 2013

McCreight R (2010) Resilience as a goal and standard in emergency management. J Homeland Secur Emerg Manag 7(1):1–7

Nakagawa Y, Shaw R (2004) Social capital: a missing link to disaster recovery. Int J Mass Emerg Disaster 22(1):5–34

Ochiai C, Shaw R (2009) Participatory urban housing reconstruction study in Aceh, Indonesia. In: Shaw R, Srinivas H, Sharma A (eds) Urban risk: an Asian perspective. Emerald Publication, Bradford, pp 233–254

Olshansky RB (2006) Planning after hurricane Katrina. J Am Plan Assoc 72(2):147–153

Shaw R (2011) Climate change adaptation research in South Asia: an overview. Asian J Environ Disaster Manag 2(4):397–426

Shaw R, Gupta M, Sharma A (2003) Community recovery and its sustainability: lessons from Gujarat earthquake of India. Austr J Emerg Manag 18(2):28–34

Sinha A, Srivastava S (2009) Essential of pre-disaster recovery planning. In: Shaw R, Krishnamurthy R (eds) Disaster management: global challenges and local solutions. University Press, Hyderabad, pp 586–610

UNEP (2005) Environmental management and disaster preparedness: lessons learnt from the Tokage Typhoon. UNEP, Osaka, 28 pp

Part I
Governance and Institutional Issues

Chapter 2
Post-Disaster Reconstruction and Institutional Mechanisms for Risk Reduction: A Comparative Study of Three Disasters in India

V. Thiruppugazh

Abstract This study examines the institutional mechanisms employed in post-disaster reconstruction programs in India after three major disasters occurring between 1993 and 2004 in three provinces—Gujarat, Maharashtra and Tamil Nadu. Extra-Ordinary Mechanisms (EOM) are usually set up in the aftermath of disasters to coordinate and speed up reconstruction process. There are many examples of EOMs that played a key role in successful reconstruction. But, only a few of them have survived as successful institutions for long-term disaster risk reduction. Several factors such as political will, availability of resources, requirements of international financial institutions, nature of bureaucratic and political leadership determine the nature and mandate of the EOMs. Based on the Indian experience, this study analyzes the challenges of sustaining the institutional arrangements for disaster management and makes an attempt to postulate the key elements needed for their effectiveness.

Keywords Extra-ordinary mechanisms • Gujarat earthquake • Institutional mechanisms • Maharashtra earthquake • Post-disaster reconstruction • Tamil Nadu tsunami

2.1 Introduction

This study, based on three major reconstruction programs undertaken in India, examines the compulsions and challenges of setting up proper institutional mechanisms not only for post-disaster reconstruction but also for long-term risk reduction. The three reconstruction programs namely, Maharashtra earthquake reconstruction after 1993 earthquake, Gujarat reconstruction program following the 2001

V. Thiruppugazh (✉)
Gujarat Institute of Disaster Management, Gandhinagar, India
e-mail: vthiruppugazh@gmail.com

R. Shaw (ed.), *Disaster Recovery: Used or Misused Development Opportunity*, Disaster Risk Reduction, DOI 10.1007/978-4-431-54255-1_2, © Springer Japan 2014

earthquake and Post-tsunami reconstruction in Tamil Nadu in the aftermath of Asian Tsunami in 2004 provide an ideal setting for the study. There are several reasons for the choice of these programs. These three were among the worst disasters in terms of death toll and destruction in post-independent India. They represent a new policy driven progression from recovery to an extensive post-disaster reconstruction effort. All the three reconstruction programs in varying degrees aimed at improvements over the pre-disaster situation to bring about better quality of life and safety standards and also use the window of opportunity created by the disaster. These programs spread over a decade also provide an opportunity to understand the evolution of the concept and formation of Extra-ordinary Mechanisms (EOM) in the aftermath of major disasters.

EOMs were set up in all three cases under investigation. While Maharashtra and Tamil Nadu set up Project Management Units (PMU), Gujarat created Gujarat State Disaster Management Authority, an agency with greater autonomy and powers than a PMU. It is instructive to examine the purpose, scope and contribution of the EOMs in the three programs. While discussions in disaster research are usually limited to the perspective of efficient and speedy reconstruction, this study examines the EOMs from a long-term perspective of efficient disaster management going beyond recovery.

To understand the institutional mechanisms set up for reconstruction, the drivers of the policy decision, and the perceptions of multiple stakeholders were examined. Primarily, qualitative methods were used for this. The data gathering included field visits to the affected villages of Maharashtra, Gujarat and Tamil Nadu for first hand verification of the ground realities. In order to understand the roles, perceptions, views and experiences of multiple stakeholders involved in the program, a careful selection of the respondents was made across different sections. Respondents were chosen across cross-sections of the society keeping in mind the social, economic, technical and political aspects of policy making and implementation processes. To enable a better understanding of the political processes, a range of political actors including Ministers, Members of Legislative Assemblies, Members of Parliament both from ruling and opposition parties, *taluka* (administrative units comprising of many villages) and village level elected representatives were interviewed.

To understand policy and implementation aspects, members of the bureaucracy at multiple levels were interviewed. In order to examine the role played by the institutional mechanisms and their effectiveness and efficiency other stakeholders such as Civil Society Organizations, professionals, financial institutions, journalists and members of the media and academics were interviewed.

2.2 Institutional Needs

Though catastrophic disasters cause large-scale death and destruction, their impact is varied on different communities, regions and nations. Not only do the impacts of the disasters differ across these but also the ability to recover. As Diamond (2005) argues, recovery is a choice made by each society driven by a number of factors. While some act decisively to reduce future losses, the others just return to *status quo ante* (National Research Council 2006). In many cases the affected areas fail to recover even to the pre-impact level. There is an increasing realization that post-disaster reconstruction should not recreate the pre-disaster vulnerabilities but aim to utilize the opportunity to build resilient communities. There are many examples of realized and missed opportunities. Reconstruction of Lisbon after the earthquake, fire and tsunami in 1755 by Marquis do Pombal is an example of realization of some of the post-disaster opportunities not only for development and future disaster mitigation (Alexander 2004) but also for political and economic consolidation (Dynes 2000, 2005). But the potential to bring about a transformation is not always realized. Reconstruction programs failing to build back better, despite aiming to do so, as in the case of 1976 Guatemalan earthquake (Bates 1982), Hurricane Mitch (Wisner 2004; Ensor 2009; Ensor et al. 2009; Telford et al. 2004) and Great Kanto earthquake in 1923 (Schencking 2006; Hein 2005) prove this point.

Many scholars have pointed out that reconstruction is essentially a major challenge for governance. "Policy makers in representative democracies are pressured to respond quickly and effectively" (Smart 2012, p. 3) and hence massive development and reconstruction need to be compressed in time and space (Olshansky et al. 2012). Reconstruction brings in a variety of stakeholders ranging from local NGOs to International Funding Agencies and hence coordination also becomes a major challenge. Hence ensuring better cooperation and collaboration among various stakeholders (Asgary et al. 2006) is one of the major governance issues in the aftermath of a disaster. The next challenge is to strike a balance between speed and quality, and consultative process and quick decision making in a compressed time frame (Olshansky et al. 2012). In addition, there is also the challenge of mobilizing and managing the flow of funds, coupled with transparency and accountability. The ultimate challenge of any incumbent government is the creation of appropriate institutional mechanisms for facing the above challenges.

The utilization of opportunities presented by the disaster depends greatly on a number of factors related to governance. Quality of leadership, planning and organization for reconstruction are considered as the major factors according to Haas et al. (1977). Rubin et al. (1985), based on the comparative study of 14 reconstruction programs argues that leadership, the ability to act and knowledge of available resources, capacity of the local officials determine the success or failure of a reconstruction program. Lack of people's participation and neglect of people's needs are cited as reasons for failures in housing recovery (Oliver-Smith 1991; Salazar and Jigyasu 2010; Barakat 2003; Barenstein 2008; Bates 1982; Jigyasu 2001; Arslan and Unlu 2006; Asgary et al. 2006). Lack of interest and coordination, lack of expertise and conflicting interest of the stakeholders (Asgary et al. 2006), and physical reconstruction becoming the main focus (Vatsa 2005) also become a hurdle for building back better.

The success or failure of a reconstruction largely depends on the ability of the State to tackle the governance issues in reconstruction (Harvey 2009), as testified by the study of post disaster reconstruction programs in different countries. States which were pro-active towards tackling governance issues through appropriate mechanisms could build back better and the others missed the window of opportunity provided by the disaster (Foley 2007; Wong 2008; Price and Bhatt 2009; Cochrane 2008; Elhawary and Castillo 2008; Fagen 2008; Willitts-King 2009). The approach towards institutional mechanisms differs, ranging from setting up stand-alone new structures to utilizing the existing structures. While India, Pakistan and Indonesia preferred setting up of new EOM (Price and Bhatt 2009; Cochrane 2008; Willitts-King 2009), El Salvador relied more on existing institutions for reconstruction (Fagen 2008).

Haas et al. (1977), in their classic study, identify the need for changing the public policy making process as the most basic governance issue among the several key issues to be addressed before disaster reconstruction is undertaken:

The first issue is fundamental: Should normal, as contrasted to extraordinary, decision-making mechanism be used in deciding how, when, and where to rebuild the heavily damaged city? At the very minimum, the question will be raised as to whether there are an adequate number of experts of the various types needed within the regular units of government? If not, shall there be one or more special task forces composed of experts, administrators, and citizens? (Haas et al. 1977, pp. 44–45).

While early debates on relevance of EOM were centered on reconstruction, the contemporary discourse goes beyond and focuses on the role of EOM in long-term risk reduction using the window of opportunity created by the disaster (Christoplos 2006). The choice of the institutional mechanism depends on a large number of factors. There are three options: (a) creating a totally new organization, (b) creating a dedicated organization drawn from existing line ministries and (c) managing the reconstruction through existing government organizations (Jha et al. 2010). Each of these has their advantages and disadvantages. The existing organizations normally have well laid out procedures, experienced manpower, horizontal and vertical linkages, and may be conversant with the problems and issues of the affected area. But the flip side is delay in decision making due to lengthy bureaucratic norms, lack of flexibility, lack of expertise, multiplicity of organizations resulting in lack of coordination (Schneider 1995; Jha et al. 2010; Olshansky et al. 2012; Neal and Phillips 1995). In addition the question remains whether normal intuitional mechanisms designed for incremental change capable of coping with the situation of recreating an entire city, or region in a short time.

The next key issue is the question of coordination. Simultaneous creation of housing and all public as well as social infrastructures need more horizontal coordination among the various government agencies. There is also the need for a single window for external players to interact. As Jha et al. (2010) point out, managing international appeals for support; arranging large credits and grants from donors and IFIs; and managing procurement, disbursement, monitoring, and evaluation also present huge challenges in the aftermath of large-scale disasters. These functions alone may require a new institutional arrangement. Creation of an EOM also sends a political message to the people that the political masters are serious about reconstruction. But, creating new institutions from a scratch and making them efficient may be time consuming (Inam 2005). As time is the essence and the "fast pace is unforgiving of mistakes and does not allow for test cases or pilot studies," adapting existing organizations and procedures and improving upon them is considered as a better option by Olshansky et al. (2012, p. 177).

There is an increasing tendency to opt for standalone EOMs in the recent past (Harvey 2009) to overcome the time consuming "bureaucratic norms" of policy making (Schneider 1995, 6). The EOMs set up after Guatemalan earthquake 1976, Orissa Super cyclone 1999, Bam Earthquake 1993, Jammu & Kashmir earthquake 1995, Haiti earthquake 2010, Tōhoku earthquake and tsunami and Fukushima Daiichi nuclear disaster in 2012, to name a few, reflect such a preference. All the governments, as explicitly stated by the Guatemalan government after 1976 earthquake (Bates 1982, p. 147), feel the need for a powerful centralized Institution with flexible speedy mechanisms to initiate, control and coordinate the rehabilitation and reconstruction. Thus EOMs are becoming more of a norm than exception due to the desire to speed up reconstruction, the insistence of the funding agencies like the World Bank, the political compulsions to display the seriousness of the government and lessons learnt from the previous disasters.

In the context of South Asia, particularly India, which is highly disaster prone, no in-depth study has been conducted on the institutional mechanisms. Many large reconstruction programs have been undertaken in India after independence. Reconstruction after cyclones in Andhra Pradesh (1977), Kandla (1998), Orissa (1999); Machhu dam breach (1979); earthquakes in Uttarkashi (1991), Maharashtra (1993), Chamoli (1999), Gujarat (2001), Jammu & Kashmir (2005); tsunami in Tamil Nadu (2004); floods in Bihar (2008), and Karnataka (2009) were all very large programs. These programs were implemented through a variety of institutional mechanisms ranging from using the existing institutions to setting up new mechanisms.

Most studies on post-disaster reconstruction are either overtly technical (Nikolic-Brzev et al. 1999; Murty et al. 2005; Jain et al. 1994; Jain et al. 1997) or sharply focused on the housing and built environment (Jigyasu 2001; Salazar 2002; Salazar and Jigyasu 2010; Barenstein 2006, 2008, 2010; Samaddar and Okada 2006; Arslan and Unlu 2006; Barakat 2003). There are also some studies on equity and gender issues in reconstruction (Winchester 2000; Krishnadas 2007; Swain et al. 2006; Pincha 2008). The research by Economic and Political Weekly Research Foundation (EPWRF 1998, 1999) provides an exhaustive documentation of the reconstruction program following the Maharashtra earthquake (1993) without making any comparative study. A detailed study by Srinivasan et al. (2005), conducted within a year of the tsunami focuses on short-term outcomes. In the Indian context comparative studies are lacking. Despite these numerous studies, research has not focused on examining the merits of these institutional arrangements vis-à-vis the success and failure of the programs.

There are several issues and challenges associated with creating an EOM for reconstruction. First is its nature and structure. Second important issue is whether the EOM set up for reconstruction should be purely temporary or be converted into a permanent one. If a permanent agency is envisaged for carrying out long term risk reduction along with post-disaster reconstruction, what powers, functions and financial arrangements are required to make it effective beyond the reconstruction period? Thirdly once a permanent body is created, can it continue to retain the extraordinary nature in the long run and be as effective as it was initially? In this empirical study I will attempt to examine the three cases from the point of view of the above challenges and questions.

2.3 Characteristics of the Three Disasters

India is among the world's ten most disaster prone countries. It is vulnerable, in varying degrees, to a large number of natural as well as man-made disasters. India's vast geography, population, geo-climatic and socio-economic conditions make it prone to various disasters (GOI 2011). Nearly 59 % of landmass is at risk of earthquakes of moderate to very high intensity. Over 12 % of the land is prone to floods and erosion. Almost 76 % of the 7,516 km long coastline is likely to experience cyclones and tsunamis. Approximately 68 % of the cultivable area experiences frequent droughts. Besides, there is the likelihood of severe industrial, chemical, biological, radiological and nuclear disasters (GOI 2009). During the last 30 years, the country has been hit by about 430 major disasters killing nearly 143,000 and affecting 1,500 million people to varying degrees. The estimated property loss due to these disasters is around US$48,000 million. The economic losses arising out of disasters has been climbing steadily from US$10,285.7 million in the period 1990–1995 to US$19,724.8 million in 2000–2005, amounting to nearly 2 % of the national GDP (GOI 2011). All the three disasters chosen for the study occurred in India, but in three different provinces or states. The political map of India (Fig. 2.1) shows the three states in which the disaster occurred.

2.3.1 Maharashtra Earthquake (1993)

An earthquake of magnitude 6.4 on the Richter scale struck the Marathwada region of the western Indian state of Maharashtra at 3.56 a.m. on September 30, 1993 (Fig. 2.2). The epicentre of the earthquake was near the village of Killari in Latur District located at a distance of about 500 km south east of Mumbai.[1] This earthquake killed 7,928, injured over 16,000 and turned out to be the worst natural disaster in the state since 1947. The disaster completely destroyed 67 villages in the districts of Latur and Osmanabad and affected nearly 2,500 villages in other districts. In total, 27,000 houses were completely destroyed and around 200,000 were partially damaged. Public buildings, roads, schools, water towers and other infrastructure were severely damaged. The total estimated damage was about US$330 million (GoM n.d.).

---

[1] Mumbai was formerly known as Bombay. The official name Mumbai is used in contemporary contexts and the old name Bombay is retained in historical allusions.

Fig. 2.1 Political map of India

2.3.2 Gujarat Earthquake (2001)

On 26th January 2001, at 8.46 a.m. on the morning of the 52nd Republic Day, one of the most destructive earthquakes to strike India occurred in Kachchh District of Gujarat (Fig. 2.3). The earthquake measuring 7.7 Mw affected about 7,600 villages, 14 towns and the mega city of Ahmedabad. Thirteen thousand and eight hundred and five people lost their lives and 167,000 were injured. About 1.2 million houses were damaged either partially or completely. The affected population was about 28 million. Six districts were severely affected and the district of Kachchh, which was

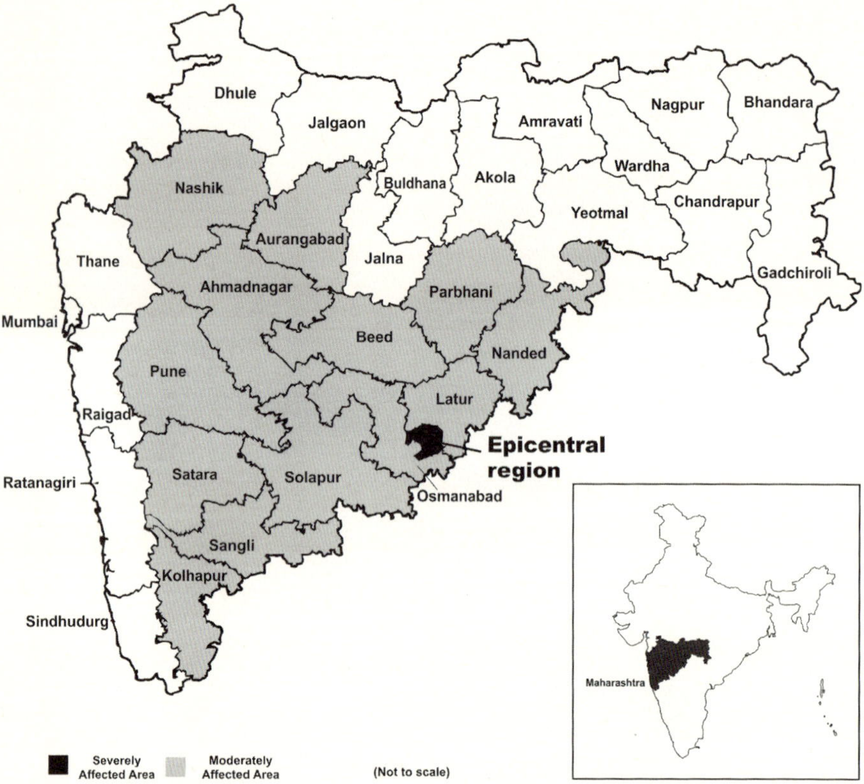

Fig. 2.2 Map of Maharashtra showing earthquake affected areas

the epicentre, was worst hit by the disaster (Mishra 2004, pp. 51–56). Out of 181 *talukas*[2] affected, 42 were declared severely affected. Around 220,000 houses fully collapsed and 917,000 were partially damaged (GSDMA 2008). The state's largest city—Ahmedabad—located 300 km from the epicentre, had the second largest death toll (752) next only to Kachchh District (12,221). In Ahmedabad more than 70 multi-storey buildings collapsed. People witnessed Reinforced Cement Concrete (RCC) buildings collapsing like a pack of cards.

Social and public infrastructure and industries also sustained severe damage. Over 50,000 school rooms and 1,500 health clinics and hospitals were damaged. The earthquake also had a severe impact on livelihoods. More than 10,000 small and medium industrial units went out of production and the livelihoods of more than 50,000 artisans were affected due to loss of workshops and tools. Kachchh District in which the epicenter of the earthquake was located was the worst affected with 12,221 people killed. One hundred and twenty four thousand houses totally

[2] The number of *Talukas* (or blocks) is 226 after reorganization.

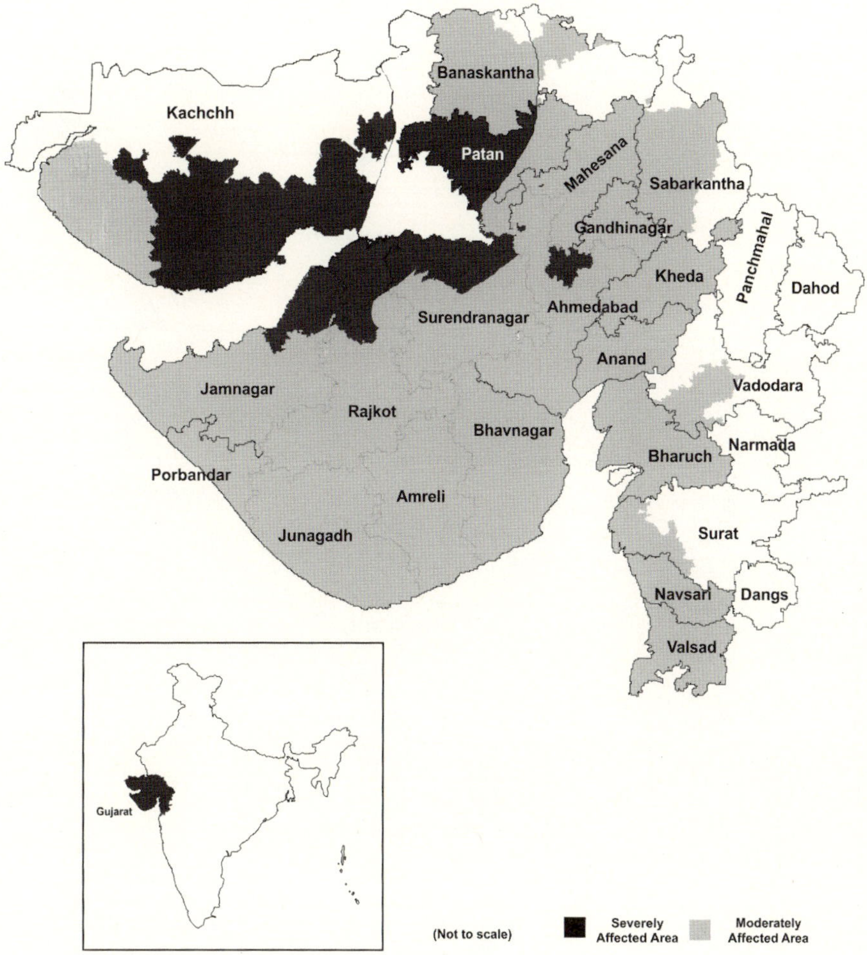

Fig. 2.3 Map of Gujarat showing earthquake affected areas

collapsed, while 210,000 were partially damaged. Four major towns of the district[3] suffered near total devastation (Mishra 2004; GSDMA 2008). The estimated total losses were about US$6 billion (World Bank and ADB 2001).

2.3.3 Tamil Nadu Tsunami (2004)

The state of Tamil Nadu was the worst affected in India when the Indian Ocean tsunami struck the east coast of India on 26th December 2004 (Fig. 2.4). It affected 238 fishing villages and 418 hamlets in the state leaving 7,997 dead and 3,625

[3] Towns are: Anjar, Bhuj, Bhachau, and Rapar.

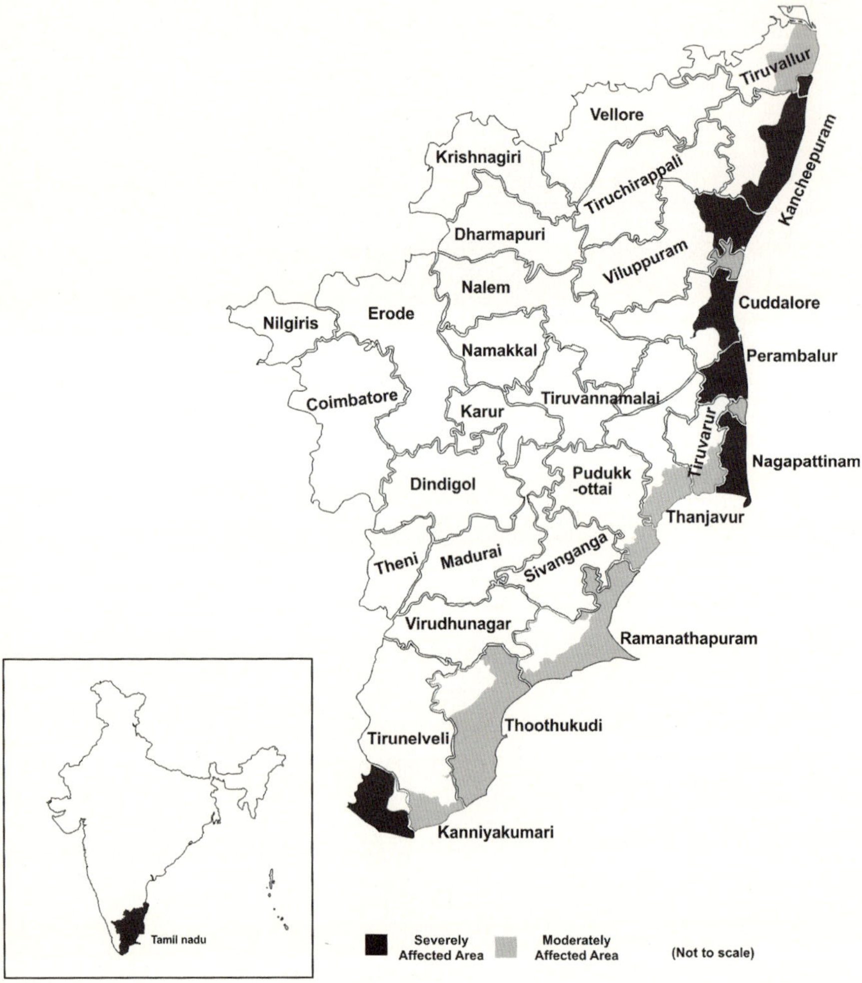

Fig. 2.4 Map of Tamil Nadu showing tsunami affected area

injured. All the 13 coastal districts were affected; but Nagappatinam with a death toll of 6,065 was the worst hit. Kanniyakumari and Cuddalore where 799 and 610 people died respectively also suffered severely. The disaster caused extensive damage to housing and livelihood assets. 53,290 houses collapsed and 11,694 were partially damaged.[4] In the fisheries sector, 51,078 fishing vessels were either lost or damaged.

[4]The accurate figures of the total number of houses damaged are not available. The reported details differ in various documents published by the Government of Tamil Nadu. The data is taken from Government of Tamil Nadu publication that gives the number of houses damaged along with breakup of fully collapsed and partially damaged (GoTN 2008).

This is one of the few disasters where the loss to income generating assets was higher than the loss to the built environment. Roads, water supply, electrical installations, minor ports, beach resorts and hotels on the beach were also affected. The preliminary loss estimate according to the joint mission of World Bank, ADB and UNDP was about US$880 million (GoTN 2005b, 2008).

2.4 Institutional Mechanisms in the Three Programs

While comparing the institutional mechanisms in three cases, it is pertinent to also keep in mind the damage profile of the three disasters. The scale of damage and destruction was much more in Gujarat when compared with the other two states (Table 2.1). Gujarat earthquake was more widespread in terms of the geographical impact and affected more population when compared with the other two. In Tamil Nadu only houses and livelihood were affected but in Maharashtra and Gujarat there was large scale damage to infrastructure. Coastal areas were affected in Tamil Nadu, and rural areas in Maharashtra. But in Gujarat both urban and rural areas were affected.

In Maharashtra, a three tier EOM was set up for reconstruction at the state level. At the first tier was the cabinet sub-committee comprising of six senior ministers for policy making, the second was Central Implementation Review Group (CIRG) for review and monitoring and the third was Project Management Unit (PMU) for implementation. The sub-committee headed by the Chief Minister was empowered to take policy decisions. A role of the high power implementation group headed by the Chief Secretary was to sort out technical problems, review and monitor. Besides these tasks, it was responsible for distributing work among various departments and to change the procedures if necessary. The policy provided for a strong and effective

Table 2.1 Comparison of the damages and losses in the three disasters

SN	Aspect	Maharashtra (1993)	Gujarat (2001)	Tamil Nadu (2004)
1	Deaths	7,928	13,805	7,997
2	Injuries	16,000	167,000	3,625
3	Population affected (million)	9.85	28.04	1.07
4	Houses—totally damaged	27,000	220,000	53,290
5	Houses—partially damaged	165,000	917,000	11,694
6	Villages affected	2,567	7,633	238
7	Towns affected	0	57	0
8	Municipal corporations	0	5	0
9	Asset losses in million US$	333	2,100	880
10	Asset loss as proportion of State Domestic Product (SDP)	0.8 %	8.9 %	2.2 %

Source: Mishra 2004; GoTN 2005c; GoTN 2008; GSDMA 2008; ADB et al. 2005; GoM n.d.; DPH 1995; World Bank and ADB 2001

PMU having adequate structure to implement the program (GoM 1994). Accordingly, a PMU headed by the Secretary and Special Commissioner, Earthquake Rehabilitation was created and sufficient powers delegated to it for project implementation (Nikolic-Brzev et al. 1999). In many aspects of implementation the PMU had the powers of the cabinet vested in it (World Bank 1999).

The creation of a PMU with full powers for implementation is considered one of the important factors that contributed to the success of the program (Nikolic-Brzev et al. 1999). The sole mandate of reconstruction given to the PMU made the difference. As one former Deputy Secretary who was in charge of the post-disaster reconstruction program in Maharashtra acknowledges, the normal mechanism would have been woefully inadequate for the task:

> In Maharashtra at that time, a PMU was very much required. The department of relief and rehabilitation did not have the capacity. And, even if we had tried to create that, the department would not have been able to focus on earthquake reconstruction in an uninterrupted way. Secondly, the simplified procedures and direct implementation galvanized a large number of staff. They could implement the program directly.

It is not just the creation of an EOM, but the support extended by the government that is crucial in making it effective. According to the then Chief Minister, it would not have been possible to meet the challenges of reconstruction without such a body:

> It was a major calamity and, in such a situation you need an exclusive set up with full authority. Unfortunately in government set up what happens is that after the bureaucracy takes some decisions in an emergency, an audit will find some fault later. He (the officer concerned) will have to then face the music for a life time. I was eager to see that such a situation is avoided. You see, ultimately members of bureaucracy are also citizens committed to the nation and society. Why not motivate them and give them authority? In a rehabilitation program, for good administration we must select right people and give them the authority. There may be a mistake here and there; but overall they get good results.

The short-term mandate given to the PMU, though highly effective in the context of reconstruction, did not help for long-term disaster management. The set up was created for reconstruction and was wound up after completion of their mandate. No permanent body was set up after the 1993 earthquake for long-term disaster management. The Mumbai floods in 2005, the next big natural disaster after the earthquake in the state, exposed the inadequacy of response capability in terms of infrastructure and institutional mechanisms (D'Souza 2005; Revi 2005). Though the need for institutional mechanisms in the form of permanent administrative structures were identified in the reconstruction program (Nikolic-Brzev et al. 1999, p. 75), they were not implemented until the state was criticised severely for mishandling the floods in 2005. The Maharashtra State Disaster Management Authority was formed in 2007 on the lines of Gujarat State Disaster Management Authority only after persistent demand from various quarters including experts (Revi 2005) and opposition parties (The Hindu 2005). Thus, the Maharashtra case highlights the need to look at EOMs in the aftermath of disasters both from short-term as well as long-term perspective.

Gujarat set up a single EOM for policy making as well as implementation. GSDMA headed by the Chief Minister was set up, within 2 weeks,[5] as a nodal agency. Besides making policy decisions, it implemented the massive reconstruction and rehabilitation program. Though a Central Implementation Group, as in the case of Maharashtra, was set up to review and monitor the progress, GSDMA had the overarching authority with regard to reconstruction. In addition, it had the mandate for long-term disaster management and risk reduction in the state. Thus in Gujarat, instead of multiple authorities there was one unified agency headed by the Chief Minister with Ministers and Secretaries of key departments as members.

GSDMA was created as a single window to deal with reconstruction activity, with rules of business different from government to facilitate quick decision making and faster implementation of the program avoiding bureaucratic hurdles and delays. GSDMA was accorded the powers of the cabinet and the decisions taken by GSDMA were implemented as if they were the decisions of the cabinet. The funds received from the World Bank and ADB were placed with GSDMA to avoid diversion of the funds for purposes other than reconstruction and rehabilitation. GSDMA, which was initially registered as an autonomous society, was subsequently converted into a permanent statutory body through the Gujarat State Disaster Management Act 2003. The act made GSDMA the apex body for disaster management in the state.

GSDMA differs considerably from the PMU set up in Maharashtra in terms of its structure. While the senior officers manning the organization were top ranking bureaucrats drawn from within state government on deputation, the junior officers, staff and experts were hired on contract basis from outside the government. In addition, it availed the services of experts and specialists by appointing them in the role of advisors and consultants. This arrangement provided for utilizing the knowledge and skills of the existing administrative machinery and at the same time to make use of expertize which does not lie in the government domain. A combination of officers well conversant with the rules business of the government and outside experts avoids the pitfalls of an entirely new agency which may commit mistakes, about which Olshansky et al. (2012) warns. The other noteworthy difference is that GSDMA has been made a very flexible organization in terms of the structure, by making it expandable and collapsible based on the situation and need. This flexibility provides for increasing or decreasing the man power and drawing on external resources at the time of crisis.

Almost all the stakeholders interviewed in the study acknowledged the important role played by GSDMA and considered it as one of the key factors for the success of the Gujarat reconstruction program. GSDMA's role as a single window mechanism enabled quick decision making, transparency, and openness. It was able to receive feedback and make mid-course corrections, and this was believed to have been one of the main reasons for better policy making and implementation. In the words of one Principal Secretary to the government, "GSDMA took to itself strong

[5] Gujarat earthquake occurred on 26th January 2001 and GSDMA was formed on 8th February 2001.

financial and managerial powers which allowed centralized planning and integration of all the various pieces of reconstruction. If we had kept it in the departments this integration would not have taken place." One of the members of the of the World Bank team associated with the program explained to me the important role played by GSDMA in Gujarat:

> I think in Gujarat's case it was necessary. It was effective in making quick decisions going outside government norms, perhaps due to the easy access to political bosses. All these are important. As a single window system, it was able to manage the whole thing in an integral fashion. It integrated various aspects like social impact and environment assessments. Otherwise, it would have resulted in departments running to each other.

The setting up of GSDMA as a two-in-one mechanism for reconstruction and disaster management is considered an innovation that set an example for other states. For many Gujarat reconstruction was special mainly due to this, as articulated by an NGO leader:

> Definitely high in my list as to why Gujarat was special, is the formation of GSDMA. I think its role was very central. It played a huge and positive role. It was an interesting innovative institutional arrangement. In my view, it performed very well.

The structure of GSDMA enabled it to achieve what other EOMs could not. Unlike the OSDMA which was constituted in Orissa after the Orissa Super Cyclone in 1999, which was headed by the Chief Secretary, the GSDMA was headed by the Chief Minister. It provided a platform to think, brainstorm and take policy decisions, thus cutting short the endless movement of files from one line department to another. The present Chief Minister and head of GSDMA, explained:

> GSDMA was not commanding other departments. That was its beauty. It became an umbrella under which frank discussions could be conducted. It also served as a forum to discuss various problems and possible alternatives freely. It provided a platform for arriving at better solutions. Due to this democratic process of decision making, there was a sense of collective responsibility to address the problems and ownership of the solutions.

The mandate given to the GSDMA for long-term disaster management is the main reason why it undertook many initiatives which were not carried out elsewhere. To sum up, setting up of the GSDMA helped post-disaster reconstruction as well as post-reconstruction activities. The Gujarat case demonstrates that the EOM created in the aftermath of disasters could be institutionalized to address long-term issues related to disaster management as well as avoid the formation of such ad-hoc bodies in future.

Unlike other states where disaster management initiatives ended with the reconstruction project, GSDMA continues with disaster management activities with budgetary support from government. The Gujarat State Disaster Management Act 2003, keeping the special purpose nature of GSDMA, mandates it to focus on long-term disaster management, capacity building and risk reduction while the routine functions of rescue and relief continue to vest with the office of the Relief Commissioner. Even today, many of the Disaster Management Authorities in other states including Maharashtra and Tamil Nadu, set up as per the mandatory requirement of the

Disaster Management Act of Government of India, remain on paper. In contrast GSDMA is a full-fledged functional organization. Continued allocation of funds to GSDMA even after the World Bank funding is over, is an example of the political interest in disaster management. This is despite the fact that development and routine administration are now being given importance over disaster management. According to many stakeholders, as articulated well by one of the former Additional Chief Secretaries to Government of Gujarat, in Gujarat long-term vision guided the reconstruction process even though it was not politically beneficial.

Institutionalizing EOMs and changing its nature and scope of work may result in dilution of its authority and make it function like any other institution in government. Many observed that, GSDMA, which has become a permanent body, now functions like any other bureaucratic arm of the government. Its business is carried out in a routine manner. The routine activities of government departments such as preparation of response plans, responding to disasters and conduct of drills are being transferred to GSDMA, thus diffusing the role of the EOM.

Tamil Nadu also set up an EOM in the aftermath of the tsunami to facilitate reconstruction. An advisory committee headed by a cabinet minister consisting of members from political and bureaucratic wings was set up to advice on finances, procurement, mobilization of human resources, to ensure inter-departmental coordination and strategic decisions (GoTN 2005c). It was created 8 months after the occurrence of the disaster and many important policy decisions were made before that. An empowered committee consisting of officers headed by the Chief Secretary was set up to provide approvals for the sub-projects, procurement and staff related matters as well as for periodic monitoring (World Bank 2005). The reconstruction was carried out through the office of the Commissioner of the Revenue Administration who is also the ex-officio State Relief Commissioner. A PMU, headed by a Project Director, was set up in the office of the State Relief Commissioner to monitor and coordinate the line departments/agencies implementing the project (GoTN 2005a).

The problems with a multiple-agency approach in Tamil Nadu shows that a unified agency approach is better than multiple agencies dealing with different aspects of reconstruction. The PMU had a limited mandate of implementation of reconstruction. While the policies were framed by the government, the procurement was approved by the empowered committee. According to one of the officials of the ADB who was heading the mission office in Chennai, the policy decisions were taken through routine procedures by the respective line departments without any sense of urgency, which many consider as the main reason for delays. Later, the state government set up a State Disaster Management Authority, as an entity separate from the PMU. This arrangement can hinder integration of disaster management with reconstruction. Separate agencies dealing with diverse aspects of disaster management, one that has a temporary mandate of reconstruction, the other with the long-term mandate of disaster management, may result in loss of institutional learning from the process of reconstruction. Above all, the EOMs in Tamil Nadu did not address the basic issue of change in policy making process which is the main purpose of choosing an EOM.

2.5 Extra-ordinary Mechanisms: An Analytical Comparison

The experience of the three states throws light on the need for extraordinary mechanisms and the will to put aside bureaucratic norms. But the mandate, scope and powers vested differed across the states. While in Maharashtra the policy decisions remained with the political executives, implementation including procurement was delegated to the PMU consisting of officers. In Tamil Nadu the powers for policy decisions were retained at the level of government and a two-tier mechanism was created that separated procurement from implementation. In Gujarat, not only were all the three powers vested in one agency (GSDMA), but that agency also brought together political executives and bureaucrats on the same platform. This coalescing of the political and bureaucratic wings within one authority enabled informed policy formulation that combined grassroots realities, the wisdom and experience of the political leaders and administrative requirements.

As discussed earlier, setting up of institutional mechanisms like GSDMA, reflected long-term perspective rather than short-term considerations. The fact that GSDMA in particular which was made a statutory authority for long-term disaster management with huge budget provisions even after the completion of reconstruction needs to be contrasted with the PMUs set up in Maharashtra and Tamil Nadu for reconstruction. Another issue with setting up EOMs is the time delay in setting up new organization (Olshansky et al. 2012). The case of GSDMA which was set up within 2 weeks of the occurrence of the disaster proves that if there is political and bureaucratic will such an organization can be set up without time delay.

Though the emergent norms following disasters warrant creation of EOMs, the mandate, scope, structure, and powers vested are often driven by factors other than the disaster. Tamil Nadu preferred to have a three-tiered set up at the state level, despite the knowledge of the success of GSDMA set up in Gujarat available to them is a case in point. Setting up EOM and delegation of powers is a political decision. The permanent mandate, powers and funds provided to GSDMA, helped in continuation of disaster management activities even after the reconstruction program was completed. This did not happen in Maharashtra and Tamil Nadu.

To sum up, the three cases examined provide answers to two important challenges in the post-disaster context: (a) when an EOM is necessary and (b) what are the desirable features for the structure and nature of the EOM. The three cases compared show that establishing an EOM does contribute to the success of the program. With regard to the structure of the EOMs, the Gujarat model based on a combination of some officers drawn from the existing bureaucracy and others from outside appear to be ideally suited to avoid pitfalls of other alternatives. This model is also more accountable while simultaneously having the requisite flexibility in approach and incorporating expertize. The strengths and weakness of the three EOMs are summarized in Table 2.2.

Learning from the EOMs constituted in the three cases, the characteristics of an effective model of EOM can be attempted. An ideal EOM should have autonomy to function freely and should have adequate powers and financial resources through budgetary support. A permanent EOM should be scalable as per the situation to

Table 2.2 Extra-ordinary mechanisms—a comparison of the strengths and weaknesses of the three programs

Program	Strengths	Weakness
Maharashtra	Special agency for reconstruction Full powers for implementation Single window for external stakeholders Could incorporate external expertize by hiring experts	PMU set up as agency set up through administrative orders Different agencies for policy making and implementation Short-term mandate of reconstruction Temporary and wound up after reconstruction
Gujarat	Autonomous agency mandated by legislation One unified agency for policy making and implementation Single window for external stakeholders Permanent body with long-term mandate of disaster management Adequate budget support from Government Could avail external expertize through hiring of experts and consultancies	Loss of importance post-reconstruction Bureaucratic power struggles and disputes over jurisdiction with regard matters related to disaster management Shifting of routine activities by other departments to the EOM Development priorities overriding long-term disaster management.
Tamil Nadu	Special agency for reconstruction Single window for external actors to interact Could incorporate external expertize by hiring experts	PMU set up as agency set up through administrative orders Different agencies for policy making, procurement and implementation Short-term mandate of reconstruction Temporary in nature and wound up after reconstruction

discharge its duties as per the mandate. It should have a high powered governing body constituted drawing the top political executives and bureaucracy to have commitment of the elected and bureaucratic wings of the government. The secretariat of the EOM should have a proper mix of manpower drawn from within and outside government to provide continuality, accountability and expertize. In order to maintain the focus of EOM on long-term risk reduction, existing functions of other agencies related to response should continue with them rather than transferring them to the new organization.

The EOM will play different roles before, during and after the occurrence of disasters. While preparedness through capacity building, training, information, education and communication and mitigation will be the major activities during normal times, coordination with external agencies, donors, lending institutions, and policy and planning for reconstruction will be the dominant activities during the time of the disaster, and reconstruction will be the main activity during post-disaster reconstruction period. The role of the EOM during the various phases of the disaster is given below (Fig. 2.5).

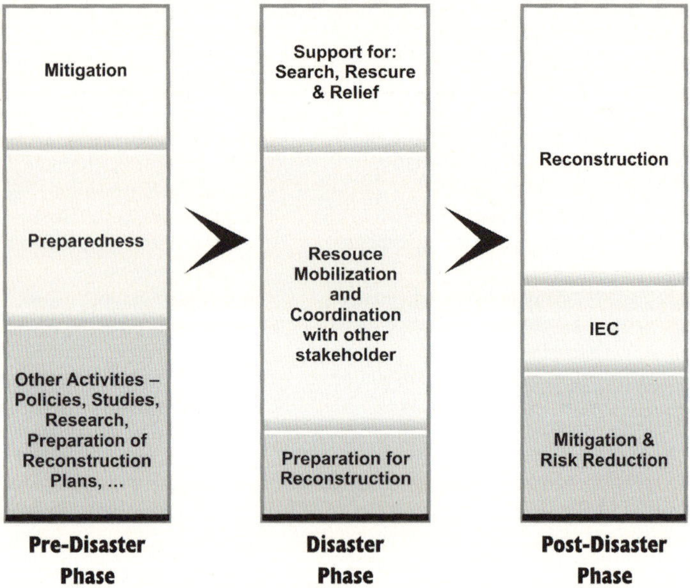

Fig. 2.5 Schematic view of functions of an EOM during before, during and after disaster

2.6 Conclusion

Creation of an extra-ordinary mechanism may have several advantages over the normal decision making process in governments. It can bring together various agencies and departments for collective decision making. It may cut down the red tape and speed up the decision making process. As the process is participatory, individual departments and agencies may own the process instead of seeing it as something thrust upon them from the above. EOMs may work as a single window for donors and lenders which will reduce their burden of having to interact with multiple government agencies. But all these depend on the nature and structure of the EOM.

The creation of an EOM may aim at overcoming the inefficiencies of a normal system, but it need not necessarily help in betterment of reconstruction. Mismatch between the routine and emergent procedures (Quarantelli 1989, pp. 10–11), lack of co-ordination, inter-departmental rivalry and fractured politics (Schencking 2006) may lead to failure of the EOMs as in the case of Japan after 1923 Kanto earthquake (Schencking 2006) or Guatemala after the 1976 earthquake (Bates 1982). A new organization created with overriding powers may not sit very well with the existing organizations. National Reconstruction Committee (NRC) created after the 1976 Guatemalan earthquake is a case in point. This EOM with special powers caused "jealousy" among other government officers. Bureaucrats and political leaders saw the emergence of NRC as a threat to their powers and tried to undercut the policies of NRC (Bates 1982). Another example is the failure of the "super reconstruction agency" in the form of a special reconstruction ministry set up in the aftermath of Great Kanto Earthquake in Japan due to political and inter-departmental rivalry (Schencking 2006). Such problems are not uncommon, hence the dilemma.

Setting up of EOMs does not guarantee better policy making or quick implementation unless, of course, they are sufficiently empowered. Failure of FEMA to respond to hurricanes Katrina and Rita due to its downgrading and dilution of authority is a case in point (Perrow 2005). GSDMA, unlike the PMUs set up in the other two states had sufficient authority and operational freedom vested with it. Besides, as emphasized by Topping (1998) in the case study of Oakland wildfire, California, GSDMA brought together representatives from relevant agencies and organizations. The success of GSDMA in integrating and coordinating diverse wings of the administration can be contrasted with the debilitating inter-departmental rivalries that ensued from the setting up of an EOM after Kanto earthquake (Schencking 2006). The success of Victorian Bushfire Reconstruction and Recovery Authority (VBRRA) set up in Australia for reconstruction after Victorian Bushfires (2009), Queensland Reconstruction Authority (QldRA) created in the aftermath of Queensland Floods (2010–2011), and formation of Canterbury Earthquake Recovery Commission (CERC) in New Zealand following Canterbury Earthquakes (2010–2012) show that if the new institutions have the capacity, coordinate with all stakeholders, engage the community and communicate with the stakeholders they can be highly successful (Smart 2012).

The structure and powers may help the EOM but its mandate is also a key factor for realizing the post-disaster opportunities. As Birkland (1997, p. 68) emphasizes, any "focusing event needs to find an advocate who will continue to press the issue on the agenda even after the immediate attention to the problem has diminished." Thus to press the issue of long-term risk reduction beyond reconstruction period a permanent agency mandated for the same may help.

Setting up new institutions or conversion of an EOM into a permanent institution to respond to disasters, as in the case of Gujarat, is not without problems. Foremost are the consequences of conversion of EOM into a permanent body. Converting an EOM into a permanent agency for long-term disaster risk reduction and using it for routine activities may result in taking away the importance of an EOM as a special purpose vehicle to respond to the disasters. It may not get the attention it got in the aftermath of disasters. The authority wielded by it is partly due to the urgency and immensity of the task and funds they had at their disposal. When both are absent, the organization loses its direction and well-defined tasks. Further, when the purpose and focus becomes broad-based, it becomes yet another agency without any special status. It is doubtful whether a permanent institution can respond as effectively in the aftermath of a catastrophic disaster, as a goal-driven EOM, as it did when it was created. While some of the EOMs were transformed into regular disaster response agencies filling in a gap, the very process of mainstreaming tends to strip it off its "special" capabilities.

Once an EOM has been transformed into a regular agency, does it make all EOMs redundant in future or would a conflict of interest arise? Given the special powers vested in some of these newly created regular agencies, it is likely that they may not be willing to recognize the requirement for EOMs, when the need arises. Several full time agencies with extra-ordinary powers to respond to disasters, including undertaking reconstruction and rehabilitation has been established in many countries including India in the last one decade or so. High powered national and regional level disaster management authorities have been set up in many countries in the aftermath of 2004, Indian Ocean Tsunami. These organizations may suffer all the perils of other institutions of government, which warrant the setting up of an EOM to respond effectively to disasters. They may resist and block the creation of an EOM if situation warrants the same. These institutions may develop vested interests and may kill the flexibility needed in the aftermath of disasters which a newly created EOM may provide in terms of openness to expert advice and formulation of procedures for speedy and effective reconstruction. The effectiveness of these institutions will be tested only when another major disaster of the similar magnitude that led to the creation of these institutions occurs again.

References

ADB, United Nations, and World Bank (2005) India post-tsunami recovery program: preliminary damage and needs assessment. ADB, New Delhi. http://www.tnrd.gov.in/externallyaidedprojects/Tsunami_rehabilation/india- assessment-full-report%5B1%5D.pdf. Accessed 24 July 2011

Alexander D (2004) Planning for post-disaster reconstruction. Paper presented at 2004 international conference and student competition on post-disaster reconstruction "planning for reconstruction." Coventry, 22–23 April. http://www.grif.umontreal.ca/pages/papers2004/Paper%20 Alexander%20D.pdf. Accessed 26 August 2010

Arslan H, Unlu A (2006) The evaluation of community participation in housing reconstruction projects after Duzce earthquake. Paper presented at 2006 international conference and student competition on post-disaster reconstruction "meeting stakeholder interests." Florence, 17–19 May. http://www.grif.umontreal.ca/pages/ARSLAN_%20Hakan.pdf. Accessed 12 June 2011

Asgary A, Badri A, Rafieian M, Hajinejad A (2006) Lost and used post-disaster development opportunities in bam earthquake and the role of stakeholders. Paper presented at 2006 international conference and student competition on post-disaster reconstruction "meeting stakeholders interests." Florence, 17–19 May. http://www.grif.umontreal.ca/pages/ASGARY_Ali.pdf. Accessed 24 December 2008

Barakat S (2003) Housing reconstruction after conflict and disaster. Humanitarian network paper no. 43. Humanitarian Practice Network, Overseas Development Institute, London

Barenstein JD (2006) Housing reconstruction in post-earthquake Gujarat: a comparative analysis. Humanitarian practice network paper 54. Humanitarian Practice Network, Overseas Development Institute, London

Barenstein JD (2008) From Gujarat to Tamil Nadu: owner-driven vs. contractor-driven housing reconstruction in India. Paper presented at 4th international i- Rec conference: building resilience; achieving effective post-disaster reconstruction. Christchurch, 30 April–2 May. http://www.resorgs.org.nz/irec2008/Papers/Duyne.pdf. Accessed 7 May 2009

Barenstein JD (2010) Who governs reconstruction? Changes and continuity in policies, practices and outcomes. In: Lizarralde G, Johnson C, Davidson C (eds) Rebuilding after disasters: from emergency to sustainability. Spon, London, pp 149–176

Bates FL (ed) (1982) Recovery, change and development: a longitudinal study of the 1976 Guatemalan earthquake. University of Georgia, Athens

Birkland TA (1997) After disaster: agenda setting, public policy, and focusing events. Georgetown University Press, Washington
Christoplos I (2006) The elusive 'window of opportunity' for risk reduction in post-disaster recovery. Paper presented at provention consortium forum 2006. Provention Consortium, Bangkok, 2–3 February
Cochrane H (2008) The role of the affected state in humanitarian action: a case study on Pakistan. HPG working paper. Humanitarian Policy Group, Overseas Development Institute, London
Diamond JM (2005) Collapse: how societies choose to fail or succeed. Viking, New York
DPH (1995) Reconstruction of earthquake affected region of Maharashtra. Charles Leopold Mayer Foundation for the Progress of Humankind. http://base.d-p-h .info/es/fiches/premierdph/fiche-premierdph-2102.html. Accessed 11 August 2011
D'Souza JB (2005) Disastrous management. Econ Political Wkly 40(36):3900–3901
Dynes RR (2000) The Lisbon earthquake in 1755: contested meanings in the first modern disaster. TsuInfo Alert 2(4):10–18
Dynes RR (2005) The Lisbon earthquake in 1755: the first modern disaster. In: Braun TEB, Radner JB (eds) The Lisbon Earthquake of 1755: representations and reflections. Studies on Voltaire and the eighteenth century. Voltaire Foundation, Oxford, pp 34–49
Elhawary S, Castillo G (2008) The role of the affected state: a case study on the Peruvian earthquake response. HPG working paper. Humanitarian Policy Group, Overseas Development Institute, London
Ensor MO (2009) Preface to the legacy of hurricane mitch: lessons from post-disaster reconstruction in Honduras. University of Arizona Press, Tuscon
Ensor MO, Ensor BE, Fuentes VE, Barrios RE (2009) The legacy of mitch: conclusions for new millennium. In: Ensor MO (ed) The legacy of hurricane mitch: lessons from post-disaster reconstruction in Honduras. The University of Arizona Press, Tuscon, pp 185–214
EPWRF (1998) Maharashtra earthquake rehabilitation documentation project. Government of Maharashtra, Mumbai
EPWRF (1999) Marathwada earthquake policy report. Government of Maharashtra, Mumbai
Fagen PW (2008) El Salvador: a case study in the role of the affected state in humanitarian action. HPG working paper. Humanitarian Policy Group, Overseas Development Institute, London
Foley C (2007) Mozambique: a case study in the role of the affected state in humanitarian action. HPG working paper. Humanitarian Policy Group, Overseas Development Institute, London
GOI (2009) National policy on disaster management 2009. National Disaster Management Authority, New Delhi
GOI (2011) Disaster management in India. Ministry of Home Affairs, Government of India, New Delhi
GoM (1994) Earthquake rehabilitation policy of Government of Maharashtra. Earthquake Rehabilitation Cell, Mantralaya, Bombay
GoM (n.d.) MEERP (Maharashtra Emergency Earthquake Rehabilitation Programme). http://mdmu.maharashtra.gov.in/pages/meerp/. Accessed 5 July 2011
GoTN (2005a) G.O.Ms. No. 172, Dated 30-3-2005. In: Nagapattinam NGO coordination and Resource Centre, Tamil Nadu (ed) Government orders related to relief and rehabilitation of tsunami affected areas and communities, 1:214–217. NGO Coordination and Resource Centre, Nagapattinam
GoTN (2005b) G.O.Ms. No. 172, Dated 30-3-2005. In: Nagapattinam NGO coordination and Resource Centre, Tamil Nadu (ed) Government orders related to relief and rehabilitation of tsunami affected areas and communities, 1:214-217. NGO Coordination and Resource Centre, Nagapattinam
GoTN (2005c) Tiding over tsunami. GoTN, Chennai
GoTN (2008) Tiding over tsunami – part 2. GoTN, Chennai
GSDMA (2008) Quarterly progress report: June–September 2008. GSDMA, Gandhinagar
Haas JE, Kates RW, Bowden MJ (eds) (1977) Reconstruction following disaster. MIT, Cambridge
Harvey P (2009) Towards a humanitarian government: the role of the affected state in disaster response. Humanitarian Practice Network, Overseas Development Institute, London

Hein C (2005) Resilient Tokyo: disaster and transformation in the Japanese City. In: Vale LJ, Campanella T (eds) The resilient city: how modern cities recover from disasters. Oxford University Press, New York, pp 213–243

Inam A (2005) Planning for the unplanned: recovering from crisis in megacities. Routledge, New York

Jain SK, Murty CVR, Chandak N (1994) The September 29, 1993, M6.4 Killari, Maharashtra earthquake in central India. EERI special earthquake report. EERI Newsl 28(1):1–8

Jain SK, Murty CVR, Jaswant NA, Sinha R, Goyal A, Jain CK (1997) Some observations on engineering aspects of the Jabalpur earthquake of 22 May 1997. EERI special earthquake report. EERI Newsl 32(2):1–18

Jha AK, Barenstein JD, Phelps PM, Pittet D, Sena S (eds) (2010) Safer homes, stronger communities: a hand book for reconstructing after natural disasters. The World Bank, Washington

Jigyasu R (2001) From "natural" to "cultural" disaster: consequences of post-earthquake rehabilitation process on cultural heritage in Marathwada region, India. Paper presented at international conference on the seismic performance of traditional buildings, Istanbul. 16–18 November. http://www.icomos.org/iiwc/seismic/Jigyasu.pdf. Accessed 31 August 2010

Krishnadas J (2007) Identities in reconstruction: from rights of recognition to reflection in post-disaster reconstruction processes. Feminist Legal Stud 15(2):137–165

Mishra PK (2004) The Kutch earthquake 2001; recollections, lessons and insights. National Institute of Disaster Management, New Delhi

Murty CVR, Greene M, Jain S, Prasad NP, Mehta V (2005) Earthquake rebuilding in Gujarat, India: an EERI recovery reconnaissance report. Earthquake Engineering Research Institute, Oakland

National Research Council (2006) Facing hazards and disasters: understanding human dimensions. The National Academies Press, Washington

Neal DM, Phillips BD (1995) Effective emergency management: reconsidering the bureaucratic approach. Disasters 19(4):327–337

Nikolic-Brzev S, Greene M, Krimgold F, Seeber L (1999) Lessons Learned over time: innovative earthquake recovery in India. Learning from earthquakes series. Earthquake Engineering Research Institute, Oakland

Oliver-Smith A (1991) Successes and failures in post-disaster resettlement. Disasters 15(1):12–23

Olshansky RB, Hopkins LD, Johnson LA (2012) Disaster recovery: processes compressed in time. Nat Hazards Rev 13(3):173–178

Perrow C (2005) Using organizations: the case of FEMA. Homeland Secur Aff 1(2)

Pincha C (2008) Indian Ocean tsunami through the gender lens: insights from Tamil Nadu, India. Oxfam America & NANBAN Trust, Mumbai

Price G, Bhatt M (2009) The role of the affected state in humanitarian action: a case study on India. HPG working paper. Humanitarian Policy Group, Overseas Development Institute, London

Quarantelli EL (1989) Disaster assistance and socioeconomic recovery: some observations. Preliminary paper # 140. Disaster Research Centre, University of Delaware, Newark

Revi A (2005) Lessons from the deluge: priorities for multi-hazard risk mitigation. Econ Political Wkly 40(36):3911–3916

Rubin CB, Saperstein MD, Barbee DG (1985) Community recovery from a major natural disaster. Monograph # 41. Institute of Behavioral Science, University of Colorado, Boulder

Salazar A (2002) The crisis and modernity of housing disasters in developing countries: participatory housing and technology after the Marathwada (1993) earthquake. Paper presented at 2002 international conference on improving post-disaster reconstruction in developing countries. Montreal, 23–25 May. http://www.grif.umontreal.ca/pages/i-rec%papers/alex.PDF. Accessed 31 August 2010

Salazar A, Jigyasu R (2010) A decade of lessons from Marathwada: earthquake vulnerability, politics and participatory housing. In: Patel SB, Revi A (eds) Recovering from earthquakes: response, reconstruction and impact mitigation in India. Routledge, New Delhi, pp 79–117

Samaddar S, Okada N (2006) Participatory approach for post-earthquake reconstruction in the villages of Kachchh, India. Ann Disaster Prev Res Inst, Kyoto Univ 49B:197–205

Schencking JC (2006) Catastrophe, opportunism, contestation: the fractured politics of reconstructing Tokyo following the Great Kanto earthquake of 1923. Mod Asian Stud 40(4):833–873

Schneider SK (1995) Flirting with disaster: public management in crisis situations. M.E. Sharpe, New York

Smart J (2012) The role of post-disaster institutions in recovery and resilience: a comparative study of three recent disasters – Victorian bushfires (2009), Queensland floods (2010–11), and Canterbury earthquakes (2010–12). Working paper 12/02. Institute for governance and policy studies, School of Government, Victoria Univesity of Wellington

Srinivasan K, Venkatesh K, Nagraj V (2005) The state and civil society in disaster response: an analysis of the Tamil Nadu tsunami experience. Tata Institute of Social Sciences, Mumbai

Swain M, Swain M, Sahoo RH (2006) Disaster and gender: impact of super cyclone on life and livelihood of women. Serials Publications, New Delhi

Telford J, Arnold M, Harth A, ASONOG (2004) Learning lessons from disaster recovery: the case of Honduras. Disaster risk management. Working paper no. 8. Hazard Management Unit, World Bank

The Hindu (2005) Setup disaster management authority: BJP, August 7

Topping KC (1998) Wildfire case study: Oakland, California. In: Schwab J, Topping KC, Eadie CC, Deyle RE, Smith RA (eds) Planning for post-disaster recovery and reconstruction. American Planning Association, Washington, pp 261–280

Vatsa KS (2005) Asset-based approach to household risk management: an analysis of selected case studies in South Asia. PhD Dissertation, The George Washington University

Willitts-King B (2009) The role of the affected state in humanitarian action: a case study on Indonesia. HPG working paper. Humanitarian Policy Group, Overseas Development Institute, London

Winchester P (2000) Cyclone mitigation, resource allocation and post-disaster reconstruction in South India: lessons from two decades of research. Disasters 24(1):18–37

Wisner B (2004) Risk and the neoliberal state: why post-mitch lessons didn't reduce El Salvador's earthquake losses. Disasters 25(3):251–268

Wong K (2008) Colombia: a case study in the role of the affected state in Humanitarian Action. HPG working paper. Humanitarian Policy Group, Overseas Development Institute, London

World Bank (1999) Implementation completion report: Maharashtra emergency earthquake rehabilitation project. Infrastructure Sector Unit, South Asia Region, World Bank

World Bank (2005) Technical annex on a proposed credit to the Government of India for an emergency tsunami reconstruction project. World Bank

World Bank and ADB (2001) Gujarat earthquake recovery program: assessment report. World Bank, New Delhi

Chapter 3
Collaborative Governance and Disaster Recovery: The National Disaster Recovery Framework (NDRF) in the U.S.

Naim Kapucu

Abstract Disasters impact communities and individuals and disrupt social-technical systems and community functions. Consequences of disasters can be minimized if communities and people reduce their vulnerabilities and increase their resilience. Disaster response received significant attention from the researchers and practitioners alike. In a sense it is easy to study short term disaster response. Investment in immediate disaster response also makes sense for policy makers. Unfortunately research on disaster recovery is very limited and it is considered the ignored phase of emergency management and existing knowledge and applied research of this phase is seriously lagging behind compared to what is required today. Disaster recovery is usually perceived and understood as a slow phase that begins after critical decisions and needs are met after a disaster. There is also a perceived notion that the government agencies at all levels have committed resources more to disaster response and relief efforts and less to recovery (and mitigation) efforts. Thus it is important to pay heed to this crucial phase of managing disasters. Recovery is a long process that offers ample opportunity to rebuild and redevelop resilient and sustainable communities. The chapter highlights that a shared effort to redevelop, restore, and rebuild a community requires effective intergovernmental and cross-sector collaboration and cooperation. Recent recovery experiences in the US, especially post-Katrina, have shown a considerable lack of coordination between different government agencies as well as political conflicts in planning and executing recovery efforts. The complex nature of recovery planning and efforts requires pre-disaster and post-disaster collaboration between different stakeholders including private, public, nonprofit organizations and citizens. This chapter focuses on collaborative governance principles applied to disaster recovery using the recent National Disaster Recovery Framework (NDRF) in the U.S. as an example. The development of NDRF is also included briefly in the chapter. This chapter is important

N. Kapucu (✉)
School of Public Administration, University of Central Florida, Orlando, FL, USA
e-mail: kapucu@ucf.edu

R. Shaw (ed.), *Disaster Recovery: Used or Misused Development Opportunity*,
Disaster Risk Reduction, DOI 10.1007/978-4-431-54255-1_3,

to gauge the usefulness of a national level recovery framework. This framework may assist in altering the general perception about the Federal government's lack of effort and planning towards disaster recovery.

Keywords Collaborative governance • Disaster recovery • Emergency management • National disaster recovery framework • NDRF

3.1 Introduction

Communities in the United States and abroad experience a variety of natural, man-made, and technological disasters. Hurricane Sandy in 2012, Hurricane Katrina in 2005, the four hurricanes that damaged portions of Florida in 2004, the earthquake in Haiti in 2010, and the earthquake, resultant tsunami and nuclear power plant accident that struck Japan in 2011, provide unfortunate reminders of the vulnerability of communities to natural disasters. These unfortunate events, like many others, illustrate how disasters impact communities and individuals, and disrupt social-technical systems and community functions. The threat of natural disasters will continue, but their consequences can be minimized if communities and people reduce their vulnerabilities and increase their resilience (The National Academies 2012).

In the field of disaster and crisis management, too much emphasis is placed on response to disasters. Unfortunately research on disaster recovery is limited. According to Rubin (2009) long-term recovery is the ignored phase of emergency management and existing knowledge and applied research of this phase is seriously lagging behind compared to what is required today. Disaster recovery has also not received much interest from administrative practice nor policy (Comfort et al. 2010). This neglect can be attributed to the fact that response compared to recovery requires immediate attention and action. Recovery is usually perceived and understood as a slow phase that begins after critical decisions and needs are met. There is also a perceived notion that the federal government has committed itself and its resources more to disaster response and relief efforts and less to recovery and mitigation efforts (Comfort et al. 2010). Thus it is important to pay heed to this crucial phase of managing disasters.

Recovery is a long process that offers ample opportunity to rebuild and redevelop communities into more resilient and sustainable communities (Comfort et al. 2010; Waugh and Smith 2006; McEntire 2006). According to Comfort et al. (2010), the recovery process is "a complex system of interacting jurisdictions, public agencies, private and nonprofit organizations, and households that are engaged in a shared effort to rebuild a community following a disaster" (pp. 669–670). A shared effort to redevelop, restore, and rebuild a community requires effective intergovernmental and cross-sector collaboration and cooperation (Kapucu 2012a, b). Recent recovery experiences in the US, especially post-Katrina, have shown a considerable lack of coordination between different government agencies as well as political conflicts in planning recovery efforts (Waugh and Smith 2006). The complex nature of recovery planning and efforts requires pre-disaster and post-disaster collaboration between different recovery stakeholders.

This chapter focuses on collaborative governance principles applied to disaster recovery. The study uses the recent National Disaster Recovery Framework (NDRF) in the U.S. as an example. The research questions for this study are: What are the key elements/principles of collaborative governance? How can the collaborative governance perspective be applied to disaster recovery? Does the NDRF fit with the collaborative governance perspective? What are some of the challenges/opportunities associated with this new framework? This study is very important to gauge the usefulness of a national level recovery framework. This framework may assist in altering the general perception about the Federal government's lack of effort and planning towards disaster recovery, which is believed to be the neglected phase of disaster management (Rubin 2009). Thus studying the framework in detail and analyzing its usefulness is critical today when communities face many threats from natural disasters.

3.2 Literature Review and Background

Emergency management entails four main stages, these include, *mitigation* to reduce threats, risks and vulnerabilities by taking specific measures such as enforcing building codes and better building practices; *preparedness* which encompasses planning, trainings and exercises; *response* activities such as search and rescue , and *recovery* which usually means "the restoration of lifelines and basic services" (Waugh and Streib 2006, p. 131). As shown in Fig. 3.1, mitigation and prevention, sometimes used interchangeably, along with preparedness are stages that exist in the disaster management cycle (pre-incident, incident, and post-incident). The nature of tasks that entail these stages is not restricted to post-incident. However, usually the tasks in these three stages are reviewed and redesigned in consideration of the

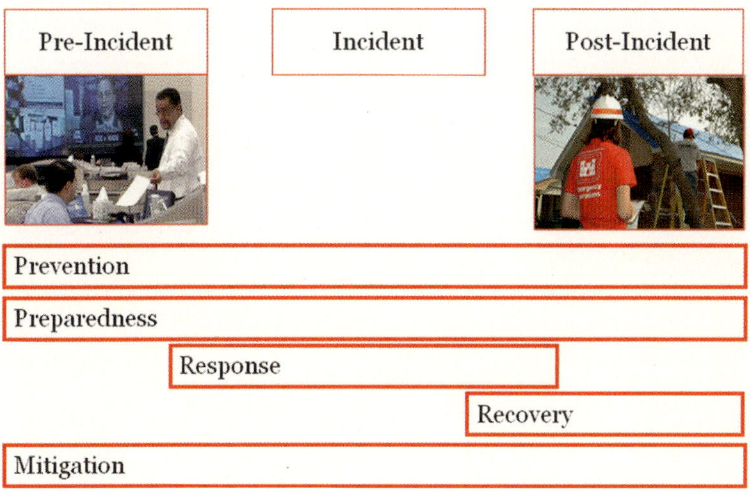

Fig. 3.1 Phases of emergency management

aftermath and consequences of disasters. Response takes effect immediately during and after the incident, while the recovery phase, as depicted in the figure, slightly overlaps the response phase and continues after the response and initial relief phase is over.

McEntire (2006) suggests that these phases should not be understood and interpreted in a linear fashion but should instead be viewed as functional areas that usually overlap each other. For instance, mitigation and preparedness both address the need to proactively prepare for disasters and apply disaster and risk reduction strategies. The strategies applied in the pre-disaster phase certainly impact the ways in which communities and agencies respond to incidents (Alesch et al. 2009; Kapucu and Ozerdem 2013; Phillips 2009).

Emergency management requires interagency coordination which is a fairly challenging task in situations where decisions have to be made in a timely manner (Derthick 2007). Response is the stage at which crucial decisions are made for immediate relief with constraints such as limited resources and time. Recovery, on the other hand, requires similar interagency coordination but constraints in recovery operations are different compared to the response stage. Although decisions are nonetheless crucial, more time and resources are available to agencies in the recovery phase. According to the National Governor's Association (NGA) (2007) "recovery is defined as the process of restoring a community to pre-disaster conditions" and "[i]t is the final phase of managing an emergency and continues until all systems return to normal or near normal. Recovery is a longer and more complex process than response, and it can take years until the entire disaster area is completely redeveloped, either as it was in the past or for entirely new purposes that are more resistant to disasters" (p. 52).

Waugh and Streib (2006) suggest that recent catastrophic disasters have changed the way long-term recovery was initially viewed. Its increased importance is reflected through an increased emphasis on pre-disaster recovery planning in overall emergency planning efforts. There has also been more emphasis in the effort to "link disaster recovery to economic development and to deal with the long-term social and economic problems exacerbated by disasters" (p. 132) which makes recovery a long, thoughtful process rather than a process that advocates quick fixes. Comfort et al. (2010) argue that disaster recovery is not a regular emergency management function since non-traditional players such as housing agencies, public works, urban planners, and a myriad of private companies dealing with infrastructure development take a front seat in recovery efforts. This makes the job of coordination and collaboration certainly more challenging but definitely necessary.

3.2.1 Collaborative Governance

Collaborative governance is a form of governing where both public and private entities are involved in collective and consensus-oriented decision-making (Ansell and Gash 2007; Emerson et al. 2012; Kapucu 2012b). Moynihan (2005) describes governance as a networked form of government that involves entities across sectors with different skills, expertise, and resources. It is a governance form that has become more popular with complex institutional structures and provides a distributed knowledge of agencies and sectors.

Ansell and Gash (2007) identify six conditions for collaborative governance. These are: this form is initiated by the government or a government agency plays the leadership role; participants include non-governmental agencies and actors; there is direct and deliberative engagement in decision-making by both state and non-state actors; formalized structures to organize, meet and engage with each other is created; decisions are made through dialog, deliberation, and consensus; and collaborative governance is aimed at improving public policy or public management. They also identify the difference between consultative techniques and collaborative techniques of engagement which reflect two-way communication and multilateral engagement and emphasize its importance in effective collaborative governance.

Stoker (1998) identifies five propositions for describing key elements of what entails governance and its structures. These are: governance structures are comprised of both state and non-state actors; boundary spanning is a common practice for dealing with public issues; power dependencies and resource dependencies exist between different agencies and entities; the structures may be self-governed networks; structures rely on the capacity and power of non-state actors in order to achieve better governance outcomes.

Along with these key principles identified in the literature, leadership is a pertinent element of collaborative and networked governance where public managers and leaders help to mobilize, facilitate, and implement collaborative and cooperative structures (Ansell and Gash 2007; Kapucu 2012b; Moynihan 2005) to achieve set goals and take responsibility to engage stakeholders in deliberative ways (Wallis and Gregory 2009). Agranoff (2006, 2007) suggests that governance can take the form of less-binding relationships like coordination and cooperation to more formal relationships that involve mandated or formal partnerships. Collaborative governance emphasizes collaboration which is beyond mere coordination and requires the achievement of shared goals and shared decision-making through both interorganizational and cross-sector efforts and relationships (Agranoff and McGuire 2003; Bardach 1998; Bryson et al. 2006; Cigler 1999). Some descriptions of governance reflect the idea that the government plays the central role in engaging other sectors while some descriptions of governance, such as those by Rhodes (1996), focus on the states having a minimal role. Another description of governance identifies the changing role of the public, private, and nonprofit sector where interdependence becomes inevitable and the blurring of sectoral boundaries becomes the norm (Stoker 1998).

Sometimes scholars do not differentiate between collaborative public management, networks and collaborative governance (Kapucu et al. 2010b). Like governance can take the form of cooperative exchange, coordinative or collaborative exchange between entities, networks can also have various forms. Brown and Keast (2003) describe cooperative networks to reflect informal and short-term relationships between entities, coordinative networks to depict joint working, decision-making and collective action for limited time action, and collaborative networks to emphasize more formal, long-term and sustainable relationships with a high level of inter-organizational trust and familiarity. However, Kapucu et al. (2009) suggest that although these terms are very similar and may follow similar characteristics and processes, in essence they are fairly different. Collaborative governance has a broader meaning compared to collaborative public management which clearly focuses on managing localities and holds public agencies and their roles in the collaborative arrangement as central and essential. Collaborative governance on the other hand has a broader, global appeal that includes collaborative public management, networks and inter-organizational and inter-jurisdictional cooperation and collaboration.

3.2.2 Collaborative Governance and Disaster Recovery

Collaborative governance can be defined as a collective effort of the stakeholders in recovering from disasters. Collaborative networks are essential in emergency management and disaster response (Drabek and McEntire 2002; Kapucu and Garayev 2012; Waugh and Streib 2006), including terrorism response (Kapucu 2012b). Collaborative emergency management aims to tackle the structural problems associated with traditionally rigid, less open command and control response and recovery systems (Birkland 2007; Kapucu 2006, 2008; Kapucu et al. 2010a; Ward and Wamsley 2007). Collaborative governance is being utilized and applied to managing disasters due to the catastrophic effects of disasters that are beyond the scope of any single jurisdiction or sector. More recent federal level changes and disaster management plans have also emphasized and supported the collaborative approach in dealing with disasters in the US. National level plans such as the National Response Framework (NRF) exude strong elements of the collaborative approach through their Emergency Support Functions (ESFs) and partnership approach.

According to Waugh and Streib (2006), "[w]hat we now call the new governance process forms the core of our national emergency response. Consensual processes are the rule" (p. 133). The field of emergency management has transformed from a top-down bureaucratic model to a more flexible network orientation that supports inter-organizational coordination, cooperation and collaboration (Waugh and Streib 2006).

The Whole Community approach has also recently been promoted by Federal Emergency Management Agency (FEMA). This is a philosophical approach on conducting better and improved emergency management by engaging the whole community (Edwards 2013). This is an approach through which citizens, government agencies, private and nonprofit organizations, community leaders and practitioners collectively decide on ways to improve their communities in order to build resiliency and a strong social structure that improves emergency management (FEMA 2011a). The Whole Community approach followed during the preparedness and mitigation phase helps to improve response and recovery stages of the emergency management cycle since partnerships and relationships built between community players and agencies have been developed and local resources and capacity have been assessed as part of this approach (FEMA 2011a). When multi-jurisdictional and cross sector partnerships have been formed in a community, then functional and resource coordination and collaboration becomes easier in the phases of emergency management (Kapucu 2012a).

The Whole Community concept has been recently applied to disasters in the US. The tornado in Joplin, Missouri in 2011 lead to the creation of a Citizens Advisory Recovery Team (CART) which is comprised of government officials, businesses, and residents. This coalition brought together community leaders to develop a collective recovery vision and develop goals for the community. CART was supported by FEMA's Long-Term Recovery Task Force, along with the Housing and Urban Development (HUD) and the Environmental Protection Agency (EPA). CART conducted community sessions and dialogue concerning recovery issues such as redevelopment of community facilities, schools, housing and infrastructure. Similarly a Recovery and Reinvestment Coordinating Team (RRCT) was developed after the 2008 flood in Iowa in Cedar Rapids. Government officials, the business community, schools, civic organizations and nonprofits acted collectively under this team to develop a plan for recovery. Goals for long-term recovery were established and discussed by involving members in open house forums and public meetings (FEMA 2011b).

It has been observed that collaborative plans and programs initiated by the government have focused more on improving the response stage of managing a disaster as compared to recovery. There has been emphasis on creating healthy relationships between different agencies and entities in the preparedness phase so that response to disasters can be improved and made more effective (Kapucu 2009). Initiatives such as the Urban Area Security Initiative (UASI) funded by the Department of Homeland Security (DHS) help urban regions develop response capabilities by investing in building and sustaining partnerships between responding agencies so that existing resources such as personnel and equipment and capabilities can be leveraged and shared (Jordan 2010).

It is important to understand the nature of recovery in the United States where communities impacted by catastrophic disasters can take many years to recover. Around 14 federal level agencies are involved in administering recovery programs and plans. To implement these programs local and state governments have to work very closely with federal level agencies, which make collaboration imperative in recovery efforts (GAO 2012).

Moreover, it is important to emphasize that federal recovery assistance is supplementary in nature and the onus of recovery efforts and activities lies with state and local governments rather than FEMA or the other federal government agencies (NGA 2007). State and local governments are tasked to help their citizens recover from disasters. However, when a disaster overburdens the capacity of local and state government's federal assistance then becomes available. Public assistance, as the name suggests is the federal assistance available to public entities such as state, local governments or non-governmental agencies that provide public services such as education, utilities, medical and rehabilitation etc. This type of assistance can finance and support the "repair, restoration, reconstruction, or replacement of a public facility or infrastructure that is damaged or destroyed" (NGA 2007, p. 48). The federal government offers disaster unemployment assistance, crisis counseling, disaster housing assistance, legal services, veteran's assistance, and tax relief along with providing low-interest loans for repairing uninsured property (NGA 2007).

Advisory committees are an example of a type of collaborative governance arrangement only when the advice of members is closely linked to the decisions that are actually made (Ansell and Gash 2007). This form of governance is popularly found in the recovery stage of managing disasters when different agencies and jurisdictions come together for revitalization and redevelopment plans and programs.

There was an impressive list of recovery plans that were developed for New Orleans' reconstruction and redevelopment after Hurricane Katrina. Although the list was impressive, the lack of coordination between those plans was evident (Comfort et al. 2010). It took around 4 years to agree on an initial strategic recovery plan for Louisiana which meant that many businesses had already decided to relocate elsewhere (Rubin 2009). There was a lack of coordination not only in the recovery planning stage but also in the implementation stage where different jurisdictions of government were not aware of each other's roles and responsibilities; this led to the utmost confusion about both disaster programs and their requirements, not to forget excessive duplication and high costs (Comfort et al. 2010). Thus, recent experiences reflect the need to improve recovery planning at the local, state and federal levels.

From 2008 to 2010, the GAO reviewed disaster recovery efforts led by FEMA and arrived at three broad conclusions and themes for successful recovery efforts. These three themes are: the need for clearly defined recovery roles and responsibilities; the importance of effective coordination and collaboration among recovery stakeholders; and the value of periodic evaluation of, and reporting on, recovery progress (GAO 2012, p. 1). There has often been confusion regarding the roles and responsibilities of various recovery agents. GAO's review of the Office of the Federal Coordinator for Gulf Coast Rebuilding (OFC) found disagreements in recovery stakeholders regarding their roles and the scope of operations which resulted in delays for recovery processes (GAO 2012).

Hurricane Katrina's scale of impact resulted in an outstripping of local and state resources with many internally displaced people and massive destruction of infrastructure and housing. The recovery efforts so far in New Orleans and the surrounding areas reflect a poor display of coordination and collaboration between the local, state and federal government (Waugh and Smith 2006). The federal government has been largely criticized for doing less for the recovery of businesses and the rebuilding of the local community and its population (Dietch and Corey 2011). The weak recovery efforts coordinated post-Katrina made Congress realize the need for a National Disaster Recovery Strategy. Under the Post-Katrina Emergency Management Reform Act of 2006, DHS was asked to develop a National Disaster Recovery Strategy (GAO 2012).

3.2.3 The National Disaster Recovery Framework

Response, compared to recovery, has been given more importance in both the research community and by the government in the planning stages of disaster management (Kapucu and Ozerdem 2013). However, experience with recent disasters and their consequences have emphasized the need for a better recovery structure and support.

In October 2009, President Obama asked DHS and HUD to establish a White House Long-term Disaster Recovery Working Group to develop a plan for the operational guidance of recovery agencies and to make recommendations for the improvement of existing recovery structure. This working group was headed by DHS and composed of representatives from more than 20 departments and agencies. As a direct result of this initiative the National Disaster Recovery Framework (NDRF) was developed for leading improved efforts for disaster recovery. The idea of this framework was not to increase the already existing recovery resource requirements but to make them more useful and effective by developing an operational structure (Carter 2011). It is a framework that is currently being utilized as a resource to help engage various stakeholders with each other in the recovery process.

The successful implementation of the framework requires inter-organizational, cross-sector, and inter-governmental coordination. NDRF "provides a flexible structure that enables disaster recovery managers to operate in a unified and collaborative manner. It also focuses on how best to restore, redevelop and revitalize the health, social, economic, natural, and environmental fabric of the community and build a more resilient Nation" (FEMA 2010, p. 1). The document identifies key principles for recovery, outlines roles and responsibilities of coordinators and stakeholders, defines a coordinating structure that supports coordination of tasks and communication flows amongst stakeholders, provides direction for both pre-and post recovery planning efforts, and explains the process through which stakeholders and communities can avail opportunities to restore and rebuild in better more sustainable ways (FEMA 2010).

The NDRF follows a similar framework to the NRF. Like the NRF, this recovery framework offers an operational structure for recovery efforts and recovery-related planning efforts. Through the implementation of the NDRF, the NRF Emergency Support function (ESF) #14 which dealt with long-term recovery has been excluded from the NRF. However, key concepts from ESF #14 have been adopted and expanded in the NDRF. Similar to ESFs that are used for response functions, the NDRF has introduced Recovery Support Functions (RSFs). "The RSFs are six groupings of core recovery capabilities that provide structure to facilitate problem solving, improve access to resources, and foster coordination among State and Federal agencies, nongovernmental partners and stakeholders" (FEMA 2010, pp. 1–2).

ESF #14 has helped to connect states and local agencies to the research community and plan for recovery. Thus, although ESFs have helped to develop avenues for improved recovery planning and efforts, its focus remains centered on the process for applying for federal assistance, etc. According to Rubin (2009) there is an imperative need to create a repository of information on recovery resources and also there is a need for technical assistance and funding to develop the capacity of local communities and states to plan recovery on their own. The NDRF addresses this concern through its newly developed RSFs and the recommended positions of recovery coordinators and managers at local, tribal, and state levels.

The NDRF has been developed to provide a common doctrine for an efficient and effective recovery. It has also been developed to complement the already existing NRF. It was created in direct response to a Congressional directive to formulate a National Recovery Strategy (FEMA 2011b). The key concepts introduced in the NDRP are: structure, leadership, and planning. Structure is conceptualized through the Recovery Support Functions (RSFs). Leadership is provided and supported through the Local Disaster Recovery Managers (LDRMs), the State and Tribal Disaster Recovery Coordinators, RSFs and their lead agencies, the Federal Disaster Recovery Coordinator and leaders belonging to private and non-profit sectors. Recovery planning is emphasized throughout the document both pre and post disaster. These three concepts, along with the involvement of relevant stakeholders and the setting of desired outcomes, represent a successful recovery (FEMA 2010).

The framework aims to transform existing recovery efforts by guiding post-disaster recovery organizations to improve their performance and coordination during the recovery phase of emergency management. Recent experiences after Hurricane Katrina reflect the lack of consensus in recovery planning. There were different and sometimes conflicting recovery plans being developed by different groups and stakeholders (Comfort et al. 2010). This framework creates and supports a strong coordination system that brings stakeholders on one platform and promotes unity of effort at different jurisdictional levels (FEMA 2010). The document also emphasizes pre-disaster recovery planning and recovery capacity building which is essential for developing community resiliency which helps in improving response and recovery from disasters which ultimately decreases recovery time and cost. The NDRF also emphasizes a shift away from recovery as being something that is limited to restoration of pre-disaster conditions to a process that reduces vulnerabilities, promotes sustainable practices and develops community resiliency (FEMA 2010).

The importance of leadership in achieving successful recovery is emphasized throughout the document. The framework prescribes and recommends that local disaster recovery managers, a state disaster recovery coordinator and a federal disaster recovery coordinator should be designated to ensure unity of effort in both the pre-disaster planning stage and post-disaster recovery efforts (FEMA 2011b). There is a separate chapter devoted to leadership describing the recommended pre- and post-recovery roles and responsibilities of recovery managers and coordinators at local, tribal, and state levels along with a detailed description of the designated position of the Federal Disaster Recovery Coordinator (FDRC). The FDRC's authority, qualifications, selection, and credentials, pre and post disaster engagement and responsibilities, and activation, transition and demobilization is described clearly in the document. However, the description within the framework remains incomplete as it will be supplemented through other supporting tools and documents later on (Carter 2011; FEMA 2011b).

There is one complete section in the framework dedicated to factors that lead to successful recovery. These factors are identified as: effective decision-making and coordination; integration of community recovery planning processes; well-managed recovery; proactive community engagement, public participation and public awareness; well-administered financial acquisition; organizational flexibility; and, resilient rebuilding. Each factor is explained albeit shortly but clearly in the chapter and provides valuable guidance for recovery stakeholders and impacted communities to think about measuring recovery goals and outcomes (FEMA 2010).

In a collaborative environment it is essential that the roles and responsibilities of each stakeholder and partners is clearly understood and defined so that exchange between partners is optimal and effective. The coordination and collaboration required in a disaster setting requires timeliness but not hastiness. This requires strong pre-disaster planning. The NDRF, in accordance with its principle of unity of effort clearly defines the roles and responsibilities of key recovery partners and stakeholders. The recommended roles and activities of individuals and families, the business community and critical infrastructure owners and operators, the nonprofit sector, and local, state, tribal and federal governments are outlined in detail (FEMA 2010).

The collaborative nature of recovery is exuded to a great extent through the RSFs. The RSFs created focus on key recovery capabilities and core recovery functions such as housing, infrastructure systems, etc. RSFs are activated post-disaster, similar to the ESFs, but they bring together stakeholders for pre-recovery planning so that roles and responsibilities of different players are identified enabling trusting relationships and strong partnerships to be developed. Although very similar to the ESF structure under the NRF, " RSFs are different from ESFs in that they have different mission objectives, partnerships, approaches, time spans and organizational structure; additionally the players and skill sets involved may be different" (FEMA 2010).

RSFs follow the same structure and overall purpose of effective coordination between different entities and stakeholders (Fig. 3.2). However, their mission and objectives, and their time spans to achieve tasks, etc. are different than ESFs. According to the FEMA 2010) "the objective of the RSFs is to facilitate the identification, coordination and delivery of Federal assistance needed to supplement recovery resources and efforts by local, state and tribal governments, as well as private and nonprofit sectors." The RSF approach helps to improve and quicken the process of recovery and redevelopment in impacted communities. The nature of recovery partnerships and the players is also fairly different than the partnerships involved and cultivated through the ESF approach. In recovery, non-traditional partners will be involved such as; private and government agencies for economic development and housing financing, and advocacy groups. Moreover, experience suggests that a more flexible, scalable and collaborative approach is utilized in forging recovery related partnerships compared to those developed in the response stage (FEMA 2010). This is because the nature of recovery involves thoughtful planning and redevelopment of a community as compared to task-oriented goals and objectives in immediate response.

The RSFs are led by federal level agencies that have been designated at the national level as depicted in Fig. 3.2. Each lead agency of an RSF will identify and

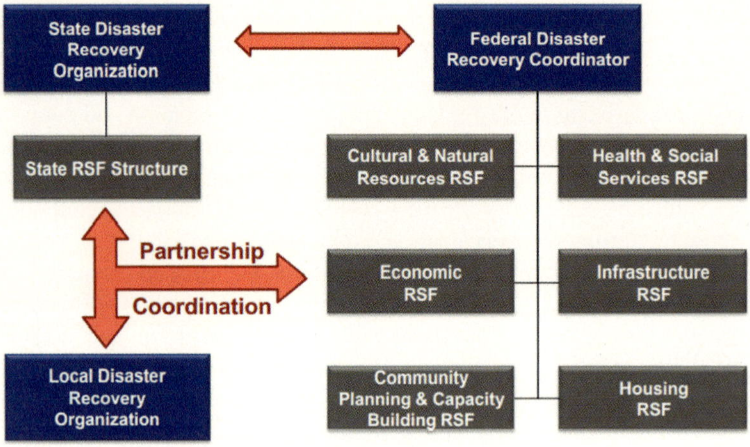

Fig. 3.2 RSF coordinating structure (FEMA 2011b)

designate a national coordinator who will oversee the recovery efforts by relevant primary and supporting agencies and ensure coordination and communication between them (FEMA 2010). The six RSFs and their lead coordinating agencies are: Community Planning and Capacity Building (DHS/FEMA), Economics (U.S. Department of Commerce), Health and Social Services (U.S. Department of Health and Human Services), Housing (U.S. Department of Housing and Urban Development), Infrastructure Systems (U.S. Army Corps of Engineers) and Natural and Cultural Resources (U.S. Department of Interior) (FEMA 2011b; Carter 2011).

This framework is line the "whole community" approach that is being adopted by FEMA to develop better prepared and resilient communities. The framework identifies the role and responsibilities of individuals and households in both pre and post disaster recovery efforts. It also identifies the various strategies and initiatives that can and should be taken during the recovery phase. For local and state governments this framework provides a guideline for improving existing recovery plans and structures as it designates coordinators for each recovery functional area. At the federal level this framework defines roles and responsibilities for agencies playing leadership and functional roles through the RSFs and also various annexes. For nonprofit organizations and the private sector this framework establishes structures that integrate their involvement in recovery efforts (FEMA 2011b).

Figure 3.3 demonstrates the expected collaborative governance networks for disaster recovery based on the RSFs in the NDRF. The figure was developed based on the functions and key organizations identified for each function in the NDRF using (see Appendix) the well-known UCINET network analysis software program (Borgatti et al. 2002). The affiliation network shows that FEMA, DHS, EPA and USDA are involved in all the RSFs. Although FEMA and DHS are playing a more primary role in the RSFs, USDA and EPA have a more supportive role to play within these RSFs and their networks. The affiliation network also reflects that the Natural and Cultural Resources RSF has most agencies that are exclusively involved

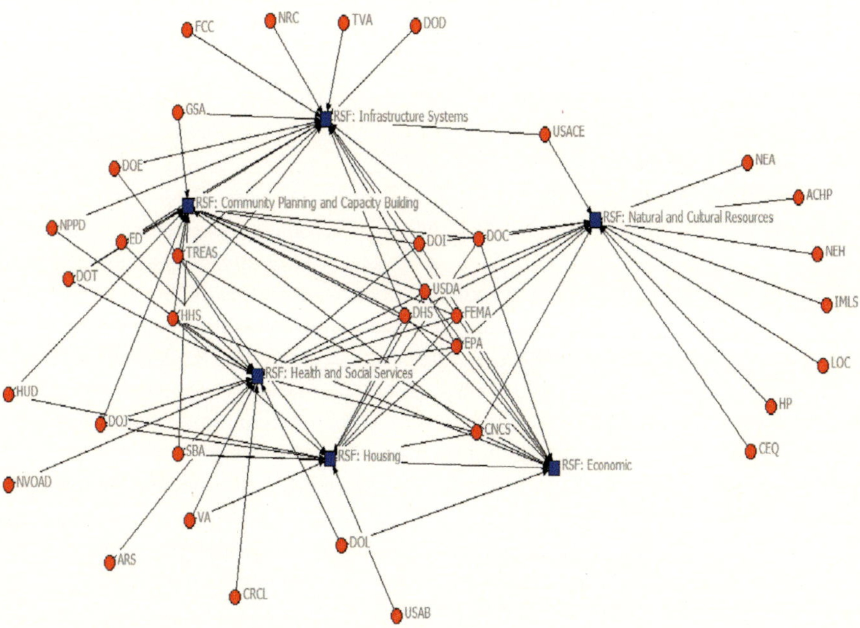

Fig. 3.3 Organizational networks in the NDRF

in this specific RSF. Agencies like NEA, ACHP, NEH, IMLS, LOC, HP and CEQ are only playing primary and supportive roles within the Natural and Cultural Resources RSF. Similarly the network shows that there are certain agencies such as TVA, DOD, NRC, and FCC that are exclusively involved in the RSF focusing on Infrastructure Systems. The agencies and departments involved in the Economic RSF and the Community Planning and Capacity Building are more integrated within the overall affiliation network depicted showing that there may be functional overlap between these RSFs or there may be agencies involved that are playing recovery roles that are not specific to a certain RSF.

Overall, the NDRF is a comprehensive document that identifies core recovery principles, roles and responsibilities of recovery leaders and coordinators at different levels of government and other sectors, and outlines a coordinating structure for different players tasked with recovery functions and operations both pre and post disaster. Alongside this framework, a National Disaster Recovery Program Database (NDRPD) has been launched in March, 2011. This tool provides a database for all programs funded by and operating at federal, state, and local government levels, as well as the private sector and nonprofit and non-governmental agencies (FEMA 2011b). It is important to emphasize that this framework is still a work in progress. This document will be revised as the National Preparedness System in developed and updated. The National Preparedness System involves prevention, protection, and mitigation frameworks along with response and recovery. Thus improvements and updates within the system will lead to the revision of existing plans (FEMA 2010).

3.3 Discussions

The NDRF in the US aims to enhance the capacity of communities to not only speed up the recovery process but to make it more effective. The emphasis on pre-disaster recovery planning throughout the framework also stresses the importance of creating a resilient community that reduces recovery costs and time (Carter 2011). Primarily the document outlines and conceptualizes the way an effective national recovery system works and outlines an interagency structure. It explains how federal agencies will organize and coordinate existing resources to improve recovery operations through the RSFs and through leaders and coordinators designated at different jurisdictional levels (FEMA 2010). The principles of leadership in collaborative governance, inter-sector partnerships and clarity of roles and responsibilities of stakeholders exist within this framework. Thus, one can argue that this framework is perfectly in line with collaborative governance principles.

Recovery experiences have reflected the importance of effective intergovernmental relationships to reach recovery goals and outcomes. Experiences in the past have shown feuds and conflicts between cities, counties, and state governments along with political rivalries deciding the future of impacted communities (Alesch et al. 2009; Rubin 2009). The literature shows that pre-disaster recovery plans at state and local levels will help to avoid conflicts regarding the roles of stakeholders. FEMA has done a worthy job of delineating the roles and responsibilities of different jurisdictions and partners in the NDRF. This helps to minimize duplication in efforts and resources and also helps to foster pre-disaster relationships and strengthen partnerships (GAO 2012).

Collaboration is one of the main objectives and goals of the NDRF. Through different strategies, the framework is supporting and encouraging collaboration between recovery stakeholders. The creation of the position of the FDRC helps to facilitate coordination of resources and efforts. The framework has also introduced recovery coordinators and managers at state and local levels which work closely with the FDRC to ensure collaboration and coordination is achieved. Along with developing leadership positions to lead collaborative recovery efforts effectively, the framework has established six RSFs that divide recovery into different functions to effectively coordinate resources and expertise (GAO 2012). There is also more useful information regarding the measurement of successful recovery and the core principles of recovery. Overall though, some of the parts of the framework lack the required detail for implementation.

The document also discusses the core principles that guide a successful recovery process. These core principles are: Individual and family empowerment, leadership and local primacy, pre-disaster recovery planning, partnerships and inclusiveness, public information, unity of effort, timeliness and flexibility, resilience and sustainability, and psychological and emotional recovery. While some of these core principles are reflected throughout the framework such as unity of effort and leadership, some core principles such as psychological and emotional recovery are only touched upon briefly throughout the plan. Moreover, while these principles point to the right direction, their implementation and the process that leads to their implementation is unclear. For instance, under the Pre-disaster Recovery Planning principle it is emphasized that the "NDRF strongly encourages innovation among the States, Tribes, localities, and the private sector in working together to identify…tools and resources…to support and sustain disaster mitigation and recovery efforts" (FEMA 2010), p. 10). The plan needs to highlight some examples of what types of innovative practices the framework does support and encourage. Moreover, it would be beneficial if these core principles are introduced in order of their importance from a national perspective or in accordance with the framework outlined. For example, unity of effort is clearly more evident as a core principle of the framework as compared to psychological and emotional recovery.

Literature shows that evaluating and monitoring recovery efforts throughout the recovery period and beyond needs to be encouraged in order to gauge which efforts and plans work and which do not. This evaluation will help to develop better recovery policies and plans. This practice was successfully followed by the city of Kobe in Japan after the 1995 earthquake for 10 years (Comfort et al. 2010; GAO 2012). The NDRF's section relating to factors and measures of successful recovery is very useful, however, what is lacking in this particular section is information on the development and implementation of specific measures and indicators for successful and effective recovery. Perhaps the inclusion of case studies and examples will provide more information and direction for communities and stakeholders. According to the GAO (2012) measuring successful and effective recovery operations is addressed in the framework, but it lacks the explanation needed regarding conceptual importance and the value of regular monitoring and evaluations.

The existing NDRF document does not seem to be a stand alone, comprehensive document. Although it provides a clear guidance and structure for recovery planning and efforts, it needs to be a supported with certain operational and training tools. The guiding document of the framework says that this existing document and structure is "supported by the ongoing development of detailed operational, management, field guidance and training tools" (FEMA 2011b).

3.4 Conclusion

Despite the efforts to create awareness that disaster management is the responsibility of local and state governments first, and not a top–down command and control operation where the federal government is always at fault, communities impacted continue to blame on federal government leadership for its ineffectiveness and lack of planning, efforts and resources. This was clearly the case in Hurricane Katrina (Waugh and Streib 2006). Thus, it makes sense for the federal government to take a lead in developing and improving existing federal level guidelines and plans. The development of the National Disaster Recovery Framework is a step in the right direction. Even with the creation of this national recovery framework, there remains the need to provide "more in-depth information regarding actual recovery experiences, successful and unsuccessful. For example; documents and guidance useful to practitioners; and much more in the way of education and training" (GAO 2012, p. 14).

There is more time needed to observe and see how this framework has been applied and implemented to more recent disasters such as Hurricane Sandy. Hurricane Sandy impacted the most populated area of the United States late October in 2012. Some significant problems observed in the disaster impacted area in terms of speedy recovery. There are some parts of the area still lacks basic public services. The NDRF was implemented first time in recovering from the disaster. It is early to collect data and evaluate the implementation of the framework during recovery operations from Sandy. Future research can focus in evaluating the disaster recovery networks from collaborative governance and integrated framework perspectives. Based on the evidence, policy makers might consider some revisions in the framework as it was the case for the National Response Framework in the US. The affiliation network (formal proposed relationships) included in the chapter can be compared with actual disaster recovery networks. It can also be determined whether this new framework needs to be complimented with additional important regulations and statutory support to ensure its implementation at local, tribal, and state levels.

Appendix: Leading Coordinating Agencies for Recovery Support Functions (RSF)

Advisory Council on Historic Preservation (ACHP)
American Red Cross (ARC)
Council on Environmental Quality (CEQ)
Corporation for National and Community Service (CNCS)
Office for Civil Rights and Civil Liberties (CRCL)
Department of Homeland Security (DHS)
Department of Commerce (DOC)
Department of Defense (DOD)
Department of Energy (DOE)
Department of the Interior (DOI)
Department of Justice (DOJ)
Department of Labor (DOL)
Department of Transportation (DOT)
Department of Education (DOE)
Environmental Protection Agency (EPA)
Federal Communications Commission (FCC)
Federal Emergency Management Agency (FEMA)
General Services Administration (GSA)
Department of Housing and Urban Development (HUD)
Department of Health and Human Services (HHS)
Heritage Preservation (HP)
Institute of Museum and Library Services (IMLS)
Library of Congress (LOC)

(continued)

(continued)

National Endowment for the Arts (NEA)
National Endowment for the Humanities (NEH)
National Protection Programs Directorate (NPPD)
Nuclear Regulatory Commission (NRC)
National Voluntary Organizations Active in Disaster (NVOAD)
Small Business Administration (SBA)
Department of the Treasury (TREAS)
Tennessee Valley Authority (TVA)
United States Access Board (U.S. ACCESS BOARD)
U.S. Army Corps of Engineers (USACE)
U.S. Department of Agriculture (USDA)
Department of Veterans Affairs (VA)

References

Agranoff R (2006) Inside collaborative networks: ten lessons for public managers. Public Adm Rev 66(s1):56–65

Agranoff R (2007) Managing within networks: adding value to public organizations. Georgetown University Press, Washington

Agranoff R, McGuire M (2003) Collaborative public management: new strategies for local governments. Georgetown University Press, Washington

Alesch DJ, Arendt LA, Holly JN (2009) Managing for long-term community recovery in the afternmath of disaster. PERI, Fairfax

Ansell C, Gash A (2007) Collaborative governance in theory and practice. J Public Adm Res Theory 18(4):543–571

Bardach E (1998) Getting agencies to work together: the practice and theory of managerial craftsmanship. Brookings, Washington

Birkland TA (2007) Lessons of disaster: policy change after catostrophic events. Georgetown University Press, Washington

Borgatti SP, Everett MG, Freeman LC (2002) UCINET for windows: software for social network analysis. Analytic Technologies, Harvard

Brown KA, Keast RL (2003) Citizen-government engagement: community connections through networked arrangements. Asian J Public Adm 25(1):107–132

Bryson JM, Crosby BC, Stone MM (2006) The design and implementation of cross-sector collaborations: propositions from the literature. Public Adm Rev 66(1):44–55

Carter T (2011) National disaster recovery framework: strengthening disaster recovery for the nation [powerpoint slides]. NCEMA fall all-hazards, Hickory, 23–28 October 2011. https://ncema.renci.org/FallConf2011/presentations/1/NDRF_.pptx

Cigler B (1999) Pre-conditions for the emergence of multi-community collaborative organizations. Policy Stud Rev 16(1):86–102

Comfort LK, Birkland TA, Cigler BA, Nance E (2010) Retrospectives and prospective on Hurricane Katrina: five years and counting. Public Adm Rev 70(5):669–678

Derthick M (2007) Where federalism didn't fail. Public Adm Rev 67(s1):36–47

Dietch EA, Corey CM (2011) Predicting long-term business recovery four years after Hurricane Katrina. Manag Res Rev 34(3):311–324

Drabek TE, McEntire DA (2002) Emergent phenomena and multiorganizational coordination in disasters: lessons from the research literature. Int J Mass Emergencies Disasters 20(2): 197–224

Edwards FL (2013) All hazards, whole community: creating resiliency. In: Kapucu N, Hawkins C, Rivera F (eds) Disaster resiliency: interdisciplinary perspectives. Routledge, New York, pp 21–47

Emerson K, Nabatchi T, Balogh S (2012) An integrative framework for collaborative governance. J Public Adm Res Theory 22(1):1–29

FEMA (2010) National disaster recovery framework. http://www.fema.gov/national-disaster-recovery-framework Accessed 2 August 2012

FEMA (2011a) A whole community approach to emergency management: principles, themes, and pathways for action, Washington. http://www.fema.gov/library/viewRecord.do?id=4941

FEMA (2011b) National disaster recovery framework [powerpoint slides]. http://www.emd.wa.gov/training/documents/NDRF_FEMA.pdf

Jordan AE (2010) Collaborative relationships resulting from the urban area security initiative. J Homeland Security and Emergency Manag 7(1):1–19

Kapucu N (2006) Interagency communication networks during emergencies: boundary spanners in multi-agency coordination. The Am Rev Public Adm 36(2):207–225

Kapucu N (2008) Collaborative emergency management: better community organizing, better public preparedness and response. Disasters 32(2):239–262

Kapucu N (2009) Interorganizational coordination in complex environments of disasters: the evolution of intergovernmental disaster response systems. J Homeland Secur Emergency Manag 6(1):1–26

Kapucu N (2012a) Disaster resilience and adaptive capacity in Central Florida, US, and in Eastern Marmara Region, Turkey. J Comp Policy Anal 14(3):202–216

Kapucu N (2012b) The network governance in response to acts of terrorism: comparative analyses. Routledge, New York

Kapucu N, Garayev V (2012) Designing, managing, and sustaining functionally collaborative emergency management networks. The Am Rev Public Adm 20(10):1–19

Kapucu N, Ozerdem A (2013) Managing emergencies and crises. Jones & Bartlett, Boston

Kapucu N, Augustin ME, Garayev V (2009) Interstate partnerships in emergency management: Emergency Management Assistance Compact (EMAC) in response to catastrophic disasters. Public Adm Rev 69(2):297–313

Kapucu N, Arslan T, Demiroz F (2010a) Collaborative emergency management and national emergency management network. Disaster Prev Manag 19(4):452–468

Kapucu N, Yuldhasev F, Bakiev E (2010b) Collaborative public management and collaborative governance: conceptual similarities and differences. Eur J Econ Political Sci 2(1):39–60

McEntire DA (2006) Disaster response and recovery: strategies and tactics for resilience. Wiley, New York

Moynihan D (2005) Leveraging collaborative networks in infrequent emergency situations . IBM Center for the Business of Government, Washington

National Governors Association (NGA) (2007) A governor's guide to homeland security. http://www.chelandem.org/documents/GGHS/GGHS.pdf

Phillips BD (2009) Disaster recovery. CRC, Boca Roton

Rhodes R (1996) The new governance: governing without government. Political Stud 44(4): 652–667

Rubin CB (2009) Long term recovery from disasters – the neglected component of emergency management. J Homeland Secur Emergency Manag 6(1):1–15

Stoker G (1998) Governance as theory: five propositions. Int Soc Sci J 50(155):17–28

The National Academies (2012) Disaster resilience: a national imperative. National Academy Press, Washington

US Government Accountability Office (GAO) (2012) Disaster recovery: selected themes for effective long-term recovery. Testimony before the subcommittee on emergency preparedness, response, and communications. Committee on Homeland Security, House of Representatives. GAO –12-813T, USGAO, Washington

Wallis J, Gregory R (2009) Leadership, accountability and public value: resolving a problem in "new governance"? Int J Public Adm 32(3):250–273

Ward R, Wamsley G (2007) From a painful past to an uncertain future. In: Rubin CB (ed) Emergency management: the American experience 1900-2005. PERI, Fairfax, pp 207–242

Waugh WL, Smith RB (2006) Economic development and reconstruction on the Gulf after Katrina. Econ Dev Q 20(3):211–218

Waugh WL, Streib G (2006) Collaboration and leadership for effective emergency management. Public Adm Rev 66(s1):131–140

Chapter 4
Typhoon Morakot and Institutional Changes in Taiwan

Jet-Chau Wen, Shao-Yang Huang, Chi-Feng Lin, Chia-Chen Hsu, and Wen-Ni Chen

Abstract Typhoon Morakot hit Taiwan in 2009 with numerous rainfalls and caused nearly 700 deaths from mudslides and huge agricultural losses that have been estimated to be reaching NT$14.59 billion. For reconstructing the damaged buildings, traffic and houses and avoiding the same problems happen again, some institutional changes were made in post-Morakot period. The major changes include amending law and establishing new organizations to reduce problems like unclear responsibilities because of the overlapping affairs. Briefly speaking, the major changes are replacing "National Fire Agency" (NFA) by "National Disaster Prevention and Protection Agency" (NDPPA) as a professional institution in dealing with future disasters and gives emphasize on the role of local governments, establishing "Special Reconstruction Regulation for typhoon Morakot" to supply more finical support for houses or industries reconstruction, amending "Disaster Prevention and Protection Act" to make the whole disaster managing system more consistently, combining the "Central Disaster Prevention and Response Council" and "National Disaster Prevention and Response Committee" to reduce the problem of affairs overlapping, and establishing "Office of Disaster Management" to supervise central and local governments to execute disaster prevention and response plans.

Keywords Disaster Prevention and Protection Act • Morakot • Taiwan

J.-C. Wen (✉) • S.-Y. Huang • C. F. Lin • C.-C. Hsu • W.-N. Chen
National Yunlin University of Science and Technology, Yunlin County, Taiwan
e-mail: wenjc@yuntech.edu.tw

R. Shaw (ed.), *Disaster Recovery: Used or Misused Development Opportunity*,
Disaster Risk Reduction, DOI 10.1007/978-4-431-54255-1_4, © Springer Japan 2014

4.1 Preface

Owing to the rapid global climate change and special geographic environment makes Taiwan encountering different disasters in past years. In past decades, there are many disasters which brought serious damages to Taiwan's society, such as the 921 earthquake, Typhoon Xangsane, Nari, and Morak etc. The issues of disaster management then become the public and media focus.

According to the study result (Natural Disaster Hotspots: A Global Risk Analysis) that provided by World Bank, Columbia University and the members of ProVention Consortium, it identifies that Taiwan may be the place on Earth most vulnerable to natural hazards, with 73 % of its land and population exposed to three or more hazards. In order to deal with disasters, the Taiwan government flung into disaster prevention institution development as the guideline for disaster prevention management.

The chapter compiled information on disaster included "921 Taiwan Chi-Chi Earthquake, 1999," "Morakot Typhoon, 2009," "Tsunami in South Asian, 2004," "Pakistan Flood, 2010," "New Zealand Earthquake, 2011," "Tohoku Earthquake and Tsunami, 2011," "Thailand Flood, 2012," and generalize the trend of global disasters from a collection of disaster instances:

- The effects of nature disasters unlike the previous that humans have met before.
- The loss and damage of nature disaster much higher.
- There are more and more complex disasters.

Furthermore, the chapter also compiled information on disaster prevention governance and institutional from England and America. The statement is as follows.

4.1.1 England

In order to deal with disasters, the England government established Civil Contingency Secretariat (CCS) to promote teamwork and cooperation with non-governmental organization and government. In addition, the government completed Civil Contingence Act and build-up and development UK Resilience to strengthen the ability of disaster management.

4.1.2 America

In order to deal with huge disasters, the America government established Federal Emergency Management Agency (FEMA) in charge of programming disaster reduction, disaster preparedness countermeasures, disaster response and recovery strategies. It combined several independent departments that relate to disaster prevention to achieve objective of centralize command and program for disaster management.

In order to deal with disaster, the government program and execute the mission of disaster prevention by National Fire Agency, Ministry of the Interior in Taiwan.

When mass casualty incidents happened, the central administrative Command Post (ACP) plays the role as a coordinator who requested resource assistance directly to the Central Emergency Operation Center (CEOC). Local government's Emergency Operation Center (EOC) is the mainly in charge of providing all sufficient rescue resources and dispatching relief manpower supplies for the local Incident Command Post (ICP).

On purpose to acquaint readers with the disaster prevention governance and institutional changes in Taiwan, the chapter passed an opinion by adopting literature survey with Morakot Typhoon as below.

4.1.3 Case Introduction

Typhoon Morakot, a devastating tropical cyclone that caused critical landfall in Taiwan during August 7–9, 2009, had reached highest recorded rainfall in southern Taiwan in the past 50 years. The path of Morakot Typhoon is plotted as shown in Fig. 4.1 and the radar echo during 00:00 7-August, 2009 to 00:00 10-August, 2009 as shown in Fig. 4.2.

August 2, 2009, Japan Meteorological Agency (JMA) reported that a tropical depression formed and about 1,000 km far from eastern Philippines. However, the depression remained weak, and was an area of convection with deep convection flaring on the western of a partially exposed low level circulation center. Later on August 4, Central Weather Bureau of Taiwan reported that the tropical depression named Morakot located at eastern Philippines turned into a tropical storm and would likely invade southern Taiwan in a nearly future. The progresses of Morakot in Taiwan are listed as below:

1. August 5, 2009 20:30: Central Weather Bureau of Taiwan issued a Sea Typhoon Alert for Morakot.
2. August 6, 2009: Severe Tropical Storm Morakot intensified into a typhoon, closed to Taiwan but moved very slowly, and embedded in a large-scale convection region with monsoon circulation.
3. August 7, 2009: Morakot caused landfall in Hualian near midnight.
4. August 8, 2009: After midnight, most of the districts in southern Taiwan recorded heavy rainfall.
5. August 9, 2009: News indicated that Xiaolin village, a mountain village with 1,300 residents in Jiaxian Township, was hit by mudslide. The entire village was nearly razed to the ground by the mudslide and only two buildings still remained. Over 600 residents were buried alive.

4.1.3.1 Rainfall

Southern Taiwan was seriously hit by the typhoon with up to 3 m of rainfall during August 7 and 8. Typhoon Morakot invaded Taiwan with record-breaking rainfall

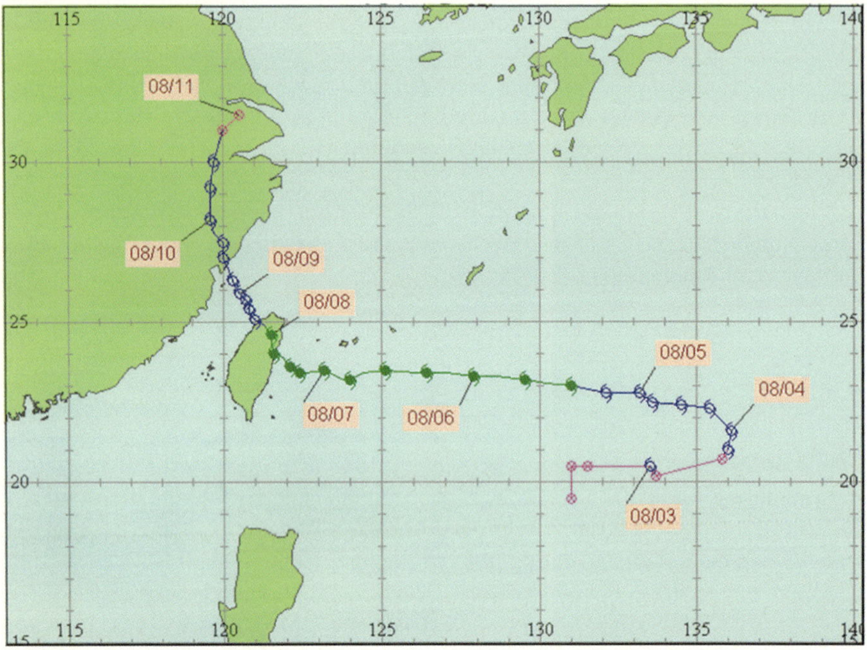

Fig. 4.1 The path of Morakot Typhoon (*Data source*: Central weather Bureau, Taiwan)

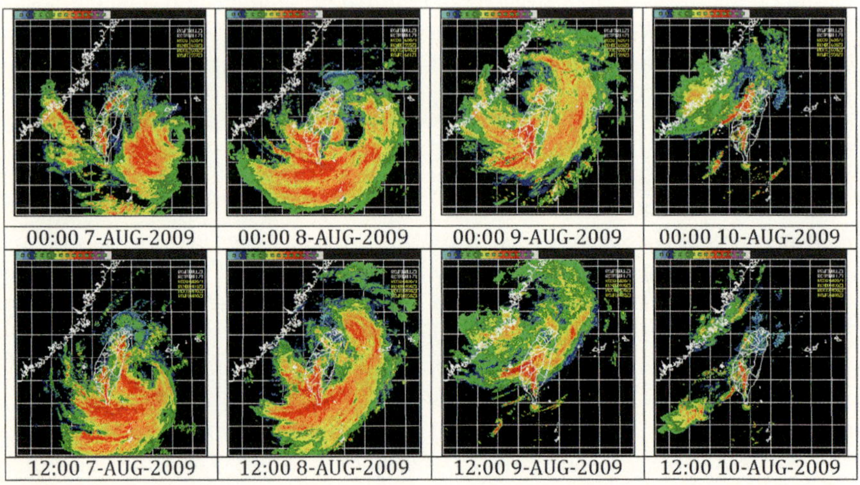

Fig. 4.2 The radar echo during 7–10-Aug (*Data source*: Central weather Bureau, Taiwan)

and strong winds which was estimated up to 13 (Beaufort scale), resulted in great loss of many lives and properties.

After Morakot landed in the midnight of August 8, almost the entire southern region of Taiwan (Chiayi County, Chiayi City, Tainan County, Tainan City, Kaohsiung County, Kaohsiung City, Pingtung County and parts of Taitung County and Nantou County) were flooded by record-breaking heavy rain. The rainfall in Pingtung County exceeded 2,600 mm (100 in), breaking all rainfall records of any single place in Taiwan induced by a single typhoon.

According to Water Resources Agency, Ministry of Economic Affairs in Taiwan, the precipitation from typhoon Morakot in these 2 days was exceeded half-yearly average in partly areas.

4.1.3.2 Impacts

During the devastated typhoon Morakot, a serious, heavy rainfall caused a deadliest mudslide in Xiaolin village, Jiasian Township of Kaohsiung County, and it resulted in nearly 700 death and thousands of people homeless.

News indicated that at least 600 people were missing in southern Taiwan. Most were residents of Xiaolin, a mountainous village with 1,300 residents in Jiasian Township. The village was destroyed by a massive mudslide and most parts were buried under a huge mudflows and landslides. It was also reported that all roads toward Namaxia Township were either blocked or washed away by severe mudslides. The deadly catastrophe lead to hundreds of residents trapped for 4 days. Survivors almost ran out of food and water. In addition, water and electricity were cut those days. It was estimated that more than 600 people died in the village due to sudden mudslide, and only two buildings remained standing in the village. A rescue helicopter, working to retrieve survivors of the mudslide crashed in early August 11, and killed three occupants. Crews were unable to reach the wreckage due to the steep terrain.

Meanwhile, around 1,000 people in Taoyuan township and 1,500 in Namasiya township were trapped, according to an emergency operation centre in Chiayi on August 14, there were still over 10,000 people trapped in the townships of Alishan, Meishan, Chuchi, Jhongpu and Fanlu. Taipei Times reported that a primary school teacher was stranded for 6 days in Jinfong township in the mountainous Taitung County, before sending a desperate email calling for help. The email explained that over 1,000 people were cut off by a massive mudslide and survivors were under starvation.

In general, typhoon Morakot caused nearly 700 deaths from mudslides and huge agricultural losses that have been estimated to be reaching NT$14.59 billion.

4.2 Institutional Changes on Disaster Prevention and Protection in Post-Morakot Period

There are five major institutional changes after typhoon Morakot happened. The differences between the two periods, pre- and post- Morakot periods are shown in Table 4.1 and the reasons and contents are explained in paragraphs below.

Table 4.1 Institutional changes after typhoon Morakot happened

Change	Period Pre-Morakot period	Post-Morakot period
Change on the government organization in charge of disaster management	Fire Agency	Disaster of Prevention and Protection Agency
Change on the act	Disaster Prevention and Protection Law	1. Disaster Prevention and Protection Law is amended 2. Special Reconstruction Regulations for Typhoon Morakot is announced
Change on the decision-making organization	Central Disaster Prevention and Response Council and National Disaster Prevention and Response Committee	Central Disaster Prevention and Response Council
Change on the staff member	None	Office of Disaster Management

4.2.1 Planned Establishing Disaster Prevention and Protection Agency

In Taiwan, National Fire Agency Ministry of Interior is the major organization for the objectives to complete a sound public safety and disaster prevention system. The missions are to improve emergency medical service, to promote professional fire prevention and fighting service, to enhance disaster rescue efficiency and to protect civilian life, property and safety. It was established on March 1, 1995 to be responsible for fire prevention, disaster rescue and emergency medical service. According to "The Organization Act of National Fire Agency, Ministry of the Interior" (2005.06.22 Amended), National Fire Agency (NFA) plans and carries out affairs of national fire prevention agency, disaster prevention and rescue; and is authorized to command and supervise national fire institutes to perform fire prevention and disaster prevention and rescue missions.

In the study of National Policy Foundation, executive institution of central disaster protection affairs should be "National Disaster Prevention and Protection Agency Ministry of Interior." But in the Organic Act of Ministry of Interior, only NFA is established for national fire administrational affairs. NFA also commands and supervises national fire institutes, and that is the reason why people misunderstand it only charges the fire affairs. Therefore the Disaster Prevention and Protection Agency (NDPPA) has to be oriented an executive institution, and Disaster Prevention and Protection Council and National Disaster Prevention and Response Committee have to be in charge of governmental rescue system coordination.

As the background of NFA still emphasized the stages of preparation and response and had not developed complete disaster prevention functions (Hsiung et al. 2010), the idea of replacing NFA by NDPPA was born after typhoon Morakot happened. In the press conference on August 18, 2010, President mentioned to establish NDPPA to substitute NFA as a professional institution in dealing with future disasters. Besides, local governments have to establish Disaster Prevention and Protection Bureaus to operate disaster prevention system effectively. Military also join disaster prevention activities as well. Before the press conference, Disaster Prevention and Protection Act (2000.7.19 announced) was amended on August 4, 2010. The amended content is including article 7 and 16. In amended article 7, it regulates "National Disaster Prevention and Protection Agency, Ministry of the Interior shall execute disaster prevention and protection business." And in amended article 16, "To deal with the response affairs of major disaster rescue, the National Disaster Prevention and Protection Agency of Ministry of the Interior shall establish the Special Search and Rescue Team and the Training center; while the municipal or county (city) government shall establish a search and rescue organization." is regulated. In these articles, the noun NDPPA is used rather than NFA. The variation also makes the Organic Act of Ministry of Interior has to be amended, and the amended draft is examined by The Legislative Yuan now (December, 2012).

In the meeting called for examining the personnel employment system after NFA substituted by NDPPA, it also concluded that in addition to the original functions of NFA, NDPPA should establish more systems in the future for raising the quantity of disaster prevention staffs, such as (1) to establish the department of mountain disaster prevention to deal with events happened in the mountain; (2) to evaluate the possibility of establishing the college of disaster prevention which belong to Central Police University to educate professional disaster prevention personnel and raise their quality.

But after NDPPA replaces NFA, some problems appear. The study of Research, Development and Evaluation Commission, Executive Yuan (2010) mentioned opinions below:

1. Although NDPPA is recharged in disaster prevention affairs, its functions overlap with Ministry of Economic Affairs, Council of Agriculture, Ministry of Transportation and Communications, and Environmental Protection Administration. Relations between each other need to be distinguished.
2. Probability of complex disasters increase year by year, so NDPPA should be responsible for communication and coordination with different institutions when a disaster happens. NDPPA is the third grade institution of Executive Yuan but it needs to lead other second grade institutions. This violates the designation of administrational organization levels and should be adjusted.
3. The central disaster prevention and protection affairs are all managed by NDPPA after Disaster Prevention and Protection Act was amended. But in professional opinion, two different fields of disaster and fire should be separated and managed by different institutions.

In this period, climate change makes more instantaneous and strong disasters happen. That is why even it is doubted the reasonability and necessity to establish NDPPA, the responsible disaster prevention and protection institution still need to exist.

4.2.2 Established Special Reconstruction Regulation for Typhoon Morakot

After typhoon Morakot hit Taiwan, it not only destroyed large parts of buildings in disaster areas but affected people's life. For the reconstructions of home country, facilities, industries, life and culture in disaster areas, "Special Reconstruction Regulations for Typhoon Morakot" was submitted by Executive Yuan on August 20, 2009. After consideration, Legislative Yuan passed the law on August 27, 2009 and declare by president on August 28, 2009.

The articles of this regulation are about reconstruction, including supplying finical support and loans for houses rebuilding; supplying low-price of health insurance for farmer and labor and bailing out the industries or enterprises which run into difficulty in business operation. In a word, there were little restrictions when implementing reconstruction of traffic, transportation and other public works.

Reconstruction should be oriented on humans and their life. Because large parts of disaster areas are aboriginal districts, the multiple cultural characters need to be respected and communities should participate in reconstruction works. Therefore reconstruction should respect people, community or tribe organization, culture and life style of disaster areas.

4.2.3 Amended the Disaster Prevention and Protection Law

After typhoon Morakot, Executive Yuan passes partly of Disaster Prevention and Protection Act on November 19, 2009 and passes by Legislative Yuan after Third Reading on July 1, 2010. Amended articles included:

- Enhance the Disaster Prevention and Protection duties of local governments, and added the local governments Autonomy of Disaster Prevention by rule.
- Integrate the "Central Disaster Prevention and Protection Council" and the "office of disaster management," fortifying disaster prevention and protection policies and measurements. National Disaster Prevention and Protection Agency, Ministry of the Interior shall execute disaster prevention and protection business.
- Amend the articles of 9 and 10: The municipal, or county (city) office of disaster management shall deal with the affairs of the municipal, or county (city) disaster prevention and protection council. The organization is enacted by the municipal, or county (city) government.
- Add the articles of 23: To insure the stream transmission of microwave communication for succeeding the responsive measures of disaster rescue, an efficient wireless transmission distance range shall be enacted and publicized duly by the Ministry of the Interior.
- Add the articles of 34: The Ministry of National Defense shall mobilize reserved servicemen's organizations to support responsive measures of major disaster

rescue as provided in the preceding paragraph. The procedures for requesting rescue support or active support for responsive measures of disaster rescue from national army, arrangement and dispatch of national defense force, commanding and coordination, negotiation and communication, education and training of disaster rescue affairs, the schedule of duty operation and other relevant operation regulations as prescribed in the fourth paragraph therein shall be duly enacted jointly by the Ministry of National Defense and the Ministry of the Interior.

4.2.4 Combined the Central Disaster Prevention and Response Council and National Disaster Prevention and Response Committee

In pre-Morakot period, there were two central disaster prevention and protection institutions which were in charge with decision making, and these are Central Disaster Prevention and Response Council and National Disaster Prevention and Response Committee. Their former functions were below:

4.2.4.1 Central Disaster Prevention and Response Council

(a) Decide basic principles of disaster prevention and response.
(b) Approve the disaster prevention and response basic plan and operation plan.
(c) Approve important disaster prevention and response policy and measure.
(d) Approve the nationwide emergency disaster contingency measures.
(e) Supervise and evaluate the central, municipal and county (city) disaster prevention and response related matters.
(f) Other matters regulated by law.

4.2.4.2 National Disaster Prevention and Response Committee

(a) Execute disaster prevention and response policies approved by Central Disaster Prevention and Response Council; promotes major disaster prevention and response tasks and measures.
(b) Plan the basic principle of the disaster prevention and response.
(c) Draw up of disaster prevention and response basic plans.
(d) Review disaster prevention and response operation plans.
(e) Coordinate unsolved contradicted matters in disaster prevention and response operation plans against regional disaster prevention and response basic plans.
(f) Coordinate with financial institutions about reconstruction fund.
(g) Supervise, evaluate, and coordinate disaster prevention and response measures of all levels of government.
(h) Other matters regulated by law.

4.2.4.3 The Reason to Combine the Two Organizations

The Central Disaster Prevention and Response Committee executive the policy of disaster prevention, but in the practically, the committee didn't stem from law. It only was a guideline. The committee assemble was by mission orients, there are so difficult to dedicate that promote disaster prevention policy.

Due to some functions of Central Disaster Prevention and Response Council and National Disaster Prevention and Response Committee overlapped, when a large scale disaster happened, those institutions couldn't integrate effectively. So it made the rescue process too complicated.

Ministry of the Interior mentioned typhoon Morakot destroyed central and southern Taiwan and revealed the severity of climate change. It showed that the challenge from natural disaster is more and more strict. The points to amend Disaster Prevention and Protection Act were how central and local governments coordinate and cooperate when facing large scale and complex disasters and how instantly and actively military support rescue work.

To solve the problems of functions overlapping, "Central Disaster Prevention and Response Council" and "National Disaster Prevention and Response Committee" were combined as Central Disaster Prevention and Response Council. For raising professional functions, the Office of Disaster Management was also established with professional personnel in Executive Yuan. It was predicted to promote the administration efficiency of disaster prevention and protection of central organizations. As Fire Agency is replaced by Disaster Prevention and Protection Agency, it will form a whole disaster prevention system. The system includes Central Disaster Prevention and Response Council to make decisions, Office of Disaster Management to supply professional advices, and Disaster Prevention and Protection Agency to execute affairs.

4.2.4.4 Functions After Combining

In accordance with the amended guidelines to set up the "Central Disaster Prevention and Response Council," the Council's missions maintained like before.

After two institutions combined, Central Disaster Prevention and Response Council held a meeting from one time per year to one time per 3 months. Including Disaster Prevention and Protection Agency, Professional Advisory Committee of disaster reduction, National Science and Technology Center for Disaster Reduction and National Search and Rescue Command Center should attend the meeting. If necessary, the Council can invite representatives of local governments, experts or scholars to attend the meeting.

4.2.5 *Established the Office of Disaster Management*

4.2.5.1 The Purpose to Establish the Office

Taiwan is located at western Pacific and Circum-Pacific Seismic Zone, which is frequently affected by typhoons. It is suffered from an average of three to four typhoons every year in this century and also has a threat of earthquake. Life and property caused by disasters grew due to the rapid growth of economic and changes of society. For this reason, improving disaster prevention and response mechanism becomes a vast goal for the government. However, it needs some more innovative thoughts to face the trend of diversified disaster with complexity. As a result, Executive Yuan of Taiwan established "Office of Disaster Management" after the disaster from typhoon Morakot.

4.2.5.2 Members of the Office

The Office of Disaster Management is under Executive Yuan of Taiwan. It has a chief who directs and supervises all affairs of the office, and has a vice chief. The office deals with issues such as typhoon, flood, landslide, earthquake, toxic chemicals, and serious explosion etc, therefore all staffs need to be transferred from other professional organizations such as fire control, country defense, information, geography, sanitation, civil engineering, water conservancy, conservation of water and soil, psychology, society, and economic.

4.2.5.3 Functions of the Office

The office deals with affairs related Central Disaster Prevention and Response Council and Central Disaster Prevention and Response Committee and those like below:

1. Draw up disaster prevention and response policy; promotes major disaster prevention and response tasks and measures.
2. Supervise each grade of governments to execute disaster prevention and response decided by Central Disaster Prevention and Response Council and Central Disaster Prevention and Response Committee.
3. Draw up disaster prevention and response basic principle and basic plan.
4. Examine elementarily disaster prevention and response operation plans and local disaster prevention and response plans.
5. Suggest laws related to disaster prevention and response to amend.
6. Assist to supervise the disaster warning, monitoring and announcing systems.
7. Assist to supervise the disaster preparing, educating, training and announcing.

8. Plan the emergency response system.
9. Assist to supervise post-disaster investigation and recovery.
10. Other matters about disaster prevention and response policy-drawing up and affair-supervising.

4.3 Conclusions

After Disaster Prevention and Protection Act amended, Office of Disaster Management presented the adjustment content of central disaster prevention and protection organizations. Article 7 of the amended act regulates that the central disaster prevention and protection organizations include (1) Central Disaster Prevention and Response Council, (2) National Disaster Prevention and Response Committee, (3) Office of Disaster Management, (4) Professional Advisory Committee of disaster reduction, (5) National Science and Technology Center for Disaster Reduction, (6) National Search and Rescue Command Center and (7) National Disaster Prevention and Protection Agency Ministry of Interior.

Each regulation to adjust central disaster prevention and protection organizations is announced on December 2, 2010 except Amended draft of Organic Act of Ministry of Interior is examined by The Legislative Yuan now. The central disaster prevention and protection system figures before and after Disaster Prevention and Protection Act amended are shown in Figs. 4.3 and 4.4. The major system changes in pre- and post-Morakot period are described below:

1. The National Disaster Prevention and Response Committee works in Fig. 4.3, but ceases working in Fig. 4.4. The original functions are executed by Central Disaster Prevention and Response Committee.
2. Professional Advisory Committee of disaster reduction, National Science and Technology Center for Disaster Reduction, National Search and Rescue Command Center and National Disaster Prevention and Protection Agency Ministry of Interior are not requested to attend the Central Disaster Prevention and Response Council in Fig. 4.3 but do in Fig. 4.4.
3. There is not a staff member in the original system to supervise the Council and Committee in Fig. 4.3. But in Fig. 4.4, Office of Disaster Management plays the role.
4. National Search and Rescue Command Center belongs to Execute Yuan in Fig. 4.3, but belongs to Central Disaster Prevention and Response Committee in Fig. 4.4 after Disaster Prevention and Protection Act was amended.

After typhoon Morakot happened, not only some acts amended but also the system changed. The key point is to reduce many affairs overlapping in different institutions. By strengthening power of regulating authorities and avoiding unnecessary coordination, major organizations of disaster prevention and protection have more energy to reduce, prepare, response and reconstruct when disaster comes. In this way, when facing future disasters, reconstruction works in Taiwan can be relied on faster help form governments and local groups such as religion groups to shorten reconstruction periods.

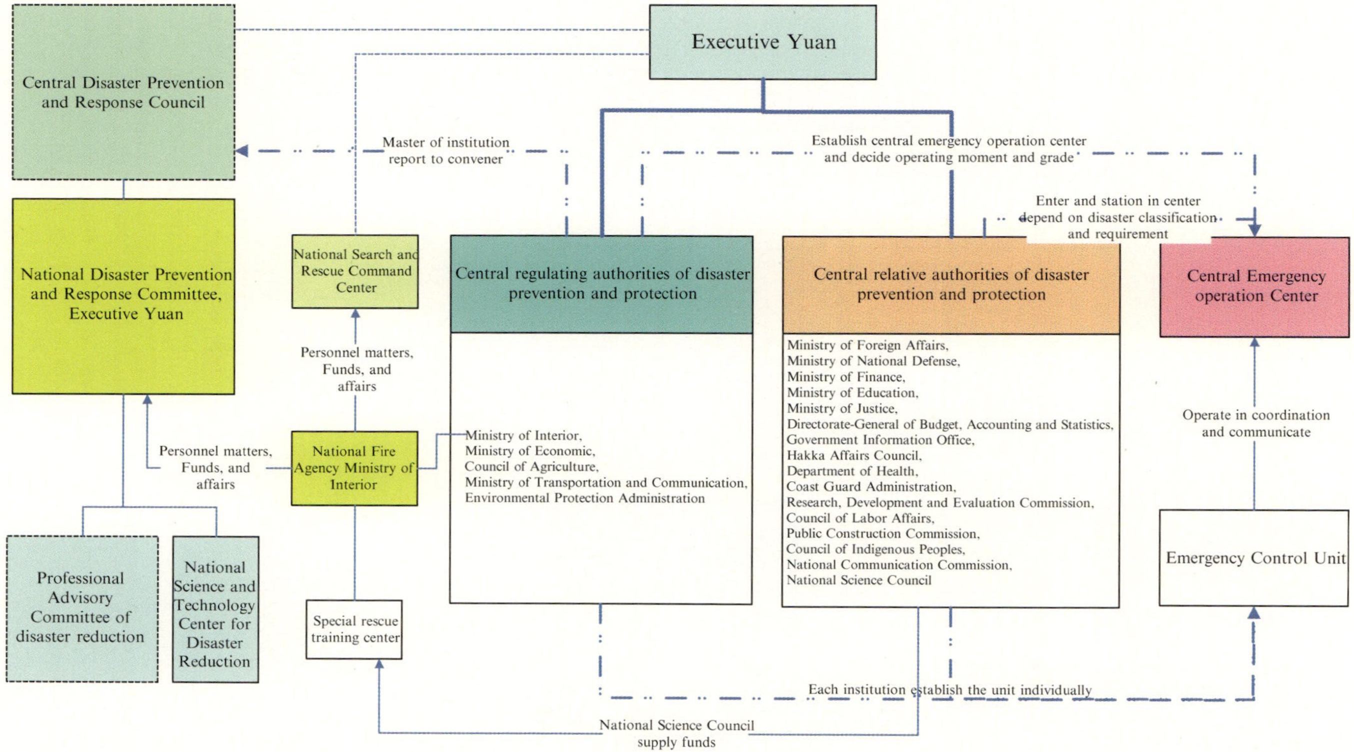

Fig. 4.3 The central disaster prevention and protection system figures before Disaster Prevention and Protection Act amended

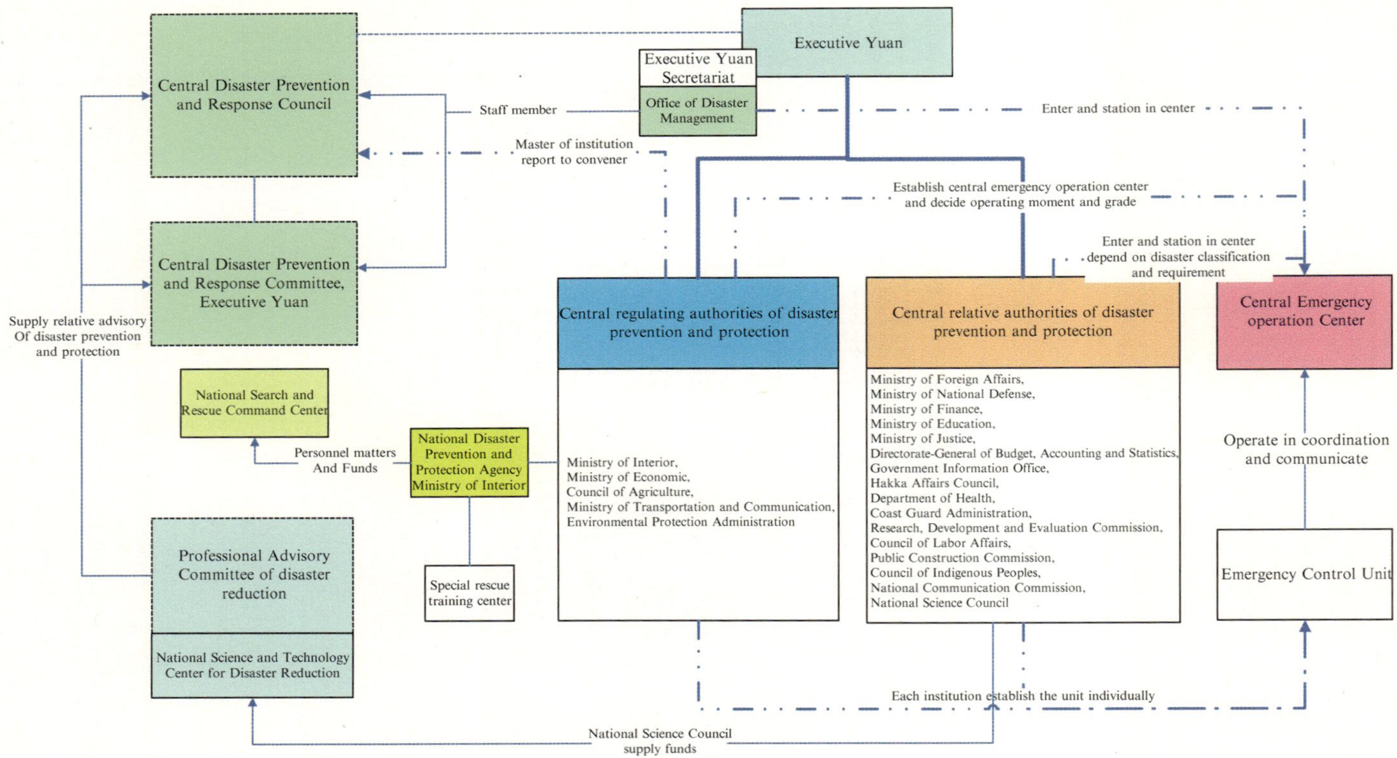

Fig. 4.4 The central disaster prevention and protection system figures after Disaster Prevention and Protection Act amended

References

Hsiung KH, Wu SSG, Yeh CH (2010) "The analysis of Taiwan disaster prevention and protection system variation". Public administration, disaster prevention and crisis management thesis, Taipei, pp 1–20

Introduction to Office of Disaster Management (in Chinese). http://www.cdprc.ey.gov.tw/cp.aspx?n=BA69C95CF86CC306

Office of Disaster Management (2010) Adjustment of central disaster prevention and protection organizations (in Chinese). http://archives.ey.gov.tw/Upload/WebArchive/01ey/20120121/www.ey.gov.tw/public/Attachment/11595943 71.pdf

Research, Development and Evaluation Commission, Executive Yuan (2010) The policy illustration of organization and work of central and local disaster prevention and protection. Research, Development and Evaluation Commission, Executive Yuan, Taipei (in Chinese)

World Bank (2005) Natural disaster hotspots: a global risk analysis. http://www.earth.columbia.edu/news/2005/story03-29-05.html. Accessed 17 November 2012

Chapter 5
Institution and Governance Related Learning from the East Japan Earthquake and Tsunami

Mikio Ishiwatari

Abstract The developing countries, which have weak governance and institution, face various difficulties for quick recovery in the aftermath of megadisasters. They have weak disaster management agencies, limited skilled and experienced staff and weak finance resources. Local governments are greatly overstretched because of a wide range of response and recovery activities. It takes some time for international development agencies to respond to disasters. This study aims to propose the practical institutional and governance mechanisms of quick recovery for developing countries and development agencies by examining practices in the East Japan Earthquake and Tsunami in 2011. The study observes that institutional and legislative features contributed to prompt recovery. For example, the governments opened roads leading to cities on devastated coasts in less than a week. Quick recovery can be attributed to robust measures prepared at normal times, which Japan have improved by learning lessons from past disasters including the Great Hanshin-Awaji Earthquake in 1995. The national government established the advance financial arrangements of subsidizing rehabilitation of public infrastructures to the local governments. Government offices have arranged prior agreements with the private sector to immediately commence recovery works. The government agencies organized to mobilize expert teams through their national networks. Structural measures of reinforcing bridges and other structures decreased recovery workloads and financial burdens. Also, the study argues challenges faced. The financial mechanizes focus on public facilities, and unfairly covers facilities owned by the private sector, such as railway, school and electricity. Local construction companies, which play crucial roles in quick recovery, are weakening capacities to respond to disasters because of budget decrease of public works.

M. Ishiwatari (✉)
Japan International Cooperation Agency, Tokyo, Japan
e-mail: ishiwatari.mikio@jica.go.jp

R. Shaw (ed.), *Disaster Recovery: Used or Misused Development Opportunity*, Disaster Risk Reduction, DOI 10.1007/978-4-431-54255-1_5, © Springer Japan 2014

Keywords East Japan earthquake and tsunami • Megadisaster • Public–private partnership • Recovery • Risk governance • Structural measures

5.1 Introduction

Rehabilitation of infrastructures is needed to quick recovery of not only economic activities but also people's daily lives (Kawata 2008). Transportation services must resume to sending search and rescuing teams, to deliver relief goods to people affected, and to start rehabilitation works. Governance and institutions, which establish collaborative actions of governments and other stakeholders, are required for quick recovery.

Enormous confusion is inevitable in the field in the immediately aftermath of megadisasters in developing countries. These countries often face difficulties in starting recovery activities, since they have weak governance and institutions, such as shortages of dedicated agencies, limited skilled and experienced staff, and weak finance resources. Also, development agencies cannot timely respond to disasters with the ordinal processes of project approval and procurement.

Activities following the East Japan Earthquake and Tsunami (EJET) in 2011 are regarded as practices of quick recovery. Japan has established institution and legislation by learning from a serious of disasters in the 1940s and 1950s and the Hanshin-Awaji (Kobe) Earthquake in 1995. The developing countries and development agencies can adapt these practices to improve their recovery activities. This study aims to propose practical institutional and governance mechanisms of quick recovery for developing countries and development agencies by examining practices in the EJET.

5.2 Issues of Disaster Recovery Governance

Governance is the overall process of deciding and implementing recovery actions by governments, organizations and populations. The governments in developing countries are forced to take hasty and reactive decisions that sacrifice relevance and sustainability, or careful but delayed decisions that sacrifices the need for urgency (IRP and UNDP 2010). This is because many developing countries do not have enough governance and institution in disaster recovery, such as specialized organizations, experienced staff, legislations, and public–private partnerships. Local governments, which are expected to play leading and coordinating roles in response and rehabilitation in the field, are greatly overstretched in post disaster situations because of a wide range of activities. Following the Indian Ocean Tsunami (IOT) in 2004, a government agency faced difficulties in supervising and managing the projects of the Asian Development Bank (ADB) at the initial stage because of limited management skills (ADB 2012). In housing reconstruction in Haiti, Davis (2012) argues that the government does not have enough leadership, strategy, policy and coordination.

Developing countries usually have limited financial capacity of recovery after a disaster (World Bank 2012c). It took three months for the Government of Lao PDR to approve a recovery fund following Typhoon Ketsana in 2009, and nearly half-year for the Indonesian Government to approve it following the IOT (World Bank 2012a). Financial protection requires a suitable legal and administrative framework that ensures to efficiently use resources in the aftermath of a disaster to mitigate the impact of natural disasters on the affected population and restore the economy of the country. A conservative fiscal policy, such as own resources and post-disaster debt instruments, contingent lines of credit available through multilateral development banks, remains the most efficient way to manage financial risk resulting from natural disasters (Ghesquiere and Mahul 2010).

Private sectors can play an important role in recovery process. In the U.S. public–private partnerships have been involved in disaster management from the late 1990s (Mitchell 2006). The Independent Evaluation Group (2011) points out through reviewing projects of the World Bank that streamlined procedures for contracting civil works and leveraging existing private sector capacity can help to avoid delays. The existing partners do not need screening for reputational risks and delivery capacity. The World Bank has applied flexible procurement measures, such as "slice and package approach". On the other hand, Gotham and Greenberg (2008) argues that market-centered approaches for recovery in New York City since 9/11 and New Orleans since the Hurricane Katrina disaster are not coherent or sustainable responses.

Communities and non-government organizations play a crucial role in recovery. Communities with strong social capital could quickly rehabilitate their towns following the Kobe earthquake in 1995 (Nakagawa and Shaw 2004). Ishiwatari (2010) reviews Japanese assistance following the IOT and stresses the importance of close collaboration between the non-government organizations and experts.

It is difficult for the development agencies to promptly response to disasters. These agencies caught up as they are in the byzantine red-tape of coordination mechanisms, conferences, planning, studies off all kinds, and contract administration, and find it hard to be nimble (Jayasuriya and McCawley 2010). The development agencies have initiated various efforts to improve the situation. ADB introduced innovation in the administration of rehabilitation grants for the IOT by delegating authority to the field mission in Banda Aceh, Indonesia, from the Headquarters in Manila, the Philippines, to promptly respond to needs on the ground (ADB 2012). The World Bank provides Development Policy Loan with Catastrophe Deferred Drawn Option, which offers a source of immediate liquidity of bridge financing after a disaster. The United Nations established the Central Emergency Response Fund to provide affected countries with more timely, reliable equitable and coordinated assistance (World Bank 2012a). ADB created the Asia Pacific Disaster Response Fund, which provides up to USD 3 million quick disbursing grant resources.

Rehabilitation of the West Coast Highway in the Sumatra Island, Indonesia following the IOT shows the importance of recovery mechanisms. The highway connecting towns on the west coast of the island was seriously damaged by the tsunami in 2004. The southern half of the highway of 122 km between Calang and Meulaboh was rehabilitated within two years by Japanese assistance. A Japanese government owned agency, which made a procurement agent contract with the Indonesian Government, promoted the quick processes of procurement, design, and construction of the project. On the other hand, rehabilitation of northern half of 150 km between Calang and Banda Aceh completed seven years after the tsunami. The project was conducted by ordinary procurement and project administration processes. Sole road design needed one and half year before starting construction works.

Japan has developed a wide range of the disaster management mechanisms based on the experience of severe disasters in the 1940s and 1950s. These mechanisms have functioned well, in particular, in rehabilitating major public infrastructures. However, support schemes to affected people's recovery, which was developed following the Kobe Earthquake in 1995, lagged behind support to the public infrastructures. The overview of these schemes is as follows:

1. Emergency response, such as providing relief goods, managing evacuation shelters and constructing transition shelters, is conducted according to the Disaster Relief Act, which Ministry of Health, Labor and Welfare manages.
2. In rehabilitation, public infrastructures, agricultural facilities, and public schools are rehabilitated according to three acts managed by Ministry of Land, Infrastructure, Tourism and Transport (MLIT), Ministry of Agriculture, Forestry and Fisheries (MAFF), and Ministry of Education, Culture, Sports, Science and Technology.[1] In case of extreme severe disasters, supports to the local governments under these measures are strengthened according to another act.[2] To support people affected, the governments provide direct assistance based on Act on Support for Reconstructing Livelihoods of Disaster Victims. This scheme was established in 1999 by learning lessons from the Kobe Earthquake, in which limited assistance was provided to the affected people to support rehabilitation of daily lives.

5.3 Rehabilitation Following EJET

Infrastructures, which were severely damaged in the EJET, were promptly rehabilitated. Damages to public utilities and social infrastructure were estimated to be about ¥1.3 trillion ($16 billion) and ¥2.2 trillion ($27 billion), respectively

[1] Act on National Treasury's sharing of expenses for project to recover public civil engineering works damaged by disaster, act on temporary measures for subsidies from National Treasury for expenses for project to recover facilities for agriculture, forestry and fisheries damaged by disaster, and act on National Treasury's sharing of expenses for recovery of public school facilities damaged by disaster.

[2] Act on special financial support to deal with the designated disaster of extreme severity.

(World Bank 2012b). Roads leading to cities on devastated coasts were opened in less than a week. Reopening roads was crucial in creating rescue routes, sending search and rescue teams, providing relief materials and starting reconstruction. Bullet train service, which connects Tokyo to Sendai within two hours, resumed within 49 days. The rehabilitation mechanisms have been improved by learning lessons from past disasters. In the Kobe Earthquake in 1995 bridges of roads and railways were severely damaged and collapsed. It took over one-and-a-half years for expressway reconstruction and 82 days for the bullet train line to be repaired.

National and local governments started rehabilitation works on the day of the earthquakes and tsunamis of the EJET. National highway 4 and the Tohoku Expressway, which are running north-south inland routes, were selected as the highest priority routes to connect to areas affected, and opened on March 12, the next day of the disaster. Fifteen east-west roads leading to coastal towns and cities devastated by the tsunamis were cleared within four days after the disaster. Finally, the coastal road running through devastated coasts, national highway 45, was opened in a week (Expert Panel on Expressway 2011; Tokuyama 2012). Because of the shape of the road networks, these works are called as "operation comb".

The Tohoku Regional Bureau, MLIT, pointed out that the following three factors contributed to prompt reopening roads:

1. The implementation of reinforcement measures before the earthquake prevented crucial damage to bridges.
2. The MLIT intensively mobilized its resources on 16 routes carefully selected in "operation comb."
3. Local construction companies immediately provided support based on pre-agreements.

Based on lessons from the EJET, a technical committee of the Central Disaster Management Council recommends as follows (Committee on Promoting Disaster Management of Central Disaster Management Council 2012):

1. To review design codes of facilities and to prepare manuals for rehabilitation
2. To formulate rehabilitation plans at normal times
3. To strengthen corporation among national and local governments, and to promote pre-agreement with private companies

Sagara and Ishiwatari (2012) recommend based on the experience that developing countries should develop the following measures:

1. To establish financial arrangement mechanisms
2. To arrange pre-disaster agreements with the private sector
3. To arrange expert teams
4. To construct disaster-resilient infrastructure
5. To prioritize key infrastructures to be rehabilitated

By summarizing these findings, quick recovery in the EJET can be attributed to the following measures prepared at normal times: (1) advance financial arrangements, (2) public–private partnership, (3) mobilization of expert teams through national networks, and (4) structural measures. Each area is discussed in the following sections.

5.3.1 Advance Financial Arrangement

In the EJET, the local governments do not share the rehabilitation costs of pubic infrastructures in practice. The national government provides the local governments with 90 % at most of the costs as subsidy. Also, the national government supplements the costs covered by the local governments as general subsidies. This 90 % of subsidy is higher than 80 % at the Kobe Earthquake in 1995. This is because the local governments in the affected areas of the EJET have weaker financial capacities than the Kobe city government.

The rehabilitation mechanisms of public infrastructure were established in the early 1950's. Japan had repeatedly suffered from typhoons and earthquakes in the 1940s and 1950s following the Second World War. During the war, Japan could not invest disaster risk management (DRM) projects. Almost every year, thousands of lives were lost, and assets and properties were seriously damaged. The rehabilitation of the public facilities imposed a considerable financial burden for governments. The national government allocated half of the budget of public works for disaster rehabilitation in 1950 (Handa 2011).

During the military occupation of Japan by the U.S., the Shoup mission studied the Japanese tax system and financial systems of national and local governments, and recommended its comprehensive reform in 1949. In disaster rehabilitation, the mission recommended that the national government should cover all costs (Kaneta 2011). If the local governments cover all costs, their budget may collapse. Disasters are unpredictable and often require a huge financial burden for the local governments. Based on this recommendation, a new law on financial support to the local governments was enacted in 1951.

The national government provides the local governments with more than two-thirds of total rehabilitation costs. Share of rehabilitation costs by the local governments are decided based on their financial capacities. Furthermore, the national government provides grand subsidy for debts of local bonds that are issued for local expenses. The local governments eventually share only 1.7 % of project costs. In case of extreme severe disasters the national government increases the support to the local governments by 10–20 % of project costs.

The program, which MLIT manages, aims at promptly and effectively rehabilitating the facilities damaged by natural disasters. Some three million facilities have been rehabilitated with some JPY 23.4 trillion, or USD 270 billion, for the last six decades under the program.

This program has the following uniqueness:

1. To cover various public facilities of river, coast, landslide protection, roads, ports, fishery ports, sewerage and parks
2. To assess project costs immediately after disasters and to promptly secure budgets by supplementary budgets
3. To start works promptly, often on the day of disaster, before cost estimation by providing subsidy retroactively

4. To rehabilitate function, not to necessarily rehabilitate original forms
5. To provide a package budget to each prefecture covering all rehabilitation works so that prefecture governments have flexibility of project implementation

MAFF has a similar program for rehabilitation of agricultural lands and facilities (MAFF 2011). This program, which covers private agricultural lands, was exceptional measures to support rehabilitation of private assets before establishing support scheme for reconstructing daily lives of affected people in 1999. For public school rehabilitation, the education ministry provides subsidy too.

The budgetary process of rehabilitation is rationally formed (Kaneta 2011). The costs of each project is decided in the presence of the officers of a local government (applicant), a line ministry (technical examiner), and a finance ministry (budget examiner) in the disaster fields.

There are some issues to be improved. The rehabilitation mechanisms have been established on an ad-hoc basis for the last over half century without a holistic perspective (Morichi 2009). For example, these mechanisms mainly cover major public facilities, such as roads and river banks, but not facilities of hospitals and schools owned by the private sector. Neither, these mechanisms cover the utilities of electricity, gas, and communication, which private companies own. Sanriku Railway, which the private sector and local governments jointly own, currently faces difficulties in rehabilitating railways damaged in the EJET. Rehabilitating the railways needs JPY 18 billion, or some USD 200 million, which is a huge burden for the railway company that had been suffering in economic difficulties in a rural depopulated area (Tsukui 2012). The national government, however, provides only 25 % of the rehabilitation costs under the existing mechanisms.

The current rehabilitation mechanisms may lead to ineffective recovery (Tajika and Miyazaki 2008; Miyazaki 2009). Local governments have limited flexibility in selecting or formulating rehabilitation projects. The mechanisms aim at rehabilitating original functions, and do not allow reconstructing them to upgraded functions. Where residents gave up returning to original places because of enormous damages, it is difficult to modify projects to match with actual situations. These mechanisms may discourage to invest DRM projects at normal times. The local governments tend to "wait for disasters" to conduct the projects, because the national government provides higher subsidies to the rehabilitation projects than ones to the projects at normal times.

5.3.2 Public–Private Partnership

The private sector played crucial roles in quick recovery. The construction companies immediately started recovery works under danger situations on the day of the earthquakes and tsunamis on March 11, 2011. Some 60 % of construction companies, which engaged in the works in the Tohoku Region, commenced the works within four hours following the earthquake. At this time, coastal areas were still at risk from tsunamis, as the Meteorological Agency was still issuing tsunami warnings. Some 70 % of the companies have damaged by the disaster (Yoneda 2012). Local construction companies utilized their advantages to reach devastated fields. They are familiar with geographical conditions, and possess heavy equipment and have operators.

Government organizations singed pre-agreements with the private sector of construction, consultant, survey and others at normal times to engage in emergency rehabilitation works following disasters. In addition to the rehabilitation works, the governments arrange pre-agreements in various areas including medical, logistics, transportation, mass media, public utilities, and waste management.

The government organizations could instruct construction companies to start the emergency works on verbal or by simple documents without ordinal procurement processes. The pre-agreements are simple and cover only principle clauses, such as guaranteeing payments by the governments and requiring private company's efforts of cooperation for rehabilitation works. Payments were settled latter based on quantities of works done.

To sustain this public–private partnership, challenge is weakening capacities of the private sector. Number of the construction companies is decreasing in recent years because of reducing government budget of constructing infrastructures. They cannot afford to own heavy equipment or to employ new engineers and technicians. The government and private sector should develop new arrangement with other partners, such as large construction companies, retired engineers and lease companies of equipment.

5.3.3 Mobilization of Expert Teams Through National Networks

The Technical Emergency Control Force (TEC-FORCE), which the MLIT established and consists of government engineers and equipped with communication facilities and heavy equipment, conducted a wide range of recovery works. These experts of the force observed damages by helicopters, assessed damages of facilities, technically supported the local government to rehabilitate transport infrastructures, provided the local governments with mobile communication equipment. In total, 18,100 person days engaged in these activities. For example, they drained inundated water from the Sendai Airport by mobile pumping equipment. The airport reopened for rescue and emergency supply four days after the earthquake, and was available for commercial services on April 13.

Government agencies mobilized expert teams through their national networks as shown in Table 5.1. These agencies have established roster systems and conducted trainings and drills at normal times. Some 3,500 experts are registered in the TEC-FORCE throughout the country. The ministry had engaged in rehabilitation works in the disasters on an ad hoc basis, but established the force in 2008 to systematically implement quick rehabilitation by mobilizing resources through national networks following resent disasters including the Kobe Earthquake.

Table 5.1 Emergency expert teams

Ministry	Expert team	Activity
Ministry of defense	Self-defense force	Search and rescue Relief goods Nuclear accident
Ministry of health, labor and welfare	Disaster medical assistance team	Emergency medical treatment
MLIT	Technical emergency control force	Damage assessment Protect secondary disasters
Coast guard agency	Coast guard	Search and rescue
Fire and disaster management agency and prefectural fire departments	Emergency fire response team	Fire fighting Search and rescue
National police agency and prefectural policy agencies	Inter-prefectural emergency rescue unit	Search and rescue Securing road transportation

Source: Osa et al. (2012)

Local governments outside the areas affected have supported these areas by their own decisions without any instruction or coordination by the national governments. Prefecture governments in the Kansai Region organized a counterpart support system, or twinning arrangement, which assign each supporting prefecture to assist only one of the prefectures affected. This arrangement could achieve efficiency, speed, continuity, and accountability as well as avoid overlapping of support and clarify responsibilities (Keicho and IRP 2012).

Since the disaster scale was larger than they had prepared for ones, all agencies faced difficulties in operating on the ground. Medical teams prepared for operations of emergency treatment for 48 hours after disasters, but had to work for ordinal treatments over months because of serious damages to medical facilities. Police and fire fighting organizations had not enough equipment to independently conduct activities in the field. They started improving their capacities for relief activities in wider areas for longer periods. The MLIT found that the chain of command of the TEC-FORCE was not enough unified, since experts were gathered from all offices throughout the country. The ministry strengthened the TEC-FORCE by creating a permanent secretariat and unifying a chain of command under the director of a regional bureau located in affected areas.

5.3.4 *Structural Measures*

Reducing damages occurred by disasters is important as same as developing recovery mechanisms. Damages by earthquake shakings were limited despite the mega-earthquake of magnitude 9 because of seismic reinforcement of infrastructure (Ishiwatari 2012). Works required to rehabilitate infrastructure were reduced. Some 1,500 bridges of national roads and 1,000 bridges of expressways in the Tohoku Region were not crucially damaged by the shaking on March 11.

Because of minor damages to roads, main works of "Operation Comb" were cleaning debris and sedimentation generated by the tsunamis and quakes. Large-scale works, such as constructing temporary bridges, were not required. If a bridge was collapsed, it took more months to open roads to affected people in devastated coastal areas. Some bridges unreinforced of local governments were severely damaged. Also, five bridges of national roads were washed away by tsunamis. It took several months to repair these bridges.

These contrast well with the damages at the Kobe Earthquake, in which bridges of expressways and railways were seriously damaged and needed enormous costs and works for recovery. In the Kobe Earthquake, 36 sections of 27 road routes including highways were closed (Ishiwatari 2011). Traffic was heavily congested at major passable routes. The congestion fatally delayed transportation of rescue teams and relief goods to devastated areas. By learning lessons from the Kobe Earthquake, the MLIT has reinforced crucial facilities and bridges throughout the country including 490 bridges in the Tohoku Region. A railway company had reinforced some 18,500 bridge piers under the bullet train lines (East Japan Railway Company 2011).

5.4 Conclusion

Japanese practices show that quick recovery from disasters requires developing institution and governance at normal times: (1) advance financial arrangements, (2) public–private partnership, (3) mobilization of expert teams through national networks, and (4) structural measures. It is too late to start considering these measures once disasters happen. Financial shares and responsibilities of local and national governments should be decided at normal times. The government should arrange and sign pre-agreement with private companies so that the companies can start works on the day of the disaster. Also, roster system of expert teams should be built up, and drills should be conducted. Retrofitting and reinforcement can decrease workloads and financial burdens of rehabilitation.

5.5 Consideration

The practices in Japan can be applied to developing countries and development agencies. The developing countries should examine shares of rehabilitation costs between the national and local governments at normal times to avoid unnecessary negotiation following disasters. The developing countries and development agencies should establish specific procurement processes of mobilizing private companies for disaster rehabilitation, which are different from ordinal processes. Government agencies can quickly utilize national networks to mobilize experts in search and rescue teams, medical teams, and engineers in larger scale than international teams.

Development agencies should enhance to support developing countries to improve institution and governance in these areas in post-disaster period.

These institutional and financial arrangements are essential, but not enough. In addition to these non-structural measures, structure measures, such as retrofitting and reinforcement works, must be implemented to mitigate damages and rehabilitation works.

Since speed is the most important factor for rehabilitation, a top-down approach is preferable in the immediately aftermath of the disasters. Consultation or coordination with communities and stakeholders is not necessarily required. As recovery proceeds, a consensus based or participatory approach is becoming to be applied to coordinate stakeholders for rehabilitating their communities.

Local contexts should be considered in improving institutions and governance. Japan has developed private–public partnerships based on ordinal relationship between governments and local construction companies. Simple agreements of a few pages are enough for them, since they have established close relationship through conducting public works at normal times. More detailed agreements including unit costs may be required in other countries.

The government can mobilize engineers and experts in the government on the ground in devastated areas. The engineering works of the governments are less outsourced in Japan than other countries. In developing countries, engineers in the private sector also should be utilized as resources in recovery.

References

Asian Development Bank (2012) Validation report: Indonesia, earthquake and tsunami emergency support project. ADB, Manila

Committee on Promoting Disaster Management of Central Disaster Management Council (2012) Final report: reestablishing resilient Japan (Yuruginai nihonno saikochikuwo mezashite (in Japanese)). Cabinet Office, Tokyo

Davis I (2012) What is the vision for sheltering and housing in Haiti? http://www.onuhabitat.org/index.php?option=com_docman&task=doc_download&gid=789&Itemid=235

East Japan Railway Company (2011) Safety report 2011 (Anzen Hokokusyo (in Japanese)). East Japan Railway Company, Tokyo

Expert Panel on Expressway (2011) Emergency recommendation based on great East Japan earthquake (Higashinihondaisinsaiwo Fumaeta Kinkyuteigen (in Japanese)). Ministry of Land, Infrastructure, Transport and Tourism, Tokyo

Ghesquiere F, Mahul O (2010) Financial protection of the state against natural disasters: a primer, policy research working paper, vol 5429. World Bank, Washington

Gotham KF, Greenberg M (2008) Post-disaster recovery and rebuilding in New York and New Orleans. Soc Forces 87(2)

Handa M (2011) Disaster rehabilitation systems from the perspective of ministry of finance (in Japanese)). River 785:23–27

IEG (Independent Evaluation Group) (2011) Natural disaster response: lessons from evaluations of the World Bank and others (Evaluation brief), vol 16. World Bank, Washington

International Recovery Platform (IRP) and United Nations Development Programme-India (2010) Governance, guidance note on recovery. IRP Secretariat, Kobe

Ishiwatari M (2010) Review of disaster rehabilitation methodologies following the Indian ocean tsunami from a human perspective. Asian J Environ Disaster Manage 2(3):275–288
Ishiwatari M (2011) Redevelopment of inland water transport for post-conflict reconstruction in Southern Sudan. In: Bruch C, Nakayama M, Coyle I (eds) Harnessing natural resources for peacebuilding: lessons from US and Japanese assistance. Environmental Law Institute, Washington
Ishiwatari M (2012) Review of countermeasures in the East Japan earthquake and tsunami. In: Shaw R, Takeuchi Y (eds) East Japan earthquake ant tsunami: evacuation, communication, education and volunteerism. Research Publishing, Singapore
Jayasuriya S, McCawley P (2010) The Asian tsunami: aid and reconstruction after a disaster. Asian Development Bank Institute, Tokyo
Kaneta K (2011) Relations with disaster rehabilitation project (Watashito saigaifukyujigyo (in Japanese)). River 785:3–4
Kawata Y (2008) Current situation and issues of disaster recovery in Japan (Wagakunino Saigaifukyuno Genjoto Kadai (in Japanese)). East Times 08(11):4–7
Keicho T, IRP (2012) Supporting and empowering municipal functions and staff, knowledge note 3-4: learning from megadisasters. World Bank, Washington
Ministry of Agriculture, Forestry and Fishery (2011) Towards prompt rehabilitation: disaster rehabilitation scheme of agricultural lands and facilities (Sumiyakana fukyunimukete: Nochi, nogyoyosisetsu saigaifukyujigyono aramashi). MAFF, Tokyo
Mitchell JK (2006) The primary of partnership: scoping a new national disaster recovery policy. The Ann Am Acad Polit Soc Sci 604:228–255
Miyazaki T (2009) Financial arrangement and roles of local government in disaster management policies (Saigaikanrensesakuniokeru Zaigensochito Chihono Yakuwari) in study on disaster-related policies from perspective of economic study (Keizaigakutekisitenwo Donyushita Saigaiseisakutaikeino Arikatani Kansuru Kenkyu (in Japanese). Economic and Social Research Institute, Cabinet Office, Tokyo
Morichi S (2009) Institutional issues of responding to risks of infrastructures (Syakaishihonno Risukutaiono Seidotekikadai (in Japanese)). JSCE Mag Civil Eng 94(8):38
Nakagawa Y, Shaw R (2004) Social capital: a missing link to disaster recovery. Int J Mass Emerg Disasters 22(1):5–34
Osa Y, Sagara J, Ishiwatari M (2012) Mobilizing and coordinating expert teams, nongovernmental organizations, nonprofit organizations, and volunteers, knowledge note 3-1: learning from megadisasters. World Bank, Washington
Sagara J, Ishiwatari M (2012) Infrastructure rehabilitation, knowledge note 4-1: learning from megadisasters. World Bank, Washington
Tajika E, Miyazaki T (2008) Financial systems in rehabilitation and reconstruction (Zaiseitekinimita Fukyu Fukono Taikei (in Japanese)). Financ Rev 91, http://www.mof.go.jp/pri/publication/financial_review/fr_list5/fr91.htm
Tohoku Regional Bureau, MLIT (undated) The reason of prompt rehabilitation of roads (Dorokeikaiga hayai riyunituite (in Japanese). http://www.thr.mlit.go.jp/road/jisinkannren-jouhou_110311/keikairiyuu.pdf
Tokuyama H (2012) Learning from Japan's ordeal. Public Roads 75(6), http://www.fhwa.dot.gov/publications/publicroads/12mayjune/04.cfm
Tsukui S (2012) Megadisaster and law (Daisaigai to Ho (in Japanese)). Iwanami Syoten, Tokyo
World Bank (2012a) Advancing disaster risk financing and insurance in ASEAN member states: framework and options for implementation. World Bank, Washington
World Bank (2012b) Executive summary of knowledge note: learning from megadisasters. World Bank, Washington
World Bank (2012c) Sendai report: managing disaster for a resilient future. World Bank, Washington
Yoneda M (2012) Local construction industry facing disaster prevention (Bosaino saizensennitatsu chikikensetugyono genjoto tenbo (in Japanese)). JSCE Mag Civil Eng 97(12):22–24

Chapter 6
Institutional Response in Education Sector in Kesennuma City

Yukihiko Oikawa

Abstract After the massive disaster of East Japan Earthquake and Tsunami on March 11, 2011, education sectors such as schools and board of education (BOE) in tsunami-affected area faced many hazards and challenges immediately and continuously. This chapter describes how educational institutions have been implementing the strategies for recovering schools and education from the disaster of earthquake and tsunami through the Cases of Kesennuma City Board of Education and Schools in Kesennuma. Kesennuma City has been promoting the Education for sustainable Development (ESD) as a part of school and community linkages. Through documenting the examples of a few specific schools, the chapter emphasizes that providing school lunch, arranging schools bus, providing scholarships to the affected children were some of the urgent tasks to restart the regular education process in the post disaster scenario. It is concluded that the ESD program has positive impact on strengthening the school community linkage, and the disaster risk reduction should be part of the future ESD programs.

Keywords Education for sustainable development • Education in emergency • Institutional response and Tohoku tsunami • Post disaster recovery

6.1 Introduction

The massive earthquake and tsunami on March 11th 2011 hit east Japan, especially huge tsunami caused terrible damages of many aspects such as personal sufferings, social, economic, environmental and cultural damages in Pacific coast line area of

Y. Oikawa (✉)
Kesennuma City Education Board, Kesennuma, Japan
e-mail: yuki812@seagreen.ocn.ne.jp

R. Shaw (ed.), *Disaster Recovery: Used or Misused Development Opportunity*,
Disaster Risk Reduction, DOI 10.1007/978-4-431-54255-1_6, © Springer Japan 2014

Tohoku. Educational Institutions and schools in tsunami affected area also suffered serious damages of not only facilities but also students and staffs.

According to the report (on Sep. 14th, 2012) of Ministry of Education, Culture, Sports, Science and Technology (MEXT), 12,150 educational institutions in 24 prefectures including 7,988 schools and universities, suffered damages by the earthquake and tsunami. Talking about personal sufferings, 659 students and faculties in educational institutions died and 74 in those are missing. 262 students and faculties were also injured.

In addition, after the East Japan Earthquake and Tsunami, as many people in affected area evacuated to the schools and educational institutions such as Kominkan (Community learning Center) and Gym. So most educational institutions were used as shelters. Especially almost all the schools in tsunami affected area functioned as a role of shelter to many evacuees after the earthquake and tsunami immediately. On March 17th, 2011, 622 schools managed the shelter in the gym or class room at its peak at as Emergency Report of MEXT says. And last shelter in school was closed on October 10th, 2011 which is 8 months later from disaster happened. Based on these tragic damages of the disaster of East Japan Earthquake and Tsunami, Educational Institutions such as board of educations had to response to recovery process of schools and education. Its response should be performed from their perspectives of missions and roles depend on the phases of recovery process and needs of schools. MEXT categorizes recovery process to restart school education into four phases from two functions as the shelter and the school on their report of "Examination about Preparation of the school facilities based on the damage of the East Japan Earthquake and Tsunami" as follows.

1. Rescue and Evacuation Period—From disaster happened to evacuation
2. Securing Lives Period—For a few days after the evacuation
3. Making Livings Period—For a few weeks after Securing Lives Period
4. Restarting School function Period—After restarting the school education

In case of Kesennuma City, the recovery process of school education was similar to MEXT's category. But it is needed to put some arrangements and additional phase or stage into the category to adjust actual recovery process of school education in Kesennuma City. Kesennuma City Board of Education categorized recovery process of school educational into five stage as follows (Table 6.1):

1. Emergency Response to manage crisis—after the disaster immediately
2. Short-term Response to sustain lives—set up and manage the shelters
3. Mid-term Response to school restarts—until school restart after set up shelter
4. Long-term Response to recover education—recover and settle schools
5. Further Response to reconstruct future—response to next disaster and future

Therefore, the response of the board of education and schools should be done according to these phases and stages. It will be described in this paper how main responses of educational institutions in Kesennuma City have been implementing so far after East Japan Earthquake and Tsunami.

Table 6.1 Main responses of educational institution according to phase and stages in Kesennuma City

Stages of responses	Situation and mission	BOE response (Board of Education)	School response
Emergency response to manage crisis	Rescue Evacuations	Collect information of educational damages	Instruct evacuation Accept evacuees
Short-term response to sustain lives	Shelter management Support students' life	Manage shelters Support to self-manage	Set up shelters in schools Check students' safety
Mid-term response to school restart	Reconstruction school system and facilities	Secure transportations Restart school lunch	Prepare for school restart
Long-term response to recover education	Support student's life and school activities	Economical and physical support to students	Psychological support for students and teachers
Further response to creative recovery	Preparation for future disaster and recovery	Record experiences Establish links(N-help)	Improve DRM, DRR, ESD and recovery education

6.1.1 Disaster of East Japan Earthquake and Tsunami in Kesennuma City

6.1.1.1 Damage of Kesennuma City by East Japan Earthquake and Tsunami

Kesennuma, a city in Miyagi Prefecture, is a region located in the northeast of Japan. With the Pacific Ocean along its eastern edge, it is a region rich in nature. Covering a total area of 333 km^2, Kesennuma City is situated at 141 East longitudes and 38 North latitudes. With a population of approximately 69,500 (July 31, 2012), the city is the center of the Kesennuma City area. The region's coast is called the Sanriku Coast. It extends from Iwate Prefecture to its north and is part of a coastline with complex inlets, also know as a rias coast line. Rivers flow from the green Mountains through the city to Bay. It is surrounded by forests, rivers and ocean, all interconnected by water. It has long been one of the leading fishing ports in Japan. Shell and seaweed farming is also very popular. The region is proud to land some of the highest catches in Japan for tuna, Pacific saury, oyster and kinds of seaweed. It is number one in Japan for bonito and shark fin.

At 2:46 p.m. on March 11, 2011, the massive earthquake of M9.0 on the Richter scale hit East Japan. The intense quake, in which we were not able to keep standing, continued 5 min or more. About 30 min later, the huge tsunami which happened once per millennium attacked the Pacific coastline of Tohoku area. This massive earthquake and tsunami cut all of lifeline, communications network, and means of transportations in pieces in a moment. More than half of city area was attacked by tsunami, and that was flooded and destroyed in no time.

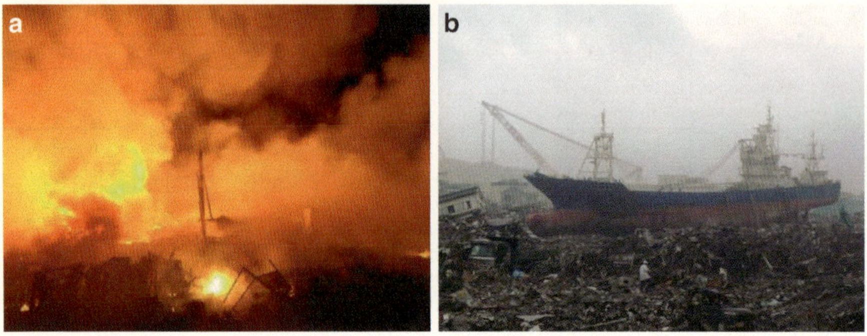

Fig. 6.1 (**a**) Fire in Shishiori, (**b**) Damaged ship transported inland

Table 6.2 Damage by East Japan earthquake and tsunami

Area	Category of damage	Actual number	Proportions (%)
Kesennuma City	Dead persons by the disaster[a]	1,038 persons	1.4
	Unidentified	34 persons	
	Dead persons relevant to disaster[b]	103 persons	0.1
	Missing persons[c]	264 persons	0.4
	Dead + missing persons[d]	1,405 persons	1.9
	Damaged company[e]	3,314 companies	80.7
	Workers who lost jobs[f]	25,236 persons	83.5

[a–d]Aug 2012 from Reports of police office
[e]Presumption of the Ministry of Internal Affairs and Communications
[f]Economic census

Especially, northern part of the city area, that is Shishiori district, was burning more than 10 days by terrible fire because of oil and gas tanks catching fire after the tsunami. Then those areas were devastated (Fig. 6.1a). Many of boats such as tuna boats, which are moored at port, clashed into the city and they were washed up on roads, rivers and buildings around bay area in Kesennuma (Fig. 6.1b). Some burning boats pushed into the city and caught the big fire like bombs.

Because of this terrible disaster, Kesennuma City had huge damages in so many aspects, such as human lives, houses, transportations, education, industries and so on. Kesennuma City had more than 1,400 victims including about 260 missing persons and 100 dead persons relevant to the disaster (Table 6.2).

The tsunami and fire has done a great deal of damage to not only fishing industries but also fishery processing industries, freezing companies, distribution industries and tourist industries in Kesennuma those are developing around bay connecting with ocean. So more than 3,300 companies destroyed by tsunami. That would correspond to 80 % of whole companies in Kesennuma. And over 25,000 workers of those affected company by tsunami lost their jobs that is 83.5 % of whole workers in Kesennuma (Table 6.2). These are unprecedented case in Kesennuma.

As a result, the population of about 5,000 of Kesennuma City decreased rapidly, and, according to this situation, about 1,000 of householders of Kesennuma also decreased comparing with the numbers before the East Japan Earthquake and Tsunami. Those are reason why many of citizens of Kesennuma City have moved to other cities or prefectures to evacuate from the disaster and to seek for new jobs for their lives.

6.1.2 Damage of Schools in Kesennuma

Almost all the schools in Kesennuma fell into alone and unaided by the disaster. Tsunami intercepted all of information, traffic, and materials and each school was surely in "a solitary island in land." The teachers of each school in Kesennuma were burdened with urgent and serious duty "how to protect children's life" in such isolated situations.

In Kesennuma, schools prepared for the earthquake and tsunami, which will be anticipated in the near future so far, and each school decided upon the disaster prevention manual, and schools have also implemented the evacuation drill in each school many times repeatedly, assuming various disaster situations. However, the scale of this earthquake disaster exceeded their assumption and a manual by far. It was certainly an "unprecedented" case. And the disaster risk management was very difficult, since the scale of damage differed from its character distinctly according to the geographical conditions in which the schools located. In addition, the directions from the board of education and connection with other schools stopped, since the communications network was cut off, so that each school had to make original decision and judgment while any information could not be informed. In each school, teachers beat their brains and courage together and they instructed last-minute refuge and evacuation action. As a result, in Kesennuma, there was no child who lost his life at school then. On the other hand, it is the deep regret that over ten students' precious lives were lost among the students who were absent on that day, who left school earlier than usual and who has come back home from school before tsunami invaded to the schools. In addition, 65 students in Kesennuma lost their both or one parent by this disaster and two teachers of elementary schools of Kesennuma were killed by tsunami at home or on the way to school (Table 6.3).

After the massive earthquake of M9.0, the tsunami reached three elementary schools, one junior high school, one high school and one kindergarten in Kesennuma City. Amongst these, Minami Kesennuma Elementary School, Shishiori Elementary School, Kesennuma Koyo High School, Ohya Kindergarten were destroyed by tsunami. As for Hashikami Elementary School and Karakuwa Kindergarten, they were seriously destroyed by earthquake (Fig. 6.2a, b). Eighteen schools located on the coastline functioned as shelters after the tsunami, while five elementary schools and junior high schools were used as posts of the Self-Defense Force, police and fire brigade. The gymnasiums of four Elementary Schools were used as mortuaries (Table 6.3).

Table 6.3 Damage of schools by earthquake and tsunami (October 2012)

Area	Category of damage	Actual number	Proportions (%)
Casualty	Victims of students(total)	13 students	0.2
	Kindergartner	1 child	
	Elementary school	7 students	
	(Missing)	(1 student)	
	Junior high school	5 students	
	Orphans by earthquake and tsunami	10 students	
	Students who lost the single parent	55 students	
	Victims of teachers	2 persons	
Damage of facilities	Serious damaged schools (total)	7 schools	17.5
	Elementary school	4 schools	19.0
	Minami Kesennuma E.S[a]		
	Shishiori E.S[a]		
	Ohya E.S[a]		
	Hashikami E.S[b]		
	Junior high school	1 schools	7.7
	Ohya J.H.S[a]		
	Kindergarten	2 schools	33.3
	Ohya kindergarten[a]		
	Karakuwa kindergarten[b]		
	School that had shelter, garrison of self-defense, firehouse, and mortuary	22 schools	64.7
	Facilities used as shelter	16 schools	47.1
	(Including classroom)	(9 schools)	(26.5)
	Facilities used as mortuary	4 schools	11.8
	Schools that have temporary houses in school yard (houses)	14 schools (1,143 houses)	41.2
	Elementary school (houses)	6 schools (143 houses)	28.5
	Junior high school (houses)	11 schools (1,000 houses)	84.6

[a]Damage by tsunami
[b]Damage by earthquake

6.2 Emergency and Short-term Response: Selected Cases of School Response

MEXT researched school damages of East Japan Earthquake and Tsunami about 3,127 schools in Iwate, Miyagi and Fukushima prefecture in January, 2012. According to the report of research named "Research of School Response in Disaster of East Japan Earthquake and Tsunami", 131 schools in three prefectures were hit by tsunami in total. And 69 schools in despite of which were not in tsunami flooded area of hazard map, were also attached by tsunami and its rate is almost half (46.3 %) of tsunami affected schools. 113 tsunami affected schools had students at schools when tsunami hit. They took evacuation depend on their locations and situations. 35 % schools evacuated to upper floors and roofs. 32 % schools evacuated to

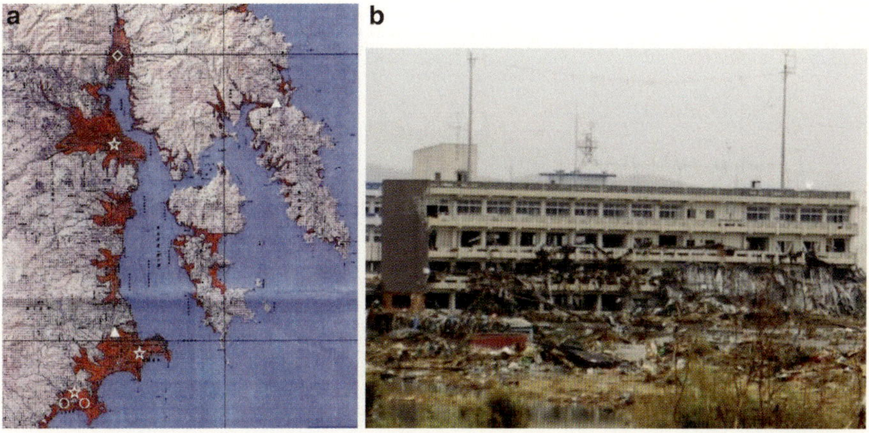

Fig. 6.2 (**a**) Damaged Schools in Kesennuma City (*Red Zone*: Tsunami affected area, Tsunami Damaged Schools: *star:* High, *diamond*: Middle, *circle*: Low, Earthquake Damaged Schools: *triangle*). (**b**) Kesennuma Koyo high school damage

high land near the schools, and 31 % schools escaped to evacuation place which were designed in advance of the disaster.

Dr. Rajib Shaw and Dr. Yukiko Takeuchi (Shaw and Takeuchi 2012) are also mentioning five types of evacuation routes of schools in tsunami affected area in Iwate and Miyagi prefecture. The route of evacuations schools took were selected depend on their situations.

In Kesennuma City, as described above, three elementary schools, one junior high school, one high school and one kindergarten were hit by tsunami. All of them took evacuations to protect students' lives according to their situations such as geographical conditions and scale of damage. Three typical cases of schools are selected and described how their school responses to evacuation, shelter management and recovery process of schools were implemented.

6.2.1 Case 1: Minami-Kesennuma Elementary School

The school, located on a riverside, was hit by tsunami, which completely destroyed the first floor. 600 students, teachers and residents evacuated to the third floor and after spending one night, the Self-Defense Force finally rescued them. The school restarted its educational activities in April 2011, using part of Kesennuma Elementary School and eventually merged with Kesennuma Elementary School in April 2012.

When the earthquake hit the school during lesson 6 at 2:46 p.m., from 3rd grade to 6th grade students who were studying in class room, escaped to school yard. First and second grade students had left school. Principle ordered teachers to take 1st and 2nd grade students back to the school. Teachers gathered other 90 % students and

Fig. 6.3 (**a**) Minami Kesennuma Elementary School hit by tsunami, (**b**) Staff Room destroyed by tsunami, (**c**) 1st grade students of Minami-Kesennuma ES studying at Kesennuma ES

kept them in school yard. School handed the parents some students whom parents came to school to take. According to the information of 6 m tsunami coming to Kesennuma from bousai musen, the principle decided that he ordered to let all of students and staff refuge to second floor of school building. The assumption of Miyagi-ken Oki Earthquake is 2 m tsunami, so manual set the evacuation place of the school and stock at gym. 600 people including 350 students, 120 local residents, 80 kids and staff escaped to central school building. Other residents also escaped to east school building.

When tsunami attacked school at 3:36, tsunami came to the school yard from the Okawa River beyond dike (Fig. 6.3a). Tsunami came up to second floor of the school, so principle directed all evacuees to move from second floor to third floor and kept the rote to the roof (Fig. 6.3b). Teachers took off the curtains and gave to students instead of blankets, and students wore them. All evacuees stayed cold night without water and foods. Staff and residents brought stock at gym to the school building for evacuees although water still covered school yard and passage to gym in early morning. Evacuees could receive supports of drinks or foods from restraint and convenience store in neighborhood. Evacuation to K-Wave (City Gymnasium) by Jeep of Self-Defense started at 3:00 p.m. next day. All evacuees of the school could move to K-Wave completely by 5:00 p.m.

In regard to the process of restarting and recovery school, first and difficult mission of school is safety check of missing students whom School handed to parents before tsunami coming and who took refuge to the shelter. And then according to direction of Kesennuma City Board of Education, teachers moved the function of school to Kujo Elementary School on 14th. The school took place the graduation and school-end ceremony at Jonan Junior High School on 23rd of March, but one student was still missing then. On April 1st the school moved to east part of Kesennuma Elementary School's building supported by staff of Kesennuma ES, parents, and volunteers. On April 21st, the school restarted with 223 students, which was estimated as 350 students before the disaster (Fig. 6.3c). On the same day, the death of missing students was confirmed. She is only victim of the school by disaster. The students studied at Kesennuma Elementary School together for one academic year. Kesennuma City and City Board of Education decided that Minami-Kesennuma Elementary School should merge with Kesennuma Elementary

Fig. 6.4 (**a**) Hallway in front of principal room, (**b**) Staff Room destroyed by tsunami, (**c**) School Garden Recovery Project with volunteers of bank

School as soon as possible, considering students' environment of learning after consulting with parents repeatedly. At last, Minami-Kesennuma Elementary School was merged with Kesennuma Elementary School on April 1st, 2012, and they started new academic year as new school.

6.2.2 Case 2: Shishiori Elementary School

Although this school is located 2 km away from Kesennuma Bay, the tsunami carried the waves up the river and caused devastating damage to the first floor. At this time, the teachers instructed evacuation to 145 students and parents. They moved to five different places for evacuation, eventually reaching a temple on a high ground and stayed there for three nights. The school activities resumed using the second and third floors. Construction of the first floor was completed at the end of August this year.

This school is located in the north of city along Shishiori River and 1.5 km far from bay area. The school had never been hit by tsunami until East Japan Earthquake happened. The new school building was established 3 month before the disaster. When earthquake hit at 2:46 p.m., 1st grade had left school before the earthquake because of four lessons. And 2nd to 6th grade students were studying in the school. Principle directed the staff to take whole students in the school to the school yard.

When tsunami attacked school at 3:36, tsunami came up the river from Kesennuma Bay and attacked the school beyond the dike. Tsunami came into first floor at 140 cm level and washed away everything in classrooms and staff room of first floor. The gym under construction (Almost finished) also destroyed by tsunami (Fig. 6.4a, b).

Staff and parents who came for their students took five steps refuges under teacher's instruction as flows.

- At first, students and staff escaped to school yard.
- Secondly, they moved to hill beside of the school and handed some students to parents. Other students, staff and parents were 145 in total,

- Thirdly, they moved to temporary office near the hill, but the space was not enough for all of evacuees to stay.
- Fourthly, they moved to the temple, that is located 1 km from the school. But they could not be accepted in the temple.
- Lastly, 145 people move to next temple where 1 km far from last temple. They stayed three nights, and handed students to their parents safely.

In regard to the process to restarting and recovery school, safety check of students was very difficult because the means of communication such as telephone, mobile phone and internet were unavailable by earthquake and tsunami, and transportations were also unavailable as teachers' cars were washed away and almost of roads were covered with rubble by tsunami and fire. Five students were missing after school handed them to parents. The death of four students was confirmed by April 2012, but one is still missing. The number of victims is worst in the school in Kesennuma City. Second is cleaning up school building. Tsunami left mud which thickness is 10 cm and lots of rubble at first floor. Staff tried to take it away to restart the school getting supports of parents, residents and staff of other school which was not affected by earthquake and tsunami. Information from school to students and parents was also difficult. Teachers walked around shelters and shops to display information such as school events and to get information of students' safety. The school also had to grasp students' changes to other school because of disaster. After the disaster, 54 students of 356 moved to other school because their houses or their parents' job were lost.

The school could restart on 21st of April 2011 using second and third floor. On 5th of September 2011, gym was reconstructed. And in August 2012, reconstruction of first floor was completed, so students are studying in whole school building now. This school also received many supports from volunteers NPO/NGOs, companies, and institutions in Japan and all over the world. BOE also supported this school through introducing donations and scholarships as well as opportunities encouraging students like concerts and sports events (Fig. 6.4c).

6.2.3 Case 3: Hashikami Junior High School

This School is very famous for being a pilot school of Disaster Risk Reduction (DRR) Education in Kesennuma. But a couple of students of this school lost their lives by tsunami on the previous day of graduation.

The school building didn't receive significant damages by the tsunami because the school is located on highland and far from the sea. However, at the time of evacuation, more than 2,000 evacuees gathered to the school for safety, making it function as a shelter for the whole community. Residents, teachers and students who had experienced DRR education/training managed the shelter by themselves.

As this School is pilot school of DRR Education in Kesennuma, it functioned as a shelter in this Disaster. More than 2,000 evacuees gather to this school then.

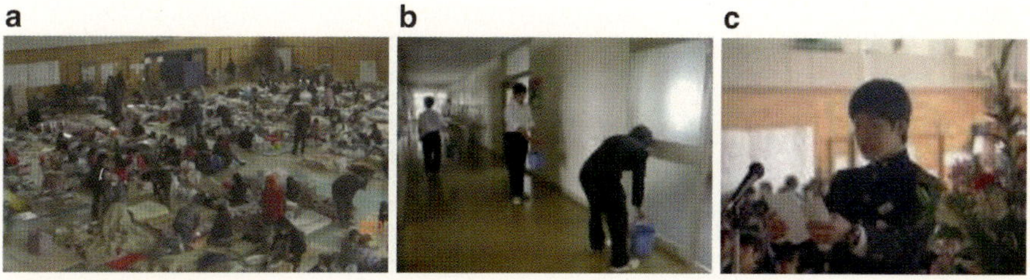

Fig. 6.5 (**a**) Shelter in the gym of Hashikami Junior High School, (**b**) Students cleaning up rest room, (**c**) Graduation ceremony with evacuees in gym

Community member, Teachers and Students as volunteer managed the shelter collaborating each other (Fig. 6.5a). Especially, students contributed to restoration of the community in the shelter, taking advantage of disaster education. Some students tried to help distribution of cooked rice and blankets to evacuees. Some students helped cleaning up rest rooms getting water from school pool because of stopping water. Some school students took care of old persons who took refuge to the shelter, with chatting and massaging (Fig. 6.5b).

One week later after the earthquake and tsunami, this school held graduation ceremony with many evacuees in gym (Fig. 6.5c). The address of representative student encouraged not only students and teacher but also evacuees and residents in gym as well as people in Kesennuma and tsunami affected area.

6.3 Mid-term Response 1: "Round School Buses Linking Shelters and School"

After the East Japan Earthquake and Tsunami, a lot of students took evacuations with their families to shelters and their relatives all over the city immediately, and they had to stay there for a long time. So school districts of schools in Kesennuma City were also destroyed by the earthquake and tsunami. In addition, the roads to schools were covered with rabbles and some parts of them were cut by tsunami and sinking under sea level, so that students could not go to their schools by walk using these roads. On the other hand, many of their parents could not use their cars because of losing them by tsunami and shortage of gasoline. And also as time goes by, amount of temporary houses were built in school yards and parks not only in Kesennuma City but also in neighboring Ichinoseki City in Iwate prefecture and students moved to the temporary houses from shelters with their family. This also made difficult for students to go to their schools (Table 6.4).

Therefore, Kesennuma City Board of Education had to provide students transportations from shelters to schools in order to restart the schools. This matter was indispensable condition to recover the schools education in Kesennuma City. Kesennuma City Board of Education tried to secure new school bus systems which

Table 6.4 Number of evacuate students in Kesennuma City after East Japan earthquake and tsunami

Living place	School level	April 15th, 2011	October 1st, 2011	April 1st, 2012	October 1st, 2012
Shelters	Elementary	288	1 (0)	0	0
	Junior high	198	4 (2)	0	0
	Total	476	5 (2)	0	0
Relatives and friends houses	Elementary	598	140 (59)	85 (29)	81 (24)
	Junior high	345	81 (27)	73 (28)	56 (22)
	Total	943	221 (86)	158 (57)	137 (46)
Temporally houses	Elementary	0	405 (143)	373 (89)	366 (90)
	Junior high	0	225 (75)	229 (76)	226 (79)
	Total	0	630 (218)	602 (165)	592 (169)
Apartment and rent houses	Elementary	–	212 (104)	154 (64)	163 (65)
	Junior high	–	131 (75)	127 (73)	113 (61)
	Total	–	343 (179)	281 (137)	276 (126)
Total	Elementary	886	758 (306)	612 (182)	610 (179)
	Junior high	543	441 (179)	429 (177)	395 (162)
	Total	1,429	1,199 (485)	1,041 (359)	1,005 (341)

Made from research of students living place by Kesennuma BOE
Number in bracket: Number of students who live outside school district

could link shelters and temporally houses to schools. But it had also many challenges and obstacles. Many of public and private buses were washed away by tsunami, so that most bus companies lost their buses, and some routs between shelters and school were unavailable by tsunami cutting. In addition, the city office and board of education could not afford budget for new school bus running. Kesennuma City Board of Education negotiated with executives of Ministry of Education and city government to get budget and asked bus companies to keep buses as "Emergency School Bus." At last, just before the school restarting, Kesennuma BOE was able to get agreements from Ministry of Education, city government and bus companies. After that immediately, Kesennuma BOE considered the round routs of school buses which link shelters to each school effectively. Finally, establishing of this new school bus system in Kesennuma could be in time for restarting school of new academic year.

After that, Kesennuma BOE has been trying to increase and adjust the routes according to the situation of students' moving and living places. This school bus system is still functioning as important transportations for students who go to their school beyond school district.

6.4 Mid-term Response 2: "Resupplying School Lunch Beyond Earthquake and Tsunami"

To restart schools and to recover education, supplying school lunch was very important immediately after the disaster of earthquake and tsunami. Because many of students evacuated to shelters or houses of relatives with families, they could not have enough meals in the morning and evening for a while. And some of students could not bring lunch boxes (bento) to the school because their families didn't have kitchens or they could not use electrical products or water for cooking as infrastructures had not recovered enough. Therefore school lunch was indispensable for students to learn at school whole day. The school lunch was life line for restarting school and recovery of education. However, after the East Japan Earthquake and Tsunami, the function of school lunch supply system in Kesennuma City stopped by damage of the earthquake and tsunami. And it had many obstacles for recovery of school lunch centers. But it was crucial that BOE restarted school lunch as soon as possible for recovery of school education.

6.5 Outline of School Lunch (Kyushoku) in Kesennuma Before the Disaster

Kesennuma City used to have three types of school lunch system in Kesennuma when it faced the disaster of East Japan Earthquake and Tsunami in March, 2011. Kesennuma City merged with Karakuwa Town in 2006 and it also merged with Motoyoshi Town in 2009, so it had complicate system of school lunch supplying.

One is full supply school lunch (Kanzen-Kyushoku) of all elementary schools and two junior high schools (Niituski Junior High Schools and Oshima Junior High Schools) in Kesennuma area, as well as all elementary schools and junior high schools in Motoyoshi Town area. Second type is semi-full supply school lunch (Hoshoku-Kyusyoku) in Karakuwa Town area. That system provides milk, soup and dishes other than staple foods for school lunch, so at lunch time, students have them with their rice as staple food which they bring by themselves from their home. Third system is milk school lunch (Milk-Kyushoku) of junior high schools in Kesennuma City area excepting Niitsuki and Oshima Junior High School. This provides just mill to students, so students at schools of the system have to bring lunch (bento) with dishes to go with the rice every day. The system of Milk Kyushoku was a long-pending problem of school lunch supplying system in Kesenuma (Table 6.5 a). Coexistence of these different systems made difficult and complicated restoration of the school lunch system in Kesennuma from the disaster of East Japan Earthquake and Tsunami.

These systems lasted until Central School Lunch Center (Chuo Kyushoku Center) established in Oct 1st. 2011 (Table 6.5b).

Table 6.5 Outline of school lunch in Kesennuma City

(a) May 1st, 2010 (before the disaster)				(b) October 1st, 2011 (after the disaster)			
Type of system	School level	Number of school	Number of students	Type of system	School level	Number of school	Number of students
Full supply	E.S.	18	3,483	Full supply	E.S.	18	3,177
	J.H.S	5	535		J.H.S	11	1,920
	Total	23	4,018		Total	29	5,097
Semi-supply	E.S.	3	320	Semi-supply	E.S.	3	302
	J.H.S	2	226		J.H.S	2	205
	Total	5	546		Total	5	507
Just milk	E.S.	0	0	Just milk	E.S.	0	0
	J.H.S	6	1,492		J.H.S	0	0
	Total	6	1,492		Total	0	0
Total	E.S.	21	3,803	Total	E.S.	21	3,479
	J.H.S	13	2,253		J.H.S	13	2,125
	Total	34	6,056		Total	34	5,604

(a) Left: May 1, 2010 (b) right: Octcber 1, 2011 (source from report of Administration Council of School Lunch in Kesennuma City, September 1st, 2012)

6.5.1 *Restart School Lunch for Recovery of Education in Kesennuma City*

Toward school restarting on April 21st, 2011, in advance Kesennuma Board of Education negotiated with all the staff of school lunch center in Kesennuma to shorten the term of temporary school lunch (bread and Milk only), and to start full supply school lunch as soon as possible by moving the steps to full supply school lunch forward, because school lunch was a life line to restart the school. But every school lunch center had many challenges to restart then, and every staff of each center said that it was impossible to start full supply school lunch for a while.

By the disaster of East Japan Earthquake and Tsunami, school lunch supply centers had damaged, and many problems and obstacles occurred at each center as flows. Therefore, the situations of each center were very difficult at that time, so that it was impossible to provide full supply school lunch to each school then (Sakuma 2012).

- Life lines such as water, electricity and gas for cooking were stopped by disaster, and recovery processes of them differed from its situation according to the conditions in which the centers faced.
- A couple of school lunch centers had to take a role as center of distributing boiled rice to evacuees in shelters.
- So staff of lunch center also had to be engaged in that mission, but not in cooking school lunch.
- It is difficult to get materials for cooking such as vegetables, meats, fishes, rice and milk because transportation and distribution system stopped.
- Some school lunch centers had troubles in cooking machine and other facilities by earthquake.
- Some trucks which carry school lunch to schools were washed away by tsunami.
- Some School lunch centers were used as bases to store provision for evacuees.
- Sanitary conditions in some lunch centers were poor, because evacuees used the lunch centers to distribute boiled rice and they entered the center with their shoes on.

However, Kesennuma City Board of Education asked staff to restart school lunch and to move to full supply school lunch as soon as possible searching for new methods and systems to overcome many challenges they faced. Staff of each school lunch center beat their brains and did their best to solve the problems for the purpose of restarting school lunch and realizing full supply school lunch as fast as possible. For example, they thought about the menu which was not needed cooking, and about the combination of foods considering supply, amount and nutrition. They also tried to seek for new supplier and supply routes which provided food materials quickly and continuously.

Thanks to these efforts of staff of each school lunch center, all of school lunch center in Kesennuma City could supply school lunches to each schools on April 25th, 2011, adjusting to school restart day. (21st was the opening ceremony and 22nd is the entering ceremony in Kesennuma, and 23rd and 24th were holiday, so these 4 day did not need school lunch.) Moreover, almost all the school lunch centers realized to provide full supply school lunch (including semi-supply) in a week after the school lunch started on 25th excepting Oshima and Motoyoshi School Lunch Center those could not use water line then. These were remarkable achievements of each school lunch center in Kesennuma City (Table 6.6).

However, problem still remained. The students at schools of milk school lunch had to bring lunch boxes (bento) to school and the students of semi-supply school lunch also had to bring rice as a staple food to school. Some of these students could not bring lunches or rice because the condition of their lives and cooking were not so good for evacuation to shelters and relatives. These students could not be rescued by school lunch supply system of Kesennuma City as ever. So Kesennuma City Board of Education considered the strategies for covering such students. At last, from April 25th until September 30th 2011, Board of Education decided on supplying box lunches (shidashi-bento) to students at school of milk school lunch who could not bring lunch, and also providing rice to some students at school of semi-school lunch who could not bring rice, utilizing National Disaster Relief Law and Study Support System of Kesennuma City communicating with federal and prefectural governments.

6.5.2 Establishing Kesennuma Central School Lunch Center

After the disaster of East Japan Earthquake and Tsunami, all of public constructions of Kesennuma City were stopped to cope with many emergency situations first. The construction of Kesennuma Central School Lunch Center which Kesennuma City Board of Education was promoting had also been stopped for a while. But City office and board of education gave priorities to establish new school lunch center in order to improve the situations of students' school lunches as soon as possible, so that BOE restart the construction of Central School Lunch Center.

Finally, Kesennuma Central School Center established on October 1st, 2011. And it worked to provide 2,700 full supply school lunches to 15 schools (6 elementary schools and 6 junior high schools). By the establishment of Kesennuma Central School Lunch Center, Kesennuma City Board of Education also could solve the problem of "milk school lunch." Central School Lunch Center started to provide full supply school lunch to 6 junior high schools those used to be milk school lunch supply (Table 6.7). As a result, all of school lunches of Kesennuma City became full supply or semi-supply school lunch. The establishment of Central School Lunch Center contributed to recovery of school lunch supplying system in Kesennuma City highly.

Table 6.6 Recovery process of school lunch centers in Kesennuma City after disaster of East Japan earthquake and tsunami

School lunch center (type of supply)	Provided schools	April 25th	April 27th	May 2nd	May 11nd	May 25th
Kesennuma center (Full supply)	6 E. schools	Temporary →		Full Supply →		
Matsuiwa center (Full supply)	3 E. schools	Temporary →		Full Supply →		
Niitsuki center (Full supply)	4 E. schools 1 J.H. school	Temporary →		Full Supply →		
Oshima center (Full supply)	1 E. school 1 J.H. school	Temporary →				Full Supply →
Motoyoshi center (Full supply)	4 E. schools 3 J.H. schools	Temporary →			Full Supply →	
Koharagi center (Semi-supply)	2 E. schools 1 J.H. school	Temporary →	Semi-Supply →			
Nakai ES kitchen (Semi-supply)	1 E. school	Temporary →	Semi-Supply →			
Karakuwa JHS kitchen (Semi-supply)	1 J.H. school	Temporary →	Semi-Supply →			

Table 6.7 Structure of Kesennuma central school lunch center

Capacity of cooking of school lunches		3,000 meals per a day
Number of schools and school lunches to supply	Elementary school	9 schools (1,350 meals)
	Junior high school	6 schools (1,350 meals)
	Total	15 schools (2,700 meals)
Number of staff and employees	Office workers	6 persons
	Cooking workers	30 persons (commissioned)
System of cooking sector		Commissioned to private sector
Type of cooking		Dry cooking (no wet)

Table 6.8 Condition of job offers in Kesennuma City after East Japan earthquake and tsunami

Month/year	Number of people seeking jobs (a)	Number of unemployment insurance (b)	People without insurance (a)–(b)	Job offers (c)	Ratio of job offers (c)/(a)
February/2011	1,778			1,019	0.57
March	1,761			923	0.52
April	4,410	1,006	3,404	838	0.19
May	6,169	5,079	1,090	1,069	0.17
June	6,325	5,511	814	1,836	0.29
July	5,417	5,008	409	1,799	0.33
August	4,835	4,660	175	1,800	0.37
September	4,627	4,210	417	1,775	0.38
October	4,321	3,760	561	1,703	0.39
November	4,355	3,523	832	1,886	0.43
December	4,287	3,335	952	1,858	0.43
January/2012	4,131	3,056	1,075	1,929	0.47
February	4,042	2,843	1,199	2,229	0.55
March	4,041	2,636	1,405	2,542	0.63
April	3,820	2,289	1,531	2,291	0.60
May	3,424	2,055	1,369	2,228	0.65
June	3,119	1,769	1,350	2,222	0.71

Date from "Hello Work" (Public Employment Security Office) of Kesennuma office

6.5.2.1 Long-term Response 1: "Economical Supports for Students and Parents"

As described above, over 80 % citizens in Kesennuma City lost their jobs by the damage of Disaster of East Japan Earthquake and Tsunami. According to this proportion, almost all the parents of students also lost their jobs. But conditions of Job offers have been very hard to seek for new jobs because most factories and business offices, more than 80 % in Kesennuma City, were devastated by tsunami and not reconstructed yet (Table 6.8). It is very serious problem not only for students' economical situations but also for school education in Kesennuma City.

6.5.2.2 Expanding Public Support System for Educational Expense

After the disaster of East Japan Earthquake and Tsunami, Kesennuma Board of Education tried to expand public support system of educational expense into students whose parents lost their jobs by the disaster of earthquake and tsunami in order to aid the students and parents who were affected by the disaster. This system covers expenses of learning materials, goods to go to school and school lunch. In addition to normal recipients of this system, Kesennuma BOE had designated 1,060 elementary school students and 654 junior high school students, 1,714 students in total who were affected by the disaster in 2011. This number of recipients including normal recipients corresponds to 41.4 % of all of elementary and junior high school students in Kesennuma City despite of under 10 % before the earthquake and tsunami. The serious situation has never been experienced in Kesennuma City so far.

6.5.2.3 New Grants for Students Affected by Disaster

On the other hand, Kesennuma Board of Education ventured to establish new grants for students who received physical and economical damages by East Japan Earthquake and Tsunami in 2011, collaborating with National Federation of UNESCO Association in Japan (NFUAJ) and donated by a bank, companies and NPOs.

One of grants is for orphaned students by East Japan Earthquake and Tsunami. This grant covers 100,000 yen as lump-sum payment and 20,000 yen per a month until their graduating from high school. In Kesennuma City, 63 orphaned students were designated in 2011. This grant is organized by National Federation for UNESCO and the budget was donated by city bank.

Another grant is for students whose parents had economical damages such as losing their jobs, houses, shops and their productive facilities. In Kesennuma City, economical damages by tsunami was so huge and serious that public supports system for educational expense was not enough to help a large number of students and parents who had serious damages by this disaster. This grant covers 20,000 yen per month for 3 years. 632 students of Kesennuma City were selected as recipients of this grant. This grant was also organized by NEUAJ and it was donated by software company, men's suits company and NPO. Both of grants were very helpful for family budget of tsunami affected families, such as expense of learning materials, excursion, club activities and so on (Table 6.9).

6.5.2.4 Long-term Response 2: "Making Substitute School Yard for School Activities"

After the earthquake and tsunami, a large number of temporary houses were built for the people who evacuated to shelters and houses of relatives in Kesennuma City. But Kesennuma has not so much flat land for building temporary houses because it

Table 6.9 Economical supports for students affected by East Japan earthquake and tsunami

Economical support	Recipient	Number of recipients	Contents of support
Public support system for educational expense	Needy parents Affected students	2,293 students (1,714[a])	Learning materials and goods Expense of school lunch
Grant for orphans by disaster	Disaster orphans	63 students[a]	Lump-sum payment: 100,000 yen Payment per month: 20,000 yen (until graduation of high school)
Grant for disaster affected students	Affected students	632 students[a]	Payment per month: 20,000 yen for 3 years

[a]Number of disaster affected students of elementary and junior high schools of Kesennuma City in 2011

Table 6.10 Temporary houses at school yard of elementary and junior high school in Kesennuma City/October 2012

School level	Number of schools	Percentage in total schools (%)	Number of temporary houses	Percentage in total temporary houses (%)
Elementary school	7	35	143	4.1
Junior high school	11	85	1,000	28.5
Total	18	55	1,143	32.6

located in Rias coastline and also massive earthquake caused subsidence of flat land of Kesennuma. So the land was short and not enough for temporary houses. These are reason why Kesennuma City government had to decide to build the temporary houses at school yards of elementary and junior high school of Kesennuma City. Beyond half of schools in Kesennuma provided their school yards to build the temporary houses. Especially, almost all the school yards of junior high schools were occupied by temporary houses for evacuees (Table 6.10).

The school which provided school yards for temporary houses could not use their school yards for physical education and club activities, so that they have been utilizing small spaces of school yards which were remained, gym and hallway. And they also borrowed school yards of near elementary schools which were not occupied with temporary houses. But Oya Elementary School and Oya Junior High School shared same schools yard, so that both of schools lost their school yards. And as Shishiori Junior High School is located far from Shishiori Elementary School, so it is impossible for the junior high school to borrow the school yard of elementary school for P.E. or club activities. These two cases are inconvenient cases. Therefore, Kesennuma City Board of Education tried to negotiate with community members and NPO/NGO to afford substitute school yards to these schools instead of their school yards.

Fig. 6.6 The well with solar panel of Kesennuma elementary school

As a result, BOE was able to borrow lands from communities and to get financial support of NPO, so that BOE could made three substitute school yards beside Oya Elementary and Junior High School as well as Shishiori Junior High School.

6.5.2.5 Further Response 1: "Constructing Electronic Power and Water Resources"

After the East Japan Earthquake and Tsunami immediately, all of power supplies in Kesennuma City were cut off because electric cables were down in many places and transformer substations were destroyed by the massive earthquake and tsunami. The electronic power supply was stopped also at every school in Kesennuma City. This caused that not only any electronic devises such as telephones, televisions, computers and stoves, but also the waterworks for drinking, cooking and toilets were not available in each schools. This situation was very serious problem for schools to protect students and evacuees and to manage the shelter at each school.

Learning from this experience and lesson, Kesennuma City Board of Education deicide to construct emergency electronic and water supply system at schools those are used as shelters in disaster for the purpose of preparing next disasters in future. Kesennuma BOE selected 16 schools (7 elementary schools and 9 junior high schools) and drilled wells and set solar panels as power resources until 2012, getting financial supports from NPO (Fig. 6.6). And Supporting from German City, Kesennuma BOE could set portable solar panels at other 11 schools in 2011 as emergency electronic resource and daily using outside. So, all of schools in Kesennuma City have solar panels as emergency electronic supply.

In addition, Kesennuma BOE is planning to set solar panels with storage battery at five schools in 2012 getting financial support from the foundation of company, and also BOE is trying to expand these same facilities to more ten schools until 2015.

Fig. 6.7 (**a**) "Move forward from East Japan Earthquake and Tsunami" published by Kesennuma City Principal's Association, Kesennuma Board of Education and Miyagi University of Education. (**b**) "Lesson from East Japan Earthquake and Tsunami" published by Kesennuma Board of Education and Ministry of Environment

6.5.2.6 Further Response 2: Improving DRM and DRR Education for Creative Recovery

After the East Japan Earthquake and Tsunami, Kesennuma City Board of Education and schools in Kesennuma City have been promoting Educational recovery as described above. In addition to these strategies, education sectors in Kesennuma City have been proceeding with physical and psychological supports for affected students fostering their resilience (Oikawa 2012).

And they are also improving and rebuilding the DRR education utilizing lessons from the disaster of East Japan Earthquake and Tsunami. For example, School Principals and BOE in Kesennuma have published the report of DRM and recovery process of school through the experience of East Japan Earthquake and Tsunami. That will be very useful as a lesson study and research of DRM of each school in Kesennuma before and after East Japan Earthquake and Tsunami. And teacher research group (Kyouiku Kenkyuin) also published a model lesson plan of DRR education from new perspective based on "Education for Sustainable Development (ESD)" (Fig. 6.7).

Moreover Kesennuma BOE is now developing "Creative Recovery Education" based on Education for Sustainable Development which Kesennuma has done so far. To realize that educational concept, Kesennuma BOE held UNESCO School International Forum in January 2012 and Kesennuma ESD/RCE Round Table

Meeting in October 30th 2012. In addition, OECD Tohoku School will be held in Kesennuma City in Mach, 2013. All of these events and practices are expected to foster future leaders who can contribute to recovery process locally and globally.

6.6 Conclusion: Lessons from Past Disaster and "N-help" for Response to Recovery

When it is analyzed that the recovery process from the disaster of East Japan Earthquake and Tsunami and it is discussed which kind of institutional response should be done according to phases or stages of recovery process, it is crucial to learn a lot of lessons from past disasters. One of huge disaster of Hanshin-Awaji Earthquake occurred in 1995 gave many lessons to the institutions and schools in affected area of East Japan Earthquake and Tsunami. Some parts of situations of Hanshin-Awaji Earthquake are similar to them of Tohoku area, so that their experiences and method are very useful for the people and institutions in Tohoku area, For example, they suggested how to check students' safeties at school, how to manage shelters with communities, how to restart schools, how to take care of students' psychological aspects, moreover, how to improve DRM and also to promote DRR Education and so on. Actualy, "EARTH" which is disaster support team dispatched by Hyogo prefecture came to Kesennuma from after the disaster immediately, and they gave lots of advices and suggestion to BOE and schools on the recovery process of school education.

As a lesson from Hanshin-Awaji Earthquake, Kobe City Board of Education established text book of DRR Education that is named "Let's carry happiness over." On the text book, it is emphasized that the importance of life, collaboration, appreciation, bonds of family and volunteer. And Kobe BOE also insists that they don't make use of their experience of Hanshin-Awaji Earthquake as negative experiences, but they tried to make the best use for educational recovery and creation of new education (Morimoto and Tahara 2012). These are completely same as concepts of people and institutions in affected area of East Japan Earthquake and Tsunami.

On the other hand, Okushiri Tsunami Disaster in Hokkaido area also gave many suggestions to tsunami affected area in Tohoku, because the disaster in Okushiri was caused by tsunami mainly, so that the character and situations of disaster are same as Tohoku. Recovery process of Okushiri Town has been done with three major strategies, one is "Reconstructing Lives," second is "Building DRR Town," last is "Enforcing Industry." They established "Disaster Recovery Foundation" and they built many resilient infrastructures against tsunami using the foundation. They also established Tsunami memorial park to memorize the experience and lessons of tsunami disaster, and to hand them down to posterity. Talking about DRR Education, Okushiri BOE established DRR education promotion committee and it made DRR manual for students based on Self-help. They disseminated the manual to whole school in town and they are performing evacuation drill repeatedly with whole residents' participation. Moreover, BOE developed Practical Guide line of DRR education, and each school put DRR lesson into their curriculum and performs systematically from elementary school to high schools collaborating community (Hokkaido Educational Research Center 2012).

Not only Kesennuma BOE but also other educational institutions and schools in disaster affected area in Tohoku are able to learn many informative lessons from two experiences of disaster and to introduce these lessons to their response effectively. In addition, institutions and schools in Japan are also able to learn globally from Indonesia and New Zeeland those experienced similar disaster to our country.

Commonalities of recovery processes of both case, Hanshin-Awaji Earthquake and Okushiri Tsunami, are practical action and response at local level, linkage with community and collaboration with diverse sectors locally and globally.

Kesennuma BOE and schools are implementing many activities collaborating with community and other institutions based on ESD and community learning (Furusato Gakusyu). Thank to these strong linkages and bonds, almost all the students were able to evacuate safely in the disaster of East Japan Earthquake and Tsunami. And then, these linkages also functioned at management of each shelter surely and will foster Mutual-help in the communities including schools furthermore (Oikawa 2012).

Finally, as described above, Kesennuma BOE are taking many responses to educational recovery depend on phase and stage. All of these responses have been promoted by not only Kesennuma BOE, but also diverse sectors such as many companies, NGO/NPO, banks and other institutions including universities collaborated and participated in these process. Namely, Kesennuma BOE has been promoting recovery making collaboration and partnerships with diverse sectors and multi-stakeholders. Kesennuma BOE calls this collaboration for recovery "N-help." "N" means NPO/NGO and Network, and also Next stage to Self-help, Mutual-help and Public help. Education sectors in Kesennuma are trying to establish recovery process through participation and collaboration among diverse actors.

References

Hokkaido Educational Research Center (2012) Hokkaido south east earthquake, Educational recovery from disaster of East Japan earthquake and tsunami. National Institute for Educational Policy Research, Yukoudo, pp# 156–160

Kesennuma Board of Education (Educational Research Group), Ministry of Environment (2012) Lesson from East Japan earthquake and tsunami. http://www.env.go.jp/jishin/park-sanriku/images/6-2.pdf. Accessed 24 July 2013

Kesennuma City Principals' Association, Kesennuma Board of Education, Miyagi University of Education (2012) To Advance After the Disaster, vol 1. http://fukkou.miyakyou.ac.jp/report/pdf/kesen1_eng.pdf. Accessed 24 July 2013

MEXT (2011) Examination about preparation of the school facilities based on the damage of the East Japan earthquake and tsunami. http://www.mext.go.jp/component/a_menu/other/detail/__icsFiles/afieldfile/2012/07/26/1323887_16.pdf. Accessed 24 July 2013

MEXT Emergency Proposal (2011) About preparation of school facilities based on disaster of East Japan earthquake and tsunami. http://www.mext.go.jp/component/b_menu/shingi/toushin/__icsFiles/afieldfile/2011/07/07/1308045_1.pdf. Accessed 24 July 2013

MEXT (2012) Research of school response in disaster of East Japan earthquake and tsunami. http://www.mext.go.jp/a_menu/kenko/anzen/__icsFiles/afieldfile/2012/07/12/1323511_1.pdf. Accessed 24 July 2013

Ministry of Education, Culture, Sports, Science and Technology (MEXT) (2011) Information of damage of disaster of East Japan earthquake and tsunami, no. 208. Press release 17 May 2011. http://www.mext.go.jp/component/a_menu/other/detail/__icsFiles/afieldfile/2012/10/30/135089_091410_1.pdf. Accessed 24 July 2013

Oikawa Y (2012) Education for sustainable development and its implication to recovery process in Kesennuma – East Japan earthquake and tsunami, Research Publishing, Singapore, pp# 165–186

Oikawa Y (2012) Education recovery based on ESD, Educational recovery from disaster of East Japan earthquake and tsunami. National Institute for Educational Policy Research, Yukoudo, pp# 175–179

Sakuma A (2012) Role of nutrition instructor and teacher in disaster of East Japan earthquake and tsunami. Report of 53th Japan nutrition instructor & teacher research meeting

Shaw R, Takeuchi Y (2012) Damage of education sector and its recovery. East Japan Earthquake and Tsunami, Research publishing, Singapore, pp# 143–163

Morimoto S, Tahara T (2012) Education recovery from disaster of Hanshin-Awaji earthquake – educational recovery from disaster of East Japan earthquake and tsunami. National Institute for Educational Policy Research, Yukoudo, pp 128–138

Part II
Education and Learning

Chapter 7
Implication of Floods—2010 on Education Sector in Pakistan

Amir Nawaz Khan and Amjad Ali

Abstract In Pakistan, like many other developing countries, floods are the recurrent and damaging phenomenon. Pakistan witnessed an exceptionally high rainfall in July and August of 2010, extending till the month of September. This rainfall proved to be one of the major contributing factors that led the country to be a victim of devastating flood. The Floods—2010 have been declared as the worst since 1929 for its extensive inundation and devastations across the country. Floods—2010 stretched to an area of 100,000 square km, which comprised of 78 districts of Pakistan. The number of the affected population is more than 20 million along with almost 2,000 reported deaths. As this ill-fated disaster traversed across the country from north to south, it seriously disintegrated many sectors that play a key role in the development of a nation. Of these moth-eaten sectors, one of the most critical one is the education sector, which already had suffered to the hands of a series of unfortunate natural and man-made disasters. Looking to social and cultural setup in the affected areas, the implication of Flood—2010 on female education is far reaching. These frequent disasters bring damages and miseries on one hand and opportunity for development on the other. The education sector has been recovered with high standards of development after the Earthquake—2005 in AJK and Khyber Pakhtunkhwa. Recently, after the Internally Displaced Persons (IDPs) Crisis—2009 in Malakand Agency of Khyber Pakhtunkhwa the education sector was found to be one of quickly developed sectors. Recovery from the Flood—2010 is under process. Physical reconstruction and teacher training programmes are the major components of the recovery strategy. Incorporating social awareness, students' counseling and focusing more on female education will have far reaching effects on overall development. This chapter gives an insight into the implications of the Floods—2010 on education sector. Section 7.1 of the chapter draws a general sketch of the floods at global level and the factors enhancing the vulnerability of the country to floods. Section 7.2 discusses

---

A.N. Khan (✉) • A. Ali
Center for Disaster Preparedness and Management, Peshawar University, Peshawar, Pakistan
e-mail: nawaz57@yahoo.com

R. Shaw (ed.), *Disaster Recovery: Used or Misused Development Opportunity*,
Disaster Risk Reduction, DOI 10.1007/978-4-431-54255-1_7, © Springer Japan 2014

the causes of Floods—2010 while the general damages of Floods—2010 is described in Sect. 7.3. The indirect and indirect impacts of the flood on education sector are explained in Sect. 7.4 whereas the way forward and conclusion are discussed in the final section.

Keywords Build back smarter • Education recovery • Pakistan flood • School education • Teacher training program

7.1 Flood Disaster: Trends Across the Globe and in Pakistan

Disasters are growing in number and ferocity (IFRC 2009a). These disasters are caused by the combination of natural hazards and vulnerability of the human population. Globally, frequency and disaster risk is increasing for weather related hazards such as floods and tropical cyclones at an alarming rate (UNISDR 2010). Floods are the dominant natural disasters in Asia Pacific region. Floods are ranked on the top in the list of top five frequently occurring natural disasters between 1998 and 2007 (IFRC 2009a, b). Every year, flooding accounts for two thirds of people affected by natural disasters (Stolton et al. 2008). This can be attributed either to a high level of vulnerability or a low level of coping capacity, or probably both. Majority of the devastating flood disasters are mostly confined to South Asia and its neighbouring country People's Republic of China (EM-DAT 2010). Floods—2010 of Pakistan is one of the top ten flood disasters as the number of people affected is more than twenty million (ADB and WB 2010).

Pakistan is a country with diverse topographical features and accumulates communities of assorted cultures. This diverse topography and culture is a blessing of nature. However, because of lack of awareness and coping mechanism the same serves to be a major factor in enhancing vulnerability of the country to natural disasters (Khan 1993, 2007). Consequently, Pakistan has been a frequent victim of floods and related disasters throughout its history. Several large scale flood disasters have struck vulnerable communities in all the four provinces of Pakistan from time to time in the recent past. Pakistan has been the hard hit by 16 major floods since its independence in 1947. These floods caused severe economic losses equivalent to Rs. 386 billion (exclusive of Floods—2010) (Akram 2010; NDMA 2010, 2011; Khan 1993, 2004; Atta-ur-Rahman and Khan 2010, 2011).

Pakistan witnessed an exceptionally high rainfall in July and August of 2010, extending till the month of September. This rainfall proved to be one of the major contributing factors that led the country to be a victim of devastating flood in the midst of the year. Monsoon rains, which started in Khyber Pakhtunkhwa on Jul 27, 2010 caused floods and damage in many areas (Akram 2010; NASA 2010; and Hanif 2010). The Floods—2010 have been assessed to be the worst since 1929. Floods—2010 stretched to an area of 100,000 km^2, which comprised of 78 districts of Pakistan. This exceptionally high floods leading to an unforgettable disaster in the history of Pakistan has affected almost one tenth population of the country

(ADB and WB 2010). The developmental gains of decades have been totally washed away in a few days. The cumulative financial loss of Floods—2010 is estimated to the tone of about PKR 765 billion (US$ 9.5 billion) which is almost one fourth of the country's financial budget outlay for the year 2010–2011. The floods 2010 eclipsed the scale of the devastating 2004 Tsunami (ABC News 2010).

7.2 Causes of Floods—2010

In Pakistan, the Monsoon season mainly occurs during June to August each year, Monsoon are actually moisture bearing winds of the Indian Ocean which rush into the region as a result of the low pressure developed because of the heating up of the land due to hot and dry summer. The winds blow till they collide with the Himalayas and rise up to form clouds to rain. This process is repeated throughout the Monsoon season. The Floods—2010 were caused by variety of factors. Some of these factors include supercharged Monsoon, westward shift of Monsoon precipitation, encroachment in rivers and torrents, deforestation and sedimentation in the water bodies. The Monsoon of summer 2010 was quite an unusual phenomenon. In early July, a strong ridge of high pressure began to develop near the Ural Mountains in Russia. This ridge became stationary and established a blocking pattern (omega block) for nearly two consecutive months. With an abnormally active jet stream riding around the periphery of the omega block into western parts of Pakistan, copious amounts of hot and moist air created a supercharged monsoon. The rapid development of a La Niña cycle in the Pacific Ocean helped to create a rarely-seen atmospheric setup that led to the flood disaster in Pakistan (Impact Forecasting 2010 and NASA 2010). Hence, the unique combination of a number of meteorological factors altered the usual path of the monsoon and shifted it further westwards. Apart from this the monsoon also thickened its concentration by many folds. As a consequence very heavier and unprecedented high rainfall occurred in the province of Khyber Pakhtunkhwa. Khyber Pakhtunkhwa received many times higher average rainfall of the province in less than 1 week. This phenomenon led to the unprecedented flash floods in the whole of northern Pakistan. These flash floods were particularly severe within the catchment areas of each and every western tributaries of Indus in the province of Khyber Pakhtunkhwa (Hanif 2010 and FAO 2010).

Lack of implementation of rules, regulation and by-laws particularly in construction and encroachments in rivers' channels and torrents in downstream areas carrying more than usual discharge was one of the important intensifying factors. Illegal construction worsened the scenario by increasing vulnerability of the communities, ultimately leaving the communities exposed to an extreme level of risk. As a result, damages caused by floods were increased to a greater extent as encroachments added to the velocity and ferocity of the floods. Indiscriminate and ruthless deforestation has also played a tremendous role in aggravating the Floods—2010. When forest cover is lost, erosion increases unprecedentedly and runoff rapidly flows into streams without any hurdle. This rapidly flowing water pick up more and more sediments leading to raising the level of river channels. Consequently, due to reduction in the channel capacity, the downstream villages, cities, agricultural fields and other infrastructure are subjected to flooding, especially during the rainy season (AAJ News 2010; Khan et al. 2010; and Atta-ur-Rahman and Khan 2010, 2011).

7.3 Damages of Floods—2010

Pakistan has a long history of flooding since its establishment. Flooding here has mainly been witnessed from the Indus River and its tributaries. Besides that damages of floods in Pakistan have been very terrible. It has been observed that the loss of human capital and social consequences as a result of disasters are particularly far reaching in developing countries, where they are often linked with or exacerbated by poverty, reflecting wider socio-economic inequalities (Khan 1993 and UNESCAP 2010). Floods—2010 have been observed as an amalgam of various types of floods across the country and have seriously disrupted the routine day to day life, natural environment and socio-economic fabric of Pakistan to an unforeseen level. In the northern mountainous areas of Khyber Pakhtunkhwa, due to its unique landscape the hazard originated as Flash Floods as a result of unprecedented high level of rainfall. However, the same was converted into a fatal river flood downstream when it got into the Indus River and inundated a greater part of Southern Khyber Pakhtunkhwa, Punjab and Sindh. Province wise Sindh holds the leading share in sufferings from 2010 devastating floods. It suffered from almost half of losses of the country as a consequence of deadly Flood—2010 (Figs. 7.1 and 7.2). The overall recovery and reconstruction cost associated with floods was estimated as approximately US$ 8.74–10.85 billion, which includes estimated costs for relief, early recovery, as well as medium- to long-term reconstruction (ADB and WB 2010).

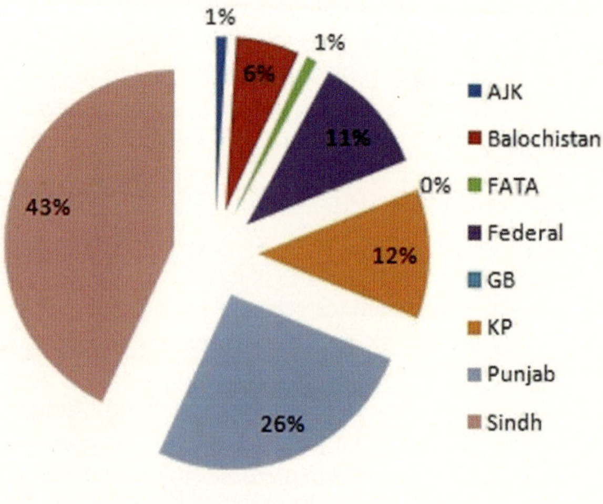

Fig. 7.1 Area wise losses of the flood: 2010 (*Source*: ADB and WB 2010)

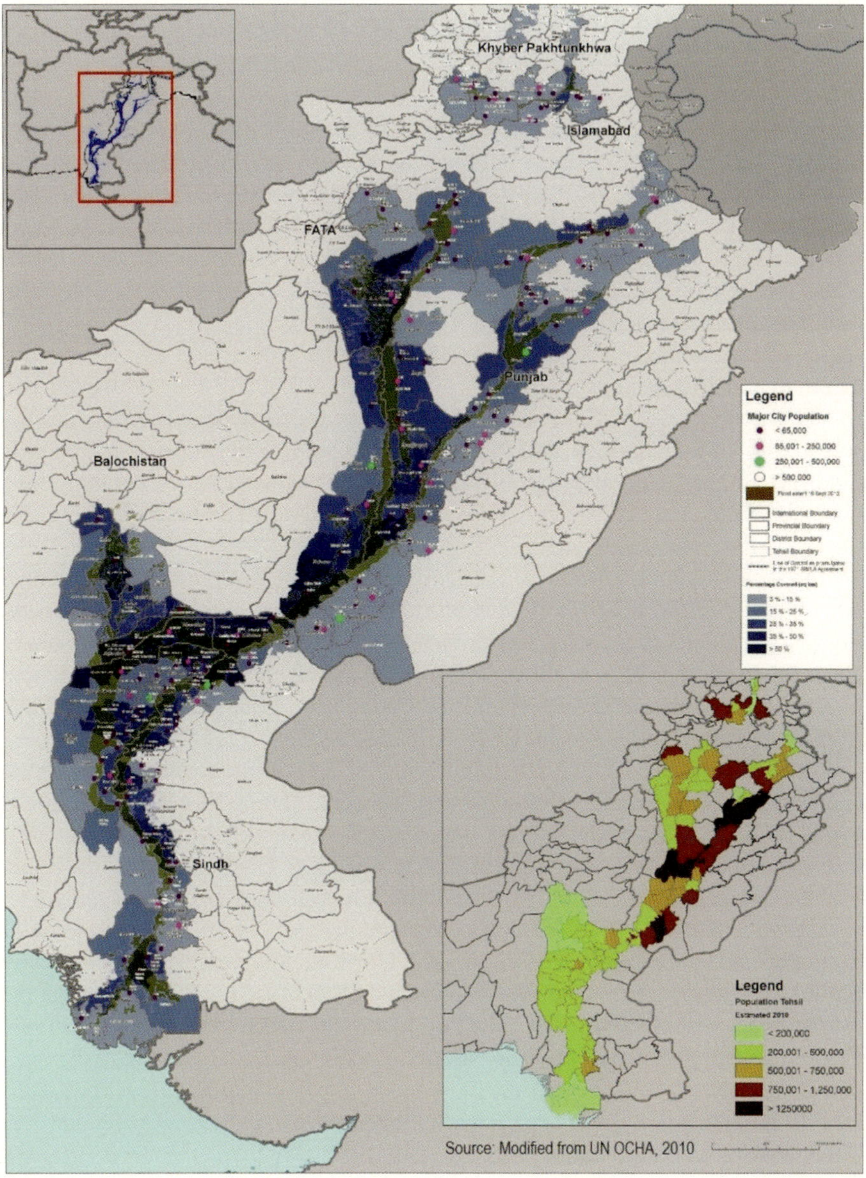

Fig. 7.2 Pakistan's flood: 2010 affected areas

7.4 Damages to Education Sector by Floods—2010

In Pakistan, natural calamities and human-induced disasters have further aggravated the already poor and inadequate access to education. The Earthquake—2005 brought destruction to a large number of educational institutions in Khyber Pakhtunkhwa and State of Azad Jammu and Kashmir (AJK). The total damaged educational institutions were 5,857 in which 4,604 were primary schools, 717 were middle schools, 479 were high and higher secondary schools and 57 were higher education institutions including one University of AJK. The total cost of the damages was estimated 545.25 million US $ (GOP 2005). Educational institutions in Khyber Pakhtunkhwa and FATA have been continuously facing terrorist threats. There were 473 educational institutions damaged in Khyber Pakhtunkhwa and two tribal agencies of FATA. The total estimated cost of the damage caused to education sector was US$ 38 million (PDMA 2009). Similarly, educational institutions in Sindh and Baluchistan were heavily hit by the cyclone and heavy rains in 2007. Gilgit Baltistan was hit by heavy rains and crisis of Ata Abad Lake in addition to damages to schools due to terrorist attacks.

The reconstruction efforts for such institutions were hardly been implemented when the July–August 2010 torrential rains and consequent floods brought devastation once again to educational institutions in majority of districts in all provinces and areas including Gilgit Baltistan, FATA and AJK. Khyber Pakhtunkhwa has been hit by all three types of disasters while in Islamabad Capital Territory educational institutions have been almost safe from all of these disasters. Damages to Universities and other institutions of higher learning occurred in the Earthquake—2005, however, fortunately no such institutions were damaged during Floods—2010 (ADB and WB 2010). It has been observed that education sector has been one of the seriously affected sectors as a consequence of Foods 2010. Damages to this sector are estimated to be US$311.3 million. The unprecedented floods have damaged a total of 10,348 educational institutions in the country. Of which 3,741 were fully destroyed and 6,666 were partially damaged. However, affected institutions are only 6.2 % of the total institutions in the country and 12 % of the total institutions, at the average, in the affected districts. The two worst affected provinces are Sindh and Punjab, where 18.5 % and 8.8 % of the pre-flood educational facilities have been damaged respectively, followed by 12.9 % and 5.6 % of pre-flood facilities damaged in Baluchistan and Khyber Pakhtunkhwa respectively (Table 7.1, 7.2 and 7.3) (ADB and WB 2010).

Besides damages to the educational buildings and services, a significant phenomenon that influenced the performance of education sector is that educational buildings in majority of the flood affected areas were used as shelter for flood survivors. Almost 7,820 schools were reported to have been fully or partially damaged in Punjab, Sindh, Baluchistan, Khyber Pakhtunkhwa, Gilgit Baltistan and AJK, and about 4,935 schools were being used as relief shelters (OCHA 2010). A decrease in the number of schools being used as shelters has been reported from Khyber Pakhtunkhwa and Sindh. The extent of damage and spilling over of vulnerability as

Table 7.1 Province/area wise damages

	Institutions damage				
Province/area	Schools	Colleges	Vocational	Rs (M)	US$ (M)
AJK	194	4		842	9.9
Baluchistan	557			776	9.1
FATA	176			415	4.9
Gilgit Baltistan	79	2		336	4.0
Khyber Pakhtunkhwa	870	13	17	2,860	33.6
Punjab	2,817	4	4	7,881	92.7
Sindh	5,655			13,355	157.1
Total	10,348	23	21	26,464	311.3

Source: DNA (2010)

Table 7.2 Education—physical damage detail—schools

Categories	AJK	Baluchistan	FATA	Gilgit/ Baltistan	Khyber Pakhtunkhwa	Punjab	Sindh	National
Higher secondary schools							20	20
Completely destroyed							10	10
Partially damaged							10	10
Secondary schools	21	25	7	8	52	141	121	375
Completely destroyed	2	2	3	1	9	19	54	90
Partially damaged	19	23	4	7	43	122	67	285
Middle schools	47	37	14	25	81	264	97	565
Completely destroyed	16	1	2	6	20	34	43	122
Partially damaged	31	36	12	19	61	230	54	443
Primary schools	126	495	155	46	737	2,412	5,417	9,388
Completely destroyed	20	25	30	12	197	604	2,627	3,515
Partially damaged	106	470	125	34	540	1,808	2,790	5,873
Grand total	194	557	176	79	870	2,817	5,655	10,348

Source: DNA (2010)

Table 7.3 Education—physical damage detail—colleges/technical and vocational institutions

Categories	AJK	Balu-chistan	FATA	Gilgit/ Baltistan	Khyber Pakhtunkhwa	Punjab	Sindh	National
Colleges	4	0	0	2	13	4	0	23
Completely destroyed	0	0	0	0	0	3	0	3
Partially damaged	4	0	0	2	13	1	0	20
Technical/ vocational lost	0	0	0	0	17	4		21
Completely destroyed	0	0	0	0	1	0	0	1
Partially damaged	0	0	0	0	16	4	0	20

Source: DNA (2010)

Fig. 7.3 Middle school affected by floods: 2010

a consequence can clearly be estimated by the fact that 1.6 million children have been affected either by the damaged schools, or because schools were being used as shelters (AAJ 2010). Recovery needs for the damages of Flood—2010 are estimated at the tone of US$ 504.8 million including US$63.7 million for short-term needs (ADB and WB 2010). A number of humanitarian organizations have been intervening in the field so as to address the immediate needs of the education sector. Prominent among the immediate needs regarding the education sector are ensuring availability of teachers and reading material, i.e. text books and tutorials etc. (Figs. 7.3, 7.4 and 7.5).

Fig. 7.4 Middle school affected by floods: 2010

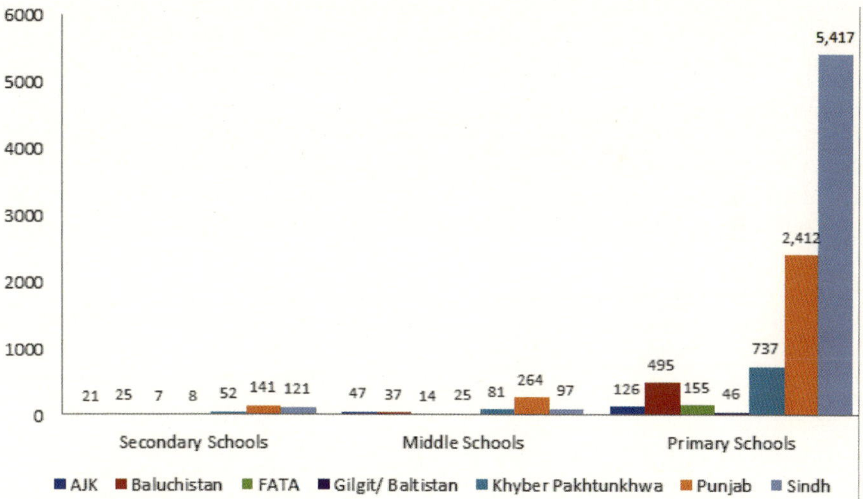

Fig. 7.5 Primary school affected by flood 2010

7.4.1 Direct and Indirect Damages to Educational Facilities

Prior to the floods, there were 172,098 educational institutions in the country (141,120 or 82 % of these being primary schools and 1,119 colleges). Among these institutions 61,500 were for girls and over 32,000 were mixed. Total enrollment in these institutions was 19,924,283 of which almost 40 % (7,865,380) were female.

A total of 10,348 institutions suffered damages from flood 2010 in around 90 affected districts. Out of these damaged institutions 3,741 were fully destroyed and 6,666 were partially damaged.

The percentage of female institutions amongst the total damaged institutions was 25 %. However, overall this represents only 6.2 % of the total institutions in the country and 12 % of the total institutions in the affected districts. The partially damaged institutions constituted around 64 % of the total damaged institutions. Hence, the data revealed that not only male but also female educational institutions were severally affected by Flood—2010. In developing country like Pakistan where female literacy ratio is already very low, the situation of female education has been further exacerbated with the damages of Floor—2010.

Taking overall scenario into focus the data revealed that damage to educational institutions has affected around 32,000 teachers and more than one million students (of which 0.42 million were girls). Province wise, the largest number of institutions affected have been reported in the Sindh province (which is 5,655 including 811 girls' schools) followed by Punjab (which is 2,817 including 1,222 girls' schools), Khyber Pakhtunkhwa (which is 915 including 626 girls' schools), Baluchistan (557), AJK (194), FATA (which is 176 including 95 girls' schools) and Gilgit-Baltistan (81). The data also revealed that more than 90 % of the affected institutions are primary schools (which are 9,368 out of total 10,348 institutions). The most heavily damaged district in the country is Jacobabad (damaged Schools 1,247) in Sindh, followed by Muzaffargarh in Punjab (damaged Schools 1,164). The other districts of the province of Sindh which were severally affected, include Thatta (damaged Schools 966), Dadu (damaged Schools 764), and Kashmore (damaged Schools 571). Rajanpur (damaged Schools 410) and Jhang (damaged Schools 311) in Punjab, and D.I. Khan (Damaged Schools 311) in Khyber Pakhtunkhwa were also among the heavily hit districts (Fig. 7.6). The total direct and indirect losses were approximately 26,464.3 million Rupees (Table 7.4). Most of the affected districts have rural population distribution, where due to social and cultural restrictions the female literacy ratio is already very low. As a matter of fact the data revealed that the women and children (the vulnerable groups) were affected severally by Flood—2010 than any other social group. However, the unfortunate aspect is that in the rehabilitation and reconstruction process these female educational institutions are at very low priority due to social and political representation. Combining all these factors the impacts of Flood—2010 on female education is far reaching (Fig. 7.7).

7.4.2 Service and Productivity Losses of Schools/Educational Institutions

Most of the affected districts with the exception of northern mountainous region are in the summer zone and the flooding occurred during the summer vacations. Hence, there was negligible loss of life among the students, teachers and staff. This also meant that floods did not result in loss of educational contact hours, the only

Fig. 7.6 Primary school affected by flood 2010

Table 7.4 Education—damage and losses

Province/area	Damage (M. Rs.)	Loss (M. Rs.)	Total (M. Rs.)
AJK	782.9	59.4	842.3
Baluchistan	608.4	167.1	755.5
FATA	361.8	52.8	414.6
G-B	311.9	24.3	336.2
Khyber Pakhtunkhwa	2,585	274.5	2,859.5
Punjab	6,383.1	1,498.2	7,881.3
Sindh	11,013.4	2,341.5	13,354.9
National	22,046.5	4,417.8	26,464.3

Source: DNA (2010)

exception being, that summer vacations were extended further for a few days. However, it was not solely because of occurrence of flood but also partly because of the coming of Muslim festival "Eid-ul-Fiter" in a few days' time. It was observed that indirect losses were multifarious. A large number of institutions have been used to provide shelter to the flood affected people. For example, in Punjab, there were 2,169 schools and in Sindh 2,372 schools were being used as shelters for the affected population.

A significant number of students and teaching and non-teaching staff suffered from emotional trauma, necessitating counseling services. This in turn required teachers to be trained for school-based psychosocial support, counseling and

Fig. 7.7 Shelter house: Govt. Girls Middle School during Floods: 2010

tutorials. Students of the schools used as temporary shelters for the survivor and displaced people were themselves being displaced with respect to (the certitude of) schooling. Similarly, those students whose schools have been destroyed or partially damaged, schooling has been interrupted unless prompt arrangements are put in place to ensure continuity. Further, in the absence of proper facilities and equipment, it is conceivable that the already poor quality of education may be impaired further. Operational costs for running the fully or partially damaged schools by using alternate means will be much higher and not feasible in the prevailing circumstances.

7.4.3 Recover and Reconstruction Needs

Pakistan, by now has the experience of the recovery and reconstruction of educational sector due to Earthquake—2005 in AJK and Khyber Pakhtunkhwa. The Earthquake Reconstruction & Rehabilitation Authority (ERRA) outlined a comprehensive policy for reconstruction of education sector. The guiding principles were based on community participation and consultation with all other stakeholders during planning, strategy formulation and implementation; donors have to fully fund package of services (building, equipment, furniture, training, capacity building) to institutions selected for support; ensuring the equity (gender, geographic areas and level of education); involvement of communities to foster ownership; and involving state agencies to manage transitional activities like semi-permanent structures etc. Some important characteristics were risk free site selection, teacher training, better equipment, high quality of civil work and community ownership. With a total cost of US $ 567.5 million of reconstruction more than 60 % recovery of the education sector has been achieved with an improved standards. This situation is far better in

Table 7.5 Education—reconstructions needs

Province/area	Short-term 6–12 months	Long-term 12–36 months	Total reconstruction (Millions) PKR	US$
AJK	61.125	1,167	1,228.125	14.4
Baluchistan	215.6	1,094	1,309.6	15.4
FATA	66.185	558	624.185	7.3
GB	33.87	507	540.87	6.4
KP	351.86	4,141	4,492.86	52.9
Punjab	1,810.305	10,060	11,870.31	139.7
Sindh	2,871.63	19,969	22,840.63	268.7
National	5,410.575	37,496	42,906.58	504.8

Source: DNA (2010)

AJK where almost 90 % recovery has been made in the education sector. It has been observed that due to resettlement of the Balakot town the overall figure of reconstruction is lower than excepted (ERRA 2012). During Internally Displaced Persons (IDPs) Crisis—2009 in Malakand Agency of Khyber Pakhtunkhwa and FATA, education was one of the worst affected sectors. With a total cost of US $ 36.7 million education has been one of the quick and highly developed sectors in Khyber Pakhtunkhwa (PDMA 2012).

Reconstruction of the educational institutions affected by Flood—2010 is under process with a two steps approach. The first step is the immediate recovery needs in the *short-term* i.e. 6–12 months; whereas the second step is simultaneous starting up of long-term reconstruction. In the short-term, various alternatives have been used to restart education work immediately. These alternatives include tents, rented buildings/borrowed buildings, temporary structures and repair of partially damaged buildings or their portions that are easily repairable. The estimated cost of such alternatives includes temporary shelters as described above, seating (*"Tat Patti," "Chowkis"* or new/repaired furniture, where ever possible) and educational materials. The teachers in such cases needed orientation, and children need to be vaccinated. The cost has been estimated to be PKR 0.3 million for each damaged institution and those being used by IDPs. *Long-term reconstruction* will entail the actual reconstruction of buildings or repair of partially damaged buildings. Overall total long-term reconstruction needs have been estimated to be PKR 42,906.58 million (US$ 504.8 million). Out of this total cost around 25 % (US$ 126.2 million) will be for girls' institutions. Beside this about PKR 5,410.575 million (US$ 63.7 million) is needed for short-term needs. The highest cost is estimated for the worst affected province of Sindh of which is PKR 22,840.63 million (US$ 268.7 million). For Punjab the reconstruction cost for long term is PKR 11,870.31 million (US$ 139.7 million), for Khyber Pakhtunkhwa, PKR 4,492.86 million (US$ 52.9 million) and for Baluchistan PKR 1,309.6 million (US$ 15.4 million), while for AJK, FATA and GB, the cost is, PKR 1,228.125 million (US$ 14.4 million), PKR 624.185 million (US$ 7.3 million) and PKR 540.87 (US$ 6.4 million) respectively (Table 7.5).

Table 7.6 Education—prioritized sector recovery framework

Activity	Short term (6–12 months)	Medium and long-term (12–36 months)
Resume teaching work	4,886.75	
Damage assessment survey	382.04	
Teacher orientation	161.705	
Reconstruction		37,496
Total (PKR million)	5,430.575	37,496
Total (US$ million)	63.89	441.13

Source: DNA (2010)

7.4.4 Proposed Sector Recovery and Reconstruction Strategies

The federal and provincial governments are in the process of formulating their recover and reconstruction strategies. To address multi-hazard risks and other common factors and requirements for reconstruction, the strategy for the education sector should form part of the overall country and province-specific strategy. The recovery and reconstruction strategy provides an opportunity that should be availed to improve access to higher quality education through improved design of physical learning spaces, social and physical access, laboratories, and capacity development of teachers and the district education offices for improved service delivery. It also provides an opportunity to ensure that building codes are enforced. Monitoring of compliance with these codes needs to be essential, and it is this issue where National, Provincial and District Disaster Management Authorities have a critical and pivotal role to play. The Education Departments at provincial and district levels will need to play a leading role in the planning and implementation of recovery and reconstruction of the education system. The capacities of administration—at the district education level and other stakeholders such as NGOs and the private sector also need to be enhanced if implementation of the reconstruction exercise is to be expedited. The proposed short term prioritized education sector recovery is in need of initial amount of Rs. 4,886.75 million for resuming teaching work, Rs. 382.04 million for damage assessment survey and Rs. 161.705 million for teacher training and orientation (Table 7.6). This strategy needs more detail prioritization and especially for female education. Similarly, students counselling and social awareness through electronic media are the other aspects that needs proper attention from all line agencies.

7.5 Conclusion

Pakistan is located in the Asia Pacific region, where 40 % of the world natural catastrophes occur. Flood is the dominant natural hazard that has been the most frequent in the region during the recent past. In Pakistan, flooding has also been very frequent and very far reaching in their effects. It was found that Pakistan has been seriously damaged in the past by a number of flooding events. However, none of them was as devastating as the Floods—2010. The Floods—2010 are declared to be the worst since 1929. These floods have eclipsed the scale of the deadly 2004 Southeast Asian Tsunami. Having triggered by a complex meteorological setup in the late summer of 2010, the floods damaged almost every part of the country, stretching from the mountainous areas of Khyber Pakhtunkhwa through Punjab and Sindh up to Arabian Sea.

Educational sector was one of the worst affected sectors. Floods—2010 have claimed around 2,000 lives, and left more than 20 million people affected. The cumulative losses from the Floods—2010 to the economy were about US$ 9.5 billion, which are almost one quarter of the financial outlay for the year 2011–12. The recovery and reconstruction cost in the aftermath of this deadly disaster is almost US$ 9–11 billion. These floods have damaged almost all the key sectors of the economy that were contributing towards the economic growth of the country. Among these important sectors, education has also been the worst affected. The total damage and loss for Pakistan's educational institutions works out to be PKR 26,464.3 million (US$ 311.3 million). The highest damage has been estimated in Sindh amounting to PKR 13,354.9 million (US$ 157.1 million) followed by Punjab with PKR 7,881.3 million (US$ 92.7 million), Khyber Pakhtunkhwa with PKR 2,859.5 (US$ 33.6 million). Damage and losses in Baluchistan were PKR 775.5 million (US$ 9.12 million), and in AJK, FATA and GB, it was PKR 842.3 million (US$ 9.9 million), PKR 414.8 million (US$ 4.88 million) and PKR 336.2 million (US$ 3.95 million) respectively. The unprecedented floods have damage a total of 10,348 educational institutions. Among theses one of the third of total institutions were fully destroyed. It was found that a considerable number of these fully damaged institutions were female institutions. The damages to these institutions not only interrupted education sector but affected the overall quality of education as well. The study recommends that the recovery strategy of educational sector should be part and parcel of the overall recovery strategy. This strategy needs to provide education across the board in an improved manner.

To foster the recovery and reconstruction of the affected areas, the approach of Build Back Smarter (BBS) is being implemented in the affected areas, which focuses mainly on the cost optimization of the multi hazard reconstruction. However, to ensure the resilience and safety of the communities, both the approaches of Build Back Better (BBB) and Build Back Smarter (BBS) need to be mixed together and more attention needs to be given to the development of resilient communities.

References

AAJ News (2010) Flood damaged schools led to education worries [online]. http://www.aaj.tv/2010/08/flood-damaged-schools-lead-to-education-worries/. Accessed 5 February 2011

ABC News (2010) Pakistan floods worse than 2004 tsunami: UN [online]. http://www.abc.net.au/news/stories/2010/08/10/2978222.htm. Accessed 20 January 2011

ADB and WB (2010) Pakistan floods 2010: preliminary damages and needs assessment. ADB and World Bank, Islamabad

Akram S (2010) Rapid assessment report of flood-affected communities in Swat district, Khyber Pakhtunnkhwa, Pakistan [unpublished report]. Save the Children, Pakistan

Atta-ur-Rahman, Khan AN (2011) Causes and environmental impacts of flood hazard in the Hindu Kush region: a case study of sample villages, Swat valley, Pakistan. Natural Hazards, Springer 2010 [published online]

Atta-Ur-Rahman, Khan AN (2010) Causes and environmental impacts of flood hazard: a case study of sample villages, Swat valley, Pakistan. In: 4th international congress of the islamic world geographers (ICIWG2010) Iran, pp 161–180. http://iciwgen.webs.com/Proceeding%20Cds/Part%203.pdf

DNA (2010) Pakistan Floods 2010: Preliminary Damage and Needs Assessment (DNA). Government of Pakistan, Asian Development Bank and World Bank. [Online]. https://www.gfdrr.org/sites/gfdrr.org/files/publication/Pakistan_DNA.pdf. Accessed 30 December 2012

EM-DAT (2010) Cred crunch [online]. http://www.cred.be/sites/default/files/CredCrunch22.pdf. Accessed 24 January 2011

ERRA (2012) Earthquake Reconstruction & Rehabilitation Authority, Education: Policy Brief. [online]. http://www.erra.pk/Reports/Education%20Policy.pdf. Accessed 30 December 2012

FAO (2010) Pakistan floods 2010 [online]. http://www.fao.org/emergencies/country_information/list/asia/pakistanfloods/en/. Accessed 6 February 2011

Government of Pakistan (GOP) (2005) Pakistan 2005 Earthquake: Preliminary Damage and Needs Assessment. Prepared by the Asian Development Bank and the World Bank, Islamabad, Pakistan. [Online]. http://www.adb.org/Documents/Reports/pakistan-damage-needs-assessment.pdf. Accessed 20 August 2010

Hanif M (2010) Latitudinal redistribution of precipitation in Pakistan (super flood in Pakistan-2010). Presentation at climate forecast application/EWS workshop of SDC IC livelihoods programme, Peshawar

IFRC (2009a) A practical guide to advocacy for disaster risk reduction. IFRC, Geneva

IFRC (2009b) Disaster: how the red cross red crescent reduces risk. IFRC, Geneva

IMPACT Forecasting (2010) Pakistan flood event recap report [online]. www.aon.com/attachments/reinsurance/201008_pakistan_flood.pdf. Accessed 25 January 2011

Khan AN et al (2010) Climate change adaptation and disaster risk reduction in Pakistan. In: Shaw R et al (eds) Climate change adaptation and disaster risk reduction: an Asian perspective. Emerald, Binglay, pp 195–215

Khan AN (1993) An evaluation of natural hazard reduction policies in developing countries with special reference to Pakistan. Pakistan J Geography 3(1 and 2):81–100

Khan AN (2004) Planning for flood hazard: a geographical approach to flood problem in district Jhelum, Punjab. Peshawar Univ Teacher's Assoc J (PUTAJ) 11:5–99

Khan MA (2007) Disaster preparedness for natural hazards: current status in Pakistan. ICIMOD, Kathmandu

NASA (2010) Flooding in Pakistan caused by higher-than-normal monsoon rainfall [online]. http://disc.sci.gsfc.nasa.gov/gesNews/pakistan_flooding_monsoon_rainfall. Accessed 10 February 2011

NDMA (2010) Flood disaster risk situations [online]. http://ndma.gov.pk/Flood%20Mitigation%20Course.html. Accessed 21 January 2011

NDMA (2011) Damages & losses. [online]. http://www.pakistanfloods.pk/en/pakistan-flood-2010. Accessed 14 February 2011

OCHA (2010) Pakistan monsoon floods situation report # 16 [online]. http://www.pakresponse.info/sitreps/OCHA_SITREP16_23_August.pdf. Accessed 5 February 2011

PaRRSA Provincial Reconstruction, Rehabilitation & Settlement Authority Provincial (2012) Government of Khyber Pakhtunkhwa [online]. http://www.pdma.gov.pk/PaRRSA/su-education.php. Accessed 30 December 2012

PDMA (2009) Damage and Needs Assessment: Immediate Restoration and Medium Term Reconstruction in Crisis Affected Areas (Education Sector). Provincial Disaster management Authority Preliminary. [Online]. http://www.pdma.gov.pk/PaRRSA/documents/DNA.pdf. Accessed 30 December 2012

Stolton S, Dudley N, Randall J (2008) Natural security-protected areas and hazard mitigation. World Wide Fund for Nature, UK

UNESCAP (2010) Protecting Development Gains-Reducing Disaster Vulnerability and Building Resilience in Asia and the Pacific-The Asia Pacific Disaster Report, 2010. UNESCAP, Bangkok, Thailand

UNISDR (2010) Natural hazards, unnatural disasters: the economics of effective prevention [online]. http://www.gfdrr.org/sites/gfdrr.org/files/nhud/files/NHUD-Report_Full.pdf. Accessed 5 February 2011

Chapter 8
Role of Civil Society and its Role in Formal and Informal Education as a Part of Disaster Recovery

Manu Gupta

Abstract In post-disaster recovery, civil society organizations (CSOs) have a critical role to pay in the education sector. The education sector is no just about rebuilding infrastructure—rather it includes restoring the confidence of young minds and through them, the society. The civil society being open in approach is able to use formal, non-formal and even informal means to reach out, especially those who are vulnerable and may have been impacted the most by disaster. Further, using participatory processes, the transition from recovery to resilience is made possible. CSOs may have a limited direct intervention in restoring formal education systems, that which public agencies are better equipped to handle, nevertheless CSOs do support through partnerships in building capacity contributing significantly to strengthening resilience of the education sector.

Keywords Civil society • Education recovery • Formal education • Informal education • Participatory process

8.1 Introduction

Recent catastrophic disaster events have wreaked havoc on the nations' educational systems impacting lives of children and teachers, disrupting education activities. Schools and school children often bear the greatest impact of disasters. In the 2001 earthquake in Gujarat (India), at least 1,884 school buildings collapsed and 5,950 classrooms were destroyed. In 2005, over 17,000 school going children perished in collapsed schools following earthquake in Pakistan. Over 8,000 schools were destroyed or damaged beyond repair. Just 3 years later, in 2008, in neighbouring

M. Gupta (✉)
SEEDS, Delhi, India
e-mail: manu@seedsindia.org

R. Shaw (ed.), *Disaster Recovery: Used or Misused Development Opportunity*,
Disaster Risk Reduction, DOI 10.1007/978-4-431-54255-1_8,

China over 9,000 children and teachers died in their classrooms when the magnitude 7.9 earthquake struck Sichuan. In the same year, Cyclone Nargis struck Myanmar. It was reported that more than 4,000 schools were destroyed and about 600,000 children were affected (UNICEF 2009). Even when loss of life is less, there are other damages that take a long time to recover. In the aftermath of East Japan Earthquake of 2011, local schools faced a shortage of teachers, affecting continuity of education; in some cases schools continued to be used as evacuation shelters up to 5–6 months as local communities refused to go back (Shaw and Takeuchi 2012).

Besides catastrophic events, there are thousands of small disasters that take place much more often adversely affecting schooling and education systems. In a recent research carried out in South Asia (ODI 2012), it was found that extreme weather events have directly impacted children's education, creating additional barriers to accessing and staying in school, as well as to the quality of schooling. In Nepal, for example, collapsed roads and bridges force children to walk further to reach school or deter them completely. Further, increased hardship and poverty linked to recurrent climate-related disasters reduced the ability of parents to pay for children's education, whether in the form of school fees or the hidden costs of replacing damaged uniforms and school supplies. Nepalese girls and boys reported that when post-disaster hardships force their parents to make trade-offs, it is usually girls who are withdrawn from schools. In both Pakistan and Bangladesh, children believed one reason for the female student drop-out rate was that the floods made school routes inaccessible or alternative routes too dangerous with parental concerns for the safety and protection of their daughters taking priority over their education.

It takes considerable time for young population to recover from such events. Living within the rubble and chaos of disaster-affected spaces can have a long term psychological impact on children's lives. In the wake of a disaster, recovery and reconstruction programmes must be initiated as quickly as possible and take into account child specific needs.

Resumption of education in post-disaster situations plays a life-saving and life-sustaining role and helps create a protective environment for all children. It instills a sense of hope and normalcy for the disaster-affected children and helps in reconstruction and building resilience. Educational continuity is crucial for both the intellectual growth and psycho-social wellbeing of children during disasters or civil strife. Schools and other learning spaces act as a safe place for children during unstable times, and can also be a point for the provision of essential services like nutrition, water, sanitation, health and protection. The return to formal education may take time and may require setting up alternative learning spaces and provision of learning and teaching materials (SAARC 2011).

Beyond restoration of education in emergencies, interventions for recovery would need to include strengthening of capacities of schools and the education sector as a whole to address the challenges of disasters.

Following emergencies, restoration of education system becomes a priority. However, while resumption of classes may provide immediate recover, the actual restoration of infrastructure may take extended time periods depending on damage. Most importantly, students and teachers need to overcome trauma and regain their confidence. Awareness, training and preparedness for future emergencies have been found as a useful contribution in the recovery process.

Civil Society Organizations (CSOs) have played an important role in partnering with local communities as well as public agencies in the education recovery process, CSOs work at the grassroots ensuring recovery programmes are contextualized and adapted to needs of the local communities. Community profiles are seldom even; There are sections within communities that are more vulnerable than others merely due exposure from a unsafe location becoming an impediment in access or special needs. CSOs are able to fill this gap in ensuring recovery programmes remain inclusive for those that are "unreachable." While increasingly governments are now ensuring education systems in the country incorporate preparedness training, there is limited translation of such policies and programmes on the ground. Primarily, the lack of capacity, at local levels is a serious limitation. CSOs have partnered with local agencies to ensure preparedness capacity is built. "School Safety" programmes have been pioneered by CSOs in regions affected by emergencies. In some cases, school safety work carried out at the grassroots has sown the seeds for a larger policy level intervention at provincial and national level.

What makes the role of CSOs are unique? Besides the fact that CSOs play an important role of "bridging the gap," they bring with them flexibility and openness in approaches that make intervention possible in short periods of time and better adapted to local needs. CSOs are able to intervene both in the formal as well as non-formal means of education. For this section, formal education has been defined as a system that exists as per government set framework and curriculum. Formal education system is considered as associated with schools, training institutions leading to recognized diplomas and qualifications. Non-formal education takes place alongside mainstream/formal systems of education and such does not usually lead to formalized certificates. Informal education is a "natural accompaniment" to everyday life. Unlike formal and non-formal education, informal education is not necessarily intentional learning (EC). In the context of education as part of disaster recovery, all three systems of education have been put to into practice mostly through humanitarian interventions. A few case examples in the subsequent sections of this chapter illustrate uniqueness in approaches that CSOs particularly are able to bring in the education sector.

8.2 "Small Change": Building Resilient Communities Through Education

In August 2010, a devastating catastrophic cloudburst in the Ladakh region of India, led to devastating flash floods in the towns and villages in the region. Over 400 lives were lost and hundreds of houses, public institutions were damaged. Along with the public agencies, several CSOs were engaged in rebuilding work. Among others, Multistory, UK and SEEDS India along with Innovations in Participatory Education

Fig. 8.1 Children using participatory tools to understand cause and effects (Photo: Sidharth Behl, Seeds)

(IPE) and CENDEP, Oxford Brookes University were involved in a innovative experiment based on the Small Change theory by Professor Nabeel Hamdi (Hamdi 2004). Based on the theory that explores how small, practical and mostly low budget interventions can act as catalysts for big and long lasting change, a unique exercise was carried out in a public girls school in Leh town in Ladakh. A team from Multistory and IPE—practitioners from the education and participatory arts sector, worked with students and teachers from Leh to address disaster risk reduction.

As part of the intervention, a 6 day workshop was carried out with Grade IX students of a local public school. The teaching module comprised distinct lessons combined with a range of creative activities. The creative activities included individual mapping exercises combined with creative writing, leaflet design and a range of participatory photography exercises. The lessons plans focused on flood hazards and risk reduction was incorporated in Geography, English and Art. The use of the arts was seen as a tool to improve and make this learning more fun and creative to enhance active and contextual learning (Fig. 8.1).

The workshop was developed on the assumption that teaching contextualized and active learning methods that relate to subjects associated with risk reduction will help strengthen community resilience to future disasters.

Further, this workshop aimed to test the assumption that participatory arts and cultural action have an important role in achieving long-term development objectives. During the workshop teachers ran a geography lesson about drainage basins. Students were asked to identify key principles of a drainage basin and apply these to the context of Leh. The students were then asked to reflect on the effects that cloudbursts would have in the terrain surrounding Leh. Findings from the workshop revealed that students' understanding about the local causes and impacts of flooding and risk reduction increased because of active, contextualized and creative teaching methods. Also, participatory arts enable people the opportunity for personal expression and empowerment both as an individual and as a group. Most important, the exercise reconfirmed that contextualized learning based on personal experiences (in this case, through arts), helped in internalizing abstract principles.

Long after the workshop intervention, the school students of this school continued to champion the cause of risk reduction and stood out as leaders in local recovery process.

Like using fine arts as form of teaching, that encouraged self-representation and expressiveness, this also recognized local indigenous knowledge. The Indian Ocean Tsunami recovery work helped uncover the age old folk song that warns people to look for signals of Tsunami in Similue Island of Indonesia. Recovery of the education sector has helped restore such traditional systems thereby helping communities understand their long standing relationship with their natural environment and in the process becoming fore-runners for community's recovery to greater resilience.

8.3 Reaching the "Unreachable"

On 2nd May, 2008, Cyclone Nargis made landfall on the western edge of the Ayeyarwady delta, in Myanmar. It caused a devastating loss of lives with an estimated 138,366 people dead or missing. The lives and livelihoods of the survivors were also severely disrupted with up to 800,000 people displaced, 450,000 housed destroyed, 3,761 schools and a substantial loss of food stocks, equipment, infrastructure and paddy fields.

As part of recovery in the education sector (Shikada et al. 2012), SEEDS implemented a school safety project to enhance the capacity of local schools to cope with natural disasters. The main problem faced during the recovery process was the lack of availability of information and resources at local level. This significantly exacerbated their risk thus undermining their capacity to "bounce back better" than before. Appropriate and timely information is crucial to reduce the vulnerability of communities, especially for people in remote areas where access to information is generally limited. To reach out to schools in such locations, and provide access to timely and appropriate knowledge and information, SEEDS conceived Mobile Knowledge Resource Centre (MKRC) (Fig. 8.2) and the Water Knowledge Resource Centre (WKRC) (Fig. 8.3). These mobile centres have been able to overcome physical and location barriers to benefit communities that are often at very high risks. The MKRC was built on a mobile truck, while the WKRC was a converted small ship. Education material of different types and topics were mounted on these vehicles. The education material was depicted through colorful decorations and local child-friendly illustrations on posters, videos, and miniature tools for understanding hazards and risks of disasters.

The Knowledge-Interest-Desire-Action tree model (Shikada et al. 2012) was adapted for developing the teaching material. Strong visuals enabled local

Fig. 8.2 Mobile knowledge resource centre, Myanmar (Photo: Seeds Asia)

Fig. 8.3 Water knowledge resource centre, Myanmar (Photo: Seeds Asia)

communities to understand easily and for long retentions. As part of the outreach program, the training of trainers was carried out for local teachers ensuring acquired skills in risk reduction and preparedness could be carried out much on a sustained basis. Like in case of Leh, causes and effects of risks and hazards have been explained through graphic presentations and models that find easier integration with existing subjects related to risk reduction.

The MKRC and the WKRC reached out more than 800 people in the Ayeyarwady and Yangon regions of Myanmar. These regions were the worst affected in Nargis cyclone, and are among the least developed in the country. The MKRC and WKRC thus helped reach communities that would have been, otherwise, left out in the long term recovery process.

8.4 Building Local Level Capacity for Preparedness

In Gujarat, India the provincial government, in partnership with CSOs introduced a pilot programme aimed at preparing schools that were located in three zones that have been affected by disasters in the past—in Ahmedabad which was affected by the 2001 earthquake, in Jamnagar affected by the 1998 cyclone and in Vadodara, an area prone to industrial disasters. A combination of formal, and non formal means aimed at complementing existing education curriculum was introduced. The programme was designed to minimal additional stress on students, while sending the message across succinctly in a manner that makes the absorption process natural. Since the region had experienced some crippling disasters, the program design had to be holistic taking into consideration the recovery needs as well as a transition into a resilient education system.

The programme design included steps to make the school buildings and building contents safer in order to avoid casualties from building collapse, falling hazards and other structural and non-structural elements. This was accompanied by both formal and non-formal systems of education—educating students on concepts through formal curriculum by introducing disaster management in formal curriculum as a subject, and through non-formal means by educating students on life skills. Informal systems were introduced to ensure learning was fun and exploratory. Informal tools such as games, activity books, hazard hunts to sensitise and train students in being able to assess risks in their immediate environment were introduced.

Training of teachers was considered critical, not only on the formal teaching of the subject of disaster management, but more importantly on transferring disaster risk reduction as a life skill. Involving parents as the link between the school and the larger community was initiated. This was felt necessary to ensure accounting for actions that will be required once the children have been secured, such as pick-up by parents and continued safety at homes and other places where risks may be encountered. Bringing in partnerships with emergency response services, public agencies and local civil society organizations to understand interdependence was a important part of the programme. Linking up school level plans with overall recovery of the community planning of the community mutually reinforced each other.

With the success and positive reception to the program, greater institutionalization was initiated by the provincial government. In partnership with local CSOs, a formal training of teachers all over the province was carried out. Text books for schools especially for Grades VIII and IX, were introduced. The approach adopted was to help students, teachers and school management to learn, reflect and to be empowered (Daisaku 2002):

To Learn: Students deepen their awareness about hazards and risks through understanding realities and knowing facts. Recent natural disasters have been well documented and shared. These serve as case studies for teachers as well as students. Wherever needed, disasters are simulated with the help of portable models. Curriculum changes strengthen the learning process.

To Reflect: Students analyze reasons that have lead to loss of life and injury in disasters. They are able to make distinction in development practices and people's actions that can cause disasters or prevent them. Students connect to their own local communities and families and share their learning with them.

To Empower: Students take concrete action toward lowering risks in the environment. Classroom and school exercises are introduced to help them take small definitive actions that can become a precursor to bigger investments for disaster risk reduction. School Management prepares school disaster management plans in which roles and responsibilities are identified and rehearsed periodically.

8.5 Teachers as Emergency Managers

Teachers play a crucial role in ensuring the safety of children in school (Fig. 8.4). The authors' experience in interacting with teachers during the implementation of school safety programmes revealed their keen interest in contemporary issues that catch the attention of the children. Stories from live events, such as most recent major catastrophes, their scientific perspective and follow up steps that children are expected to take, are areas of interest which teachers like to incorporate in their lessons.

Further, teachers act as guardians to children in school. In the event of emergency evacuation, the teachers take on the role of "Emergency Managers" guiding actions that children need to take.

Teachers also serve as links between parents and children. Parents rely on teachers for the welfare and upbringing of their children. There is, hence, an implicit trust reposed on teachers. A trust that can potentially be reinforced if teachers can also provide a guarantee to parents that children are "safe" under their teachers' care.

In formal school setup, however, school level disaster management does not figure in the "job task" of the teachers. Teachers find it as an "additional responsibility" and sometimes, even as burden to be learning skills and practicing disaster management.

The approach to school safety therefore needs to take a "complementary" approach where teachers' are able to view the subject as an extension of their

Fig. 8.4 Teachers play a critical role in recovery of education sector (Photo: Mamta Joshi)

existing curricula. Much work needs to be done in interpreting disaster reduction through mainline subjects such as science, mathematics, physical education and social studies. Currently, in most cases, disaster management is integrated with geography, however viewed from a larger perspective this may not be adequate. Practical lessons in life saving skills add value to existing classes on physical education. Similarly, the subject of science can provide a useful framework for understanding cause-effect relationships, which would help students reflect better on the inherent links between ourselves and our natural environments.

8.6 Building Resilience: From Action to Policy

In disaster affected areas, interventions in schools is an opportunity for learning and applying lessons from experience in which students may have lost friends and family members. The mutually reinforcing efforts of government, civil society organizations, school community and parents have resulted in school based learning gaining significant momentum and appreciation. In certain cases, such initiatives have served as precursors to larger policy level shifts and investments. CSOs such as the Families for School Seismic Safety in British Columbia, Canada through successful grassroots advocacy have ensured a US$1.3 billion commitment by the State Government to ensuring all schools are brought up to acceptable life safety standards by 2019 (Monk 2007). Likewise, in India, civil society work contributed to initiative by Government in form of a—National School Safety Programme (NDMA 2012)—a project to be implemented in 8,600 schools in India by the National Disaster Management Authority in partnership with Ministry of Human Resource Development.

As an outcome and follow-up of the successful safety programme in schools in the earthquake affected areas of Gujarat province, the Ahmedabad Agenda of Action for School Safety (Prevention Web 2007) was evolved by Governments, UN-ISDR and CSOs working in education recovery. The Ahmedabad Action Agenda while reaffirming the Priority for Action 3 of the Hyogo Framework for Action 2005–2015—to use knowledge, innovation and education to build a culture of safety and resilience at all levels, also recognized the UN Millennium Development Goal (Goal 2) to Achieve Universal Primary Education by the year 2015. It sets the goal of achieving—Zero Mortality of Children in Schools from Preventable Disaster by the year 2015. Towards achieving this goal, it lays out the following principal action points

- *To include disaster risk reduction in the formal curriculum at both primary as well as secondary levels.*
- *To promote disaster risk reduction through co-curricular activities in schools acknowledging that school children need to develop "survival skills" first, along with "life skills" and "academic inputs."*
- *Complete risk assessment and safety measures must be undertaken to ensure zero potential damage to new school buildings.*
- *Mandatory safety audit of all existing school buildings with respect to their location, design and quality of construction and prioritizing them for demolition, retrofit or repair.*
- *Mobilize parent, student, local community and school staff to champion school safety.*

The Ahmedabad Action Agenda for School Safety and many subsequent declarations such as the Bangkok Action Agenda and the Islamabad are policy initiatives born out of devastating disasters that have caused huge education losses in the Asian region. In a way, the disasters become turning points for policy shifts in ensuring education sector in countries becomes much more resilient to risks of disasters. These declarations have found universal acceptability and guided policies and programmes in many countries and regions since.

8.7 The Emerging Approach: Achieving Resilience of Education Sector Through Recovery

As the case examples described herein illustrate, CSOs have played a proactive and catalyzing role in ensuring smooth recovery of education sector transitioning into resilience building. Disasters have, in fact, served as windows of opportunity for building capacity of the education sector to be prepared better for future disasters while addressing underlying vulnerabilities. The case examples illustrate how various systems of education put to use effectively in the early as well as long term recovery of disaster affected areas (Table 8.1).

Table 8.1 Education systems used in disaster recovery

	Formal education	Non-formal education	Informal education
In early recovery	Training of teacher	Use of arts Use of games	Experiential learning—cause and effect
In long terms recovery linked to development	Curriculum level changes Policy level changes	Preparedness training First aid, search and rescue, systems and support	Participatory risk assessment—town watching etc

Central to the work carried out by CSOs is the people centered approach. Involving children, teachers and their parents in the recovery process has made the transition process least traumatic translating into a "built back better" state. With a wide array of tools, approaches and institutional strengths, disaster recovery was inclusive, that which built local capacity and helped in spreading a culture of preparedness.

References

Daisaku I (2002) The challenge of global empowerment: education for a sustainable future. Soka Gakkai International

Monk T (2007) School seismic safety in British Columbia. Reg Dev Dialogue 28(2), United Nations Centre for Regional Development

Multistory, Innovations in Participatory Education (2013) SEEDS India, CENDEP, Oxford Brookes University. The small change forum: building resilient community – creative learning workshop report, Leh. http://www.multistory.org.uk/downloads/projectstories/Multistory_Creative_Teaching_Workshop_Report.pdf. Accessed January 2013

Nabeel Hamdi (2004) Small change: about the art of practice and the limits of planning in cities. Routledge, London

NDMA (2012) National school safety project. http://ndma.gov.in/ndma/nssp-projects/Final_NSSP_brochure[2]%20English.pdf. Accessed January 2013

ODI (2012) Climate extremes and child rights in South Asia: a neglected priority. Katie Harris and Kelly Hawrylyshyn. http://www.odi.org.uk/publications/6838-climate-extremes-child-rights-south-asia-neglected-priority. Accessed December 2013

Preventionweb (2007) Ahmedabad action agenda for school safety, UNISDR. http://www.preventionweb.net/english/professional/trainings-events/edu-materials/v.php?id=5146. Accessed January 2013

SAARC (2011) Framework for care, protection and participation of children in disasters. A memorandum on lifelong learning, unit E-3. South Asian Association for Regional Cooperation European Commission, Kathmandu

Shaw R, Takeuchi Y (2012) East Japan Earthquake and Tsunami: Key lessons for the education sector. Kyoto University & UNESCO Bangkok

Shikada M, Than Myint U, Ko Ko Gyi U, Nakagawa Y, Shaw R (2012) Reaching the unreachable: Myanmar experiences of community based disaster risk reduction. Bingley, Emerald

UNICEF (2009) Best practices and lessons learnt: UNICEF Myanmar's response following Nargis. UNICEF Myanmar, Myanmar. http://www.unicef.org/infobycountry/files/Myanmar_Nargis_Evaluation.pdf. Accessed January 2013

Chapter 9
New Insights of Education Sector from East Japan Earthquake and Tsunami

Yukiko Takeuchi and Rajib Shaw

Abstract This chapter discuss about education sector's role and management through experience of East Japan Earthquake and Tsunami. Education sector's main role is to provide proper education to the future citizens. Public schools are located in the community, and therefore, community people have certain level of recognition to school in the regular time as well as in the emergency time. It is often considered to be contradicted that if school becomes a shelter, how to continue the education. School as the core vital community infrastructure becomes the shelter for natural reasons. However, how to continue education during the post disaster is also important. Thus, the education in emergency becomes of utmost importance, and the school manual should be modified to reflect these lessons. Locations of the school, its structure, function, layout all become very important for countries like Japan, which has long coast line and relatively less flat areas. Thus, it becomes very important that the E-HFA (Hyogo framework for Action for Education sector) is properly realized and practiced in local governments in the high-risk areas.

Keywords Disaster education • Education governance • E-HFA • School facility • Teachers role and linkage with community

---

Y. Takeuchi
Global Survivability Studies (GSS), Kyoto University, Kyoto, Japan

R. Shaw (✉)
International Environment and Disaster Management, Graduate School of Global Environmental Studies, Kyoto University, Kyoto, Japan
e-mail: shaw.rajib.5u@kyoto-u.ac.jp

R. Shaw (ed.), *Disaster Recovery: Used or Misused Development Opportunity*, Disaster Risk Reduction, DOI 10.1007/978-4-431-54255-1_9,

9.1 Introduction

The importance of education in disaster risk reduction has been emphasized in several international governance and policy agenda, frameworks, conferences as well as UN programs. Chapter 36 of Agenda 21, on "Education, Awareness and Training" stated "Education, including formal education, public awareness and training, should be recognized as a process by which human beings and societies can reach their fullest potential" (UNEP 1992). Furthermore, the UN/ISDR System Thematic Cluster/Platform on Knowledge and Education argued that "Education for disaster risk reduction is an interactive process of mutual learning among people and institutions. It encompasses far more than formal education at schools and universities, and involves the recognition and use of traditional wisdom and local knowledge for protection from natural hazard" (UN/ISDR 2005). In the 2006 the Review of the Role of Education and Knowledge in Disaster Risk Reduction pointed out that "Education, knowledge and awareness are critical to building the ability to reduce losses from natural hazards, as well as the capacity to respond to and recover effectively from extreme natural events when they do, inevitably, occur" (Wisner 2006). The Second Asian Ministerial Conference on Disaster Risk Reduction (2007, India) urged governments to make school safety and the integration of disaster risk reduction into school curricula a priority on the national agenda (UN/ISDR 2007). The Third Asian Ministerial Conference on Disaster Risk Reduction (2008, Malaysia) recognized education as an essential contribution to effective implementation of disaster risk reduction and concrete impact tin terms of shifts in behaviors art the local level, where communities are most vulnerable to disasters (UN/ISDR 2008). Last but not least, the UNESCO Education for Sustainable Development (ESD) program emphasized that "Education is the primary agent of transformation towards sustainable development, increasing people's capacities to transform their visions for society into reality" (UNESCO 2005).

Based on the above context, this chapter discuss about education sector's role and management through the experience of East Japan Earthquake and Tsunami. Education sector's main role is to educate students. However, public schools are located in the community, and the community has some recognition to school in the regular time as well as emergency time. This chapter tries to discuss following five contents: context of school centered disaster risk reduction in Japan, situation of school based damages in Tohoku region, how the schools played an important role in post disaster situation, and how this lesson can be transferred to future community based, school centered disaster risk reduction in Japan and other Asian region.

9.2 School Centered Disaster Risk Reduction in Japan

Generally, in Japan, the schools have manual to prepare for disaster. Those standards are to focus only to the student's safety during their time within school facility. Most of the manuals do not confirm with the school students outside the school boundary. However, a real education needs to give proper direction on how to behave within or outside the schools. The Basic Disaster Management Plan was revised in 1995 based on the experiences incurred at the time of the Great Hanshin-Awaji Earthquake. The plan clarifies the duties assigned to the governments, public corporations and local government in implementing disaster risk reduction measures. For easy reference to countermeasures, the plan also describes the sequence of disaster countermeasures such as preparation, emergency response, recovery and reconstruction according to the type of disaster. This plan was developed in 1963, and was revised afterwards in 1971, 1995, 2000, 2002, 2004, 2005, 2007 and 2008 (Takeuchi and Shaw 2009). Of course, the school management was also changed based by The Basic Disaster Management Plan. The Great Hanshin-Awaji Earthquake affected area's local government and school try to prepare for next disaster with local community people at school. One example is "Bousai Fukushi Community (Bokomi)" (Takeuchi et al. 2012), which is described in the next section. There are several examples of school-centered disaster risk reduction in Japan. Following part gives two specific examples of this from the past disasters in Japan, one in 1995 and the other in 2004.

9.2.1 Welfare Based School-Centered Model in Kobe

After the devastating Great Hanshin-Awaji earthquake on 17 January 1995, which left heavy traces of destruction on various infrastructures and resulted in the loss of more than 6,400 human lives, the voluntary support of many surviving residents in Kobe in rescue operations highlighted the potential of well-functioning communities. The inherent characteristics of communities were to provide a network among people, and strengthen their social capital and ability to respond to potential disasters (Shaw 2012a, b; Shaw and Goda 2004; Nakagawa and Shaw 2004).

From the Hanshin Awaji-Earthquake, it is obvious that the age group of over 65 year-old is one of the most vulnerable groups in particular living in the density area with old houses. Learning from these findings, the Kobe City Government has been conducting an important initiative for developing "BOKOMI" with the aim to building resilience of its communities against disasters. BOKOMI is the short term of "Bousai Fukushi Community" (Disaster preparedness and welfare community), where the disaster preparedness is linked to daily welfare of the people within the same school district. It was understood that the needs of the aged community is daily welfare. Therefore, to continue to the disaster preparedness activities, it is required to link this to welfare activities.

After the pilot phase in 11 districts, the BOKOMI concept was formalized in 1997 according to the Mayor's decision, and mainstreamed in all the school districts of Kobe City. BOKOMIs are established based on municipal elementary schools districts in Kobe City (Matsuoka et al. 2012; Takeuchi et al. 2012). The total number of municipal elementary schools is 191. The number of BOKOMI steadily increased and reached 100 % coverage in 2008. The reason why BOKOMIs are based in elementary school districts is because "welfare-community" groups were already established in each elementary should district and thus, disaster prevention activities could be integrated into these existing groups. In addition, elementary schools are designated as evacuation sites for communities in emergencies in Japan. These are the key reasons why BOKOMIs are established in each elementary school district.

The process of establishing BOKOMI in a district is based on multi-stakeholder consultation in the district. Firstly, the establishment of a BOKOMI is discussed and agreed by local government organizations, including the local city office (ward office) and fire station, together with leaders of local residents and other local multi-stakeholders. BOKOMI is a community-based organization comprised of local residents' associations, women's associations, elderly associations, child committee member, youth associations, PTA, local fire station, and local business entities. In order to support activities of BOKOMIs, the Kobe City Government provides various support measures like small funding, materials for community activities, rescue tools, training by fire professionals etc.

Main activities by BOKOMIs have two perspectives; disaster prevention and risk reduction activities and welfare related activities in the school district. These activities are combined and carried out together. Disaster-prevention and risk reduction activities by BOKOMI:

- Disaster drills and training
- DRR education program with schools
- BOKOMI junior team (fostering children's teams to lead and work on DRR activities)
- Public awareness event
- First-aid seminar, checking emergency materials and equipment
- Town watching and preparation of community safety map, risk reduction activities with rescue workers and fire fighters (identification evacuation root, removal of object blocking these roots, fixing furniture etc.)

Combining with welfare activity:

- Regular communication within communities to form their unity, so that they can take action, when emergency/disaster happens, considering needs of vulnerable groups such as elderly and disabled people
- Learning how to support the people with special needs during disasters (elderly people and handicapped people).

The key aspect of this example is to link the community based disaster risk management activities to local daily needs like health and welfare activities. The example shows that BOKOMI was formed in all 191 school districts in Kobe, and is still continued, even after 18 years after the disaster.

9.2.2 School Based Community Preparedness Model in Saijo

At the time of the typhoon no. 21 and 23 in 2004, mountainous area of Saijo City of Ehime prefecture (Yoshida et al. 2009) was especially damaged. Land condition and concentrated heavy rain were the major factors, but there were other reasons concerning concentrated aged population in the region. So, some elderly people had difficulty in evacuating and needed help of young people. Low awareness of disaster prevention is also a problem. Plain area is rather urban in nature, and there are many young people. So, it is necessary to make "disaster prevention network" between the plain area and the mountainous area, so as to help elderly people in the mountainous area in case of a disaster. As the driftwood stuck with bridge pier caused flood to the plain area, disaster in the mountainous area have bearings with that in the plain area. Both residents have to know each other about the circumstances.

After the disaster, Kyoto University started a program with Saijo city education board to enhance the linkage of school and communities through participatory neighborhood watching. For these reasons, mountain watching was implemented in Saijo City. Mountain watching is just like town watching and it is conducted in the mountainous area. Main target was the children, and also residents in the mountain, teachers, municipal officials and forest workers were involved. The working field was upper area of a river along school. Participants watch the site damaged by the typhoon in 2004 and hear the story from victims.

At the same time, town watching was implemented in the plain area. The main target was students and teachers, parents, Jichikai (resident or neighborhood association) and municipal officers. They walked around the school zone and search for dangerous places, useful facilities in case of disasters and favorite places, which they don't notice otherwise in daily life. At first, town watching was implemented in five elementary schools and mountain watching in three junior high schools as "disaster education program," which was an activity of 12-year-old education project. The project started in 2005, and continued till 2007. From 2008 onward, the city education board took leading role in continuing the process in the implemented schools, and also expanded it to all primary and secondary schools in the city.

The above two examples show that to institutionalize disaster education it is very important to work with the local education board. The start needs to be as pilot project with selected schools. The involvement of the teachers in the process should be high, so as to ensure ownership. The linkage with the education board needs to be strong. The second part of the process is to disseminate the experiences widely to all the schools through the education system in the city or municipality. The process described here is to incorporate local DRM into education system.

9.3 Situation of School During Tohoku Disaster

In the East Japan Earthquake and Tsunami of Tohoku, the education sector experienced major damages, like other sectors of housing, infrastructures, energy and livelihoods. As explained above, education sector not only enhances the capacity of

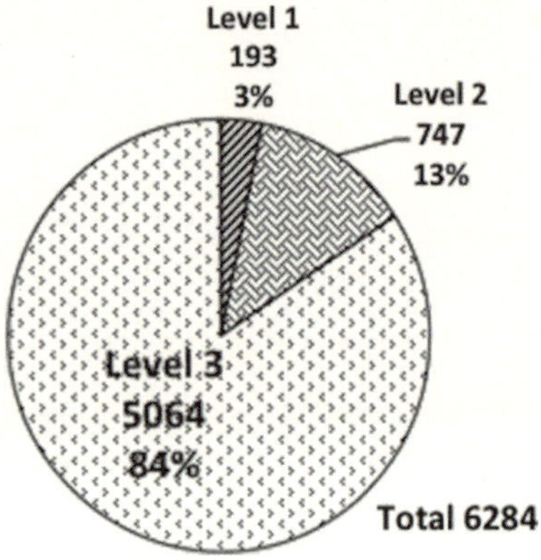

Fig. 9.1 School number of damage level by Tohoku earthquake and tsunami

future generation, it has other role in the community in usual time and also during disaster. For example, it becomes an evacuation center and also helps in developing relationship with communities through communication. Stability of social and environment of education sector is key elements for recoveries from this disaster. In total, 6,284 public schools received damage and 733 school students/teachers died or missing by Tohoku Earthquake and Tsunami, 2011 (MEXT 2011). MEXT classified damages into three levels. Figure 9.1 shows the breakdown of school number in the three damage levels. One hundred and ninety three schools belong to damage level 1, which was total destruction, and school cannot be used. For level 2, it was significant damages, and also need some amount of structural and non-structural measures, and for level 3, it was minor damages, mostly non-structural.

For the new constriction, education sector, school and community people have to discuss and decide about grand design of new school. In this disaster, many schools and students/teachers received damage. One of the key reasons for the damage was the proximity of the schools to the coastlines. Not all the coastal schools had human damages. Some other elements, for example, size of school, school structure, link with the community, disaster risk reduction education etc. affected the human damage level. This section provides five examples of school situation, which are derived from detailed interview and field survey with the school principals and education board of the city government (Takeuchi and Shaw 2012). Figure 9.2 shows the flow of different types of schools presented in this chapter. The major differences were whether the school area received tsunami damage or not. When the schools were hit by tsunami, there were two possibilities: move to other school (because of the level of destruction), or the school became the evacuation center for the people. There were some schools, which did not have any damage, and accepted other schools in its location. Four schools (two elementary schools and two junior high schools) are explained in this section.

Arahama elementary school is located Sendai plane, within 200 m from the coastline. The Chile Tsunami affected the area on 27th February 2010. After this

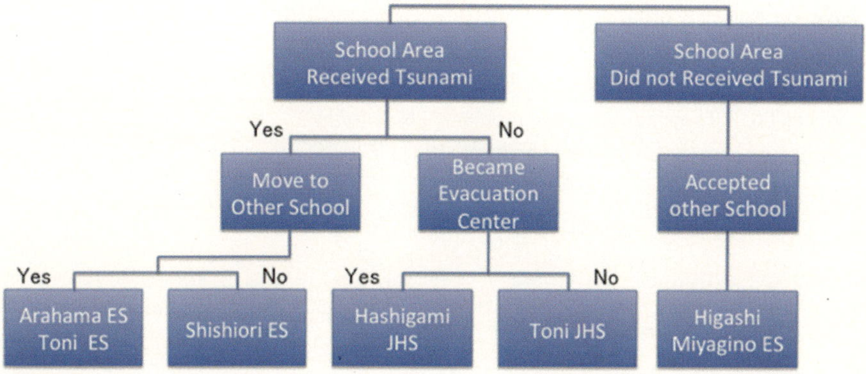

Fig. 9.2 Types of school situation

Tsunami, the school principal revised the disaster management plan by increasing the storing capacity of emergency food and utilities and moving the evacuation area from the gymnasium to the third floor of school building. In consideration of the time required to take shelter in another elementary school, it was decided the students would be confined to this school during the disaster. After disaster, Arahama elementary school has is temporarily relocation to other school.

Toni Elementary School is located in Sanriku mountain area. Characteristic of Sanriku mountain area is narrow flat area and steep slope. Eleven meter concrete dyke was developed along the coast, and Toni ES was located near this dike. At the 11th March 2011, school students and teachers evacuated to height place than school. But school building completely broken by Tsunami. After disaster, Toni elementary school also has is temporarily relocation to other school and move again to temporary school building with Toni junior high school after 9 months later. Unlike the Toni elementary school, the Toni junior high school (JHS) was located in higher ground. Therefore, it was not directly affected by tsunami. But, since the building was quite old, it had significant earthquake damages, and could not be used for the evacuation center. The school also re-started in the gymnasium, and from January 2012, the school started in the temporary building built in the playground.

Hashikami junior high school was no damage to the school building due to earthquake or tsunami. Since the school is located in a relatively higher area, after the disaster more than 2,000 people came to school for evacuation. The school is close to the Route 45, and therefore the evacuees included both the local community as well as the passerby from the Route 45. However, the gymnasium hosted around 800 people for the first several nights. Although the local community leader took the charge of the gymnasium, the school principal was also responsible for its operation. Since many people who evacuated to the school lost their houses, the evacuees stayed in the gymnasium for more than 8 months, till the temporary houses were constructed. Table 9.1 shows the summary of the observations in each school.

Table 9.1 Summery of each school

Name of the school	Damage of building	During tsunami	After 3.11	Current situation
Arahama ES	Tsunami reach to second floor	320 people evacuate school building over fourth floor	Rescued by helicopter or move to other evacuation place by oneself	Temporary usage in Higashi Miyagino ES
Toni ES	Tsunami reach to third floor	Move to height area	Stay 1 day in community hall	Temporary usage in Heita ES. Temporary school building was developed
Hashikami JHS	No damage	Many people evacuate to school	Established evacuation center for over 8 month and developed temporary house in ground	Can use own facility. But without ground. Because, temporary house build on the ground
Toni JHS	Building broken by earthquake	Move to height area	Stay 3 days in company office	Temporary usage if gymnasium. Temporary school building was developed

9.4 What was the Role of School Throughout Disaster?

Most of the schools were designated as evacuation center by local government. Therefore, when disasters happen, community people evacuate to school. As described in the previous section and also in Takeuchi and Shaw (2012), when small disasters happen, evacuation time scale is just 1–2 days. But, in case of the East Japan Earthquake and Tsunami disaster, many people evacuated to school for about 8–10 months until the temporary houses were built. To be an effective shelter, the schools need not only to keep some emergency goods, but also some facility for long term evacuation. Especially, focusing to the current disaster, higher school building is important evacuation place in the community. There needs to have a balance between the school facility in the normal time and appropriate preparedness for recovery time for new school in the affected areas (Fig. 9.3).

As explained by Takeuchi and Shaw (2012), the school functions as evacuation site in all disasters. However, the length of time the school is used as evacuation site is different according to the kind of disaster and region of occurrence. This section shows the qualitative classification of duration of use of school as evacuation site, which is classified into five levels with five indicating long duration of use of school as evacuation site and one indicating short duration of use of school as evacuation site. Each indication of the five levels is shown as follows.

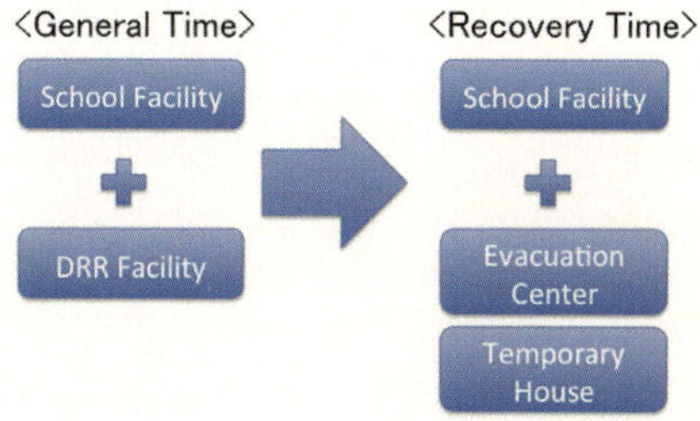

Fig. 9.3 How to make the balance with schools and others?

- Level 1: About 2–3 days
- Level 2: About within 1 month
- Level 3: About 1–2 months
- Level 4: About 3–6 months
- Level 5: About more than 6 months

Iwate, Miyagi and Fukushima prefectures experienced heavy damage because of the tsunami following the Great East Japan Earthquake. Most of the houses and infrastructure in the coastal area in the three prefectures received serious damage. As a result, so many people were evacuated. However, in the coastal area of Iwate prefecture, evacuation sites were closed within 5 months after the Tsunami disaster because the construction of the temporary houses went smoothly. The reasons were that the community size is small because it is located in a deeply–indented coastline as well as there was an upland area which was not affected by the tsunami. On the other hand, in the coastal area in Miyagi prefecture, the construction of the temporary houses became a challenge. There are two reasons for this. One is that Miyagi prefecture is a plain area. A wide area was affected by the tsunami and many people suffered. As a result, the number of temporary houses required was many. Second is that it was difficult to secure open spaces for temporary houses because it took a lot of time to remove the rubbles from a wide area. It was the same in the case of the Fukushima prefecture. The construction of the temporary houses became difficult for a different reason. There are nuclear evacuation zones in the Fukushima prefecture. Because of this, it is difficult to construct temporary houses in several areas. It is, thus, reasonable to calculate that tsunami disaster in metropolis counts for level 5, local area counts for level 4, and coastal area where population is small counts for level 3.

9.5 What to Learn on School from Tohoku Disaster?

For disaster risk reduction power to be cultivated through the class in the normal section, it is required to combine the power to live for, the power to think about to survive a disaster; and judge oneself situation not belief at the time of a disaster, and is connected for power to act. This is root and the base of the concept of "disaster prevention education to protect one's life by oneself" demanded from the disaster prevention education in the school. On the basis of influence and the effect that the picture which education tool used with the disaster prevention education teaching materials by the report of Kamaishi city gave a student, it is demanded those picture as the teaching materials positively. Students review all subjects and annual curriculum from a viewpoint of the disaster prevention education to raise a disaster prevention power through the class in the section, and it is necessary for each subject to educate disaster prevention taken in the daily life that cooperated.

The situation at the time of the disaster is changing depending on a local characteristic. Therefore it becomes basic to always know the area in carrying out disaster prevention education. Not only people know the visible science of inanimate nature, but also it is included in knowing the social connection in the area and the local constitution and history to know the area. When not only they lay the foundation of self-act, mutual assistance at the time of the disaster, but also can tie it to "N" assistance (power of the network) which is acceptance of network and NGO/NPO to tie an area and the local outside demanded relief, restoration, the revival in case of the East Japan great earthquake disaster from to, as for this social connection, it is with important human infrastructure.

Because the disasters are related to various fields like environment, welfare, peace, it is necessary for it to be able to enter the field of vision to take in sustainable education (ESD = Education for Sustainable Development), which is the action technique of these problems. Students become able to think about disaster prevention more deeply on carrying out these actions by cooperating with crisis control section or the university of the City. A comparison between the Kamaishi city (where an intensive disaster education program was carried out) and Kesennuma city (where the intensive ESD program was carried out with specific focus on community network) shows that in both the cities, the student survival rate was 99.8 %. This is quite significant that in Kamiashi, the success rate is attributed to disaster education (often known as Kamaishi Miracle). In contrast, for Kesennuma, it is attributed to the power of community network, and school-community linkages, which was developed through ESD. Thus, both the models are equally important and found to the effective. In future, possibly a combination of both these models would be desirable.

It is necessary that the staff of a school know the area like what showed in a part of the school education to ensure the security of students. Some of the important factors are: delivery and the safety confirmation of a student, protector correspondence and the NGO, NPO, the correspondence to the outside supporting group such as volunteers, media correspondence, the administration of the refuge and the delivery to the administration person in charge, the care of the heart to a students, a student are demanded afterwards. The training to the staff of a school is required to solve these problems, but it is necessary for making a constant rule for transfers of clarifying a role of the staff of a school, the delivery of a students, the student and the refuge administration to the administration, acceptance of the outside support and the correspondence to the media to make a manual.

Disaster prevention education to the staff of a school is carried out by incorporating the staff of a school in a teacher training process and the staff of a school training and ensures the security of a students, the student and, as an important talented person binding a school and an area together, is brought up as a leading figure of the disaster prevention education again. Those analyze it with the accumulation of experience and the lesson at the time of the disaster of the staff of a school in the East Japan great earthquake disaster, and it is thought that connected for disaster prevention education to the staff of a school who is a more practical effect by sharing them with the staff of a school of other areas.

9.6 Key Lessons of Education Sectors

School damages in the affected areas need further detailed investigations to understand the reasons for the damages and their potential future remedy. Broadly, the key lessons can be categorized into:

- *Structure, Location, Layout,*
- *Function and Educational Continuity,*
- *Human Resources and Training, and*
- *Effectiveness of Disaster Education*
- *New role of school and multi-stakeholder dialogue*

For an integrated approach that incorporates disaster risk reduction (DRR) into the education sector, Gwee, Shaw and Takeuchi (2011) identified and modified 16 out of the original 22 tasks suggested for the implementation of the five priorities for action of the Hyogo Framework for Action (HFA) to adapt them for use in the education sector. The 16 tasks are referred to as E-HFA or Education in the HFA as shown in Table 9.2. The following analysis tries to link the lessons to the E-HFA framework. This is to be noted that these are qualitative analysis, and should not be taken as the granted as the key lesson, rather, these should be taken as key initial learning.

In several school buildings structure was one of the key issues. In spite of a magnitude 9 earthquake, there were no major damages to most of the school buildings in terms of partial or total collapse. However, the damages due to tsunami were observed to be widespread. Several schools were retrofitted for seismic safety, as part of national- and prefecture-led programs. This shows a high level of education governance, linked to the E-HFA Priority 1 (Task 3). However, in many cases, multi-hazard approach was not incorporated, which is linked to E-HFA priority 4 (Task 13). In terms of structure, many of the school buildings had curved roof, instead of flat roofs, which prevented the students and community members to take shelter on the rooftop. The curved roof was promoted to reduce the water logging on the roof in the heavy rainfall areas, however, it became a barrier for evacuation during tsunami. The need of multi-hazard assessment is highlighted in this case also.

Table 9.2 Sixteen tasks relevant to the education sector

Priority 1: Developing institutional basis for disaster risk reduction (DRR) in education	
Task 1.	Engage in multi-stakeholder dialogue to establish the foundation for DRR education
Task 2.	Create or strengthen mechanism for systematic coordination for DRR education
Task 3.	Assess and develop the institutional basis for DRR education
Task 4.	Prioritize DRR and allocate appropriate resources for DRR education
Priority 2: Identifying, assessing, and monitoring disaster risks in the education sector	
Task 5.	Establish risk assessments for the education sector
Task 6.	Strengthen early warning in the education sector through effective communication and dissemination mechanism
Priority 3: Building a culture of safety through DRR education	
Task 7.	Develop program to raise public awareness of DRR
Task 8.	Include DRR in the education system
Task 9.	Develop DRR training and learning at the school and community
Task 10.	Enhance dissemination of DRR information
Priority 4: Reducing the underlying risk factors in the education sector	
Task 11.	Environment: understand sustainable ecosystem, environment, and natural resources management
Task 12.	Establish measures to incorporate DRR in urban and land-use planning
Task 13.	Structures: strengthen mechanisms for improved building safety and protection of critical facilities in the education sector
Task 14.	Disaster recovery: develop a recovery planning process that incorporates DRR
Priority 5: Preparing for effective emergency response and recovery in education	
Task 15.	Build on disaster preparedness capacities and mechanisms in the education sector
Task 16.	Assess disaster response preparedness capacities and mechanisms through strengthened planning

The other point of the school building is linked to its layout. It was found that the schools with a layout perpendicular to the coastline had less damages, as against those parallel to the coastline. In several cases, the schools were placed to the south facing, especially in the Sendai plain, which had relatively large flat area, and north south coast line. Therefore, the tsunami hit the side of the school buildings (Fig. 9.4). In case of Sanriku areas, due to lack of availability of land areas, the school were placed parallel to the coast, where the whole school building faced tsunami waves, and had significant damages. The location of the school was another crucial issue. Many of the damaged schools were located very close to the coastline, within 100 or 200 m only (e.g., the Arahama elementary school, Toni elementary school etc.). This is related to Priority 2 (Task 5, in terms of risk assessment), and Priority 4 (Task 12, in terms of land use planning). Schools, being a social infrastructure, need to be considered with appropriate risk assessment, as well to have an in-depth understanding on the underlying risk factors, including proper land use planning.

School function is an important issue during emergency period, as exemplified by E-HFA priority 5 (Task 15 and 16). The school became temporary shelter for several non-damaged school, like Hashikami JHS, where people lived in the school gymnasium for several months. This also posed a serious problem in the education continuity, which is often discussed as "Education in Emergencies." Several of the schools started its new academic year after 1 month, and therefore, several schools

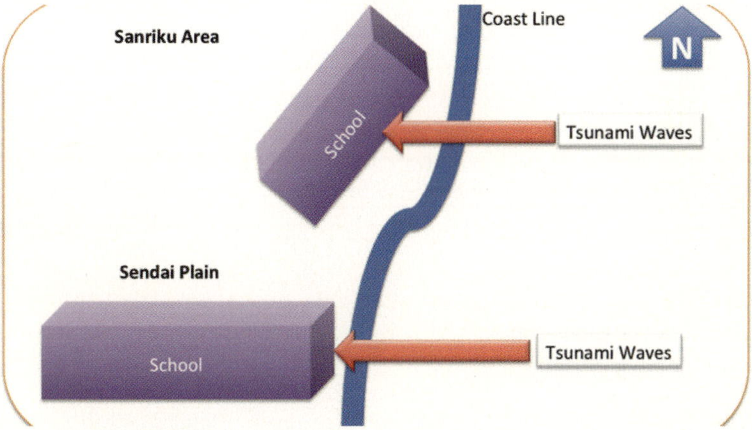

Fig. 9.4 Layout of school buildings contributing to damages

need to shorten its length of summer vacation to recover the lost days. In case of Arahama ES, students and local communities spent one night in the school building, and therefore the emergency goods and food were very important, which is linked to E-HFA Priority 5 (task 16, emergency response capacity).

Human resources, teachers' training, and emergency manual were other key important issues. In many cases, the school principal had to take decisive actions for the evacuation of the students, where there was no clear information on the tsunami timing and height. Although in the evacuation or emergency manual, one evacuation site was identified and designated, the teacher had to take their own decision based on the local situation, and in some cases, they moved with the students to four or five different places for safer evacuation (in case of Shishiori ES). Therefore, it shows the need of teacher training and need of decision making during emergency. This is related to E-HFA Priority 3 (Task 9: training). Also, proper management plan for education in emergency is required, which point to the need of E-HFA 5 (Task 15: disaster response capacity).

In several cities disaster education helped the students to take proper evacuation behavior, not only in the school but when they were outside the school. A classic example is the "*Kamaishi Miracle*," where many students evacuated spontaneously as part of their disaster education programs. During evacuation, the Kamaishi Higashi JHS took with them the Unotsumai ES students, both of the schools were located near the coast, next to each other, had disaster education and joint emergency drills. Thus, E-HFA Priority 3 (Task 8: include DRR in education system) is found to be extremely important.

During the school recovery programs, in several cities, multi-stakeholder collaboration is established in cooperation with the local residential association, school principals, education board, academics and other related stakeholders. As explained above, in many parts of the affected areas, aged population has a relatively higher percentage, and number of school going children are gradually decreasing.

Table 9.3 Key initial lessons of EJET education sectors in terms of E-HFA

Key lessons and issues

E-HFA	Task	Structure, location, layout	Function and educational continuity	Human resources and training	Effectiveness of disaster education	New role of school and multi-stakeholder dialogue
E-HFA 1	Task 1					High performance
	Task 2					
	Task 3					
	Task 4					
E-HFA 2	Task 5	Low performance, need improvements				
	Task 6	Low performance, need improvements				
E-HFA 3	Task 7					
	Task 8				High performance	
	Task 9			High performance		
	Task 10					
E-HFA 4	Task 11					
	Task 12					
	Task 13	Low performance, need improvements				
	Task 14					
E-HFA 5	Task 15		Moderate performance			
	Task 16		Moderate performance			

Low performance, need improvements
Moderate performance
High performance

Therefore, new roles of schools are required to link to the local communities. Thus, the community needs and priorities are reflected in the future developed of school through multi-stakeholder dialogues. The MEXT has started the new concept of school centered resilient community development in the affected areas. Therefore E-HFA priority 1 (Task 1: multi-stakeholder dialogue) has a strong significance and importance. Table 9.3 shows a tentative evaluation of the key lessons and issues from the current disaster.

9.7 Implications to Asian Context

Fernandez, Shaw and Takeuchi (2012) made an analysis from 25 specific cases of school damages from 11 different Asian countries for 6 types of hazards. Most commonly implemented recovery actions relate to Task 14 (disaster recovery), Task 7 (public awareness of DRR), Task 13 (physical structures, i.e., building codes, retrofitting, protection of critical facilities, etc.), Task 15 (disaster preparedness, i.e., drills, standby funds, etc.), and Task 12 (land-use planning, i.e., safe location for schools). Tasks under Priority 4 (reducing underlying risks) are performed 50 % of the time.

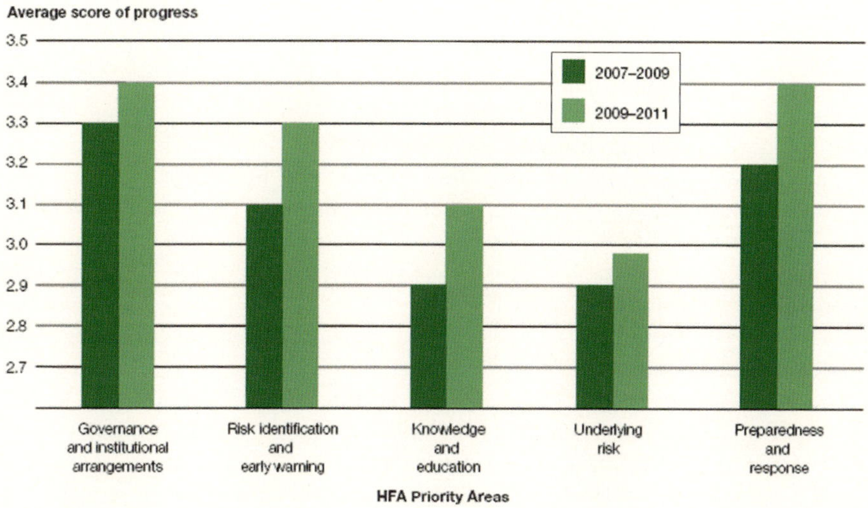

Fig. 9.5 Global progress against the HFA

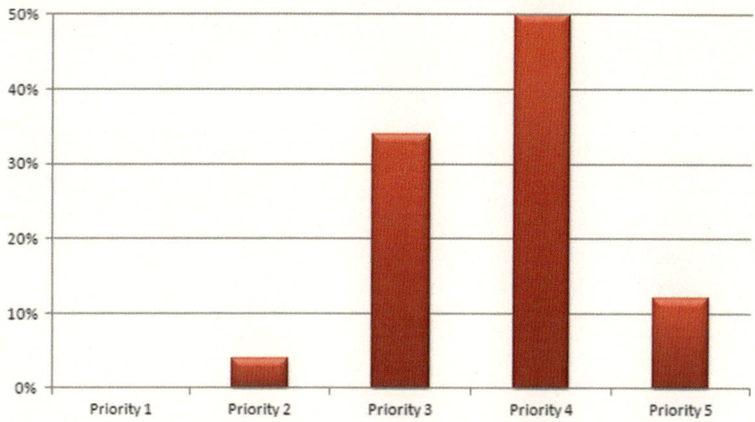

Fig. 9.6 Classification of the Asian case studies according to E-HFA

Tasks not done in any of the 25 case studies are all four tasks in Priority 1 (institutional basis for DRR in education), Task 6 (early warning), and Task 16 (assessment of disaster preparedness). The case study lessons, categorized according to the E-HFA priority area they belong to, are then graphed and compared to the result reported in the 2011 Global Assessment Report on Disaster Risk Reduction (GAR 2011).

According to GAR 2011, whereas substantial progress is being made globally against the HFA priority targets in early warning, disaster preparedness, and emergency response, countries are still struggling to address the underlying risk drivers (Fig. 9.5). Interestingly, reflecting the lessons from the 25 case studies, Fig. 9.5 shows almost the opposite tendency. In Fig. 9.6 (IEDM 2012), Priorities 4 and 3 are performed on more occasions than the other priorities. Figure 9.5 seems to suggest that perspectives change after one has experienced a disaster, hence the difference on the focused priorities. This is interesting in that although the cases are more on post-disaster response and recovery, HFA 4, which focuses on underlying risk factors are incorporated into the recovery process. This is significant in the sense that it encompasses the future risk reduction perspectives. For obvious reasons, the schools are focused, and HFA 3 has gotten more emphasis. It should also be noted that the examples presented here are rather randomly selected from different Asian countries on different types of disasters. A more systematic analysis may provide more insight in the progress of E-HFA.

The analysis from the current cases of school damages in the East Japan Earthquake and Tsunami (EJET) shows that all the E-HFA priorities are equally important, however, needs to be performed at different levels by different sets of stakeholders. While the Asian examples have more focus on E-HFA priority 3 and 4, the EJET shows the importance of HFA 1 and 3 as successful cases, and E-HFA 5 needs improvements. However, E-HFA 2 and 4 will need future improvements, and can be considered as one of the key lessons.

School is beyond only education. School damages are not only restricted to the educational damages. School is directly linked to the community, and school recovery is linked to community recovery. The 16 E-HFA tasks are to be performed at all levels, i.e., national, local, community, and school, to achieve sustainable implementation. When considering disaster risk reduction (DRR) education, it should not be limited to the education curriculum only, but should also include related issues such as structural and non-structural safety measures; legislative measures supporting the integration, implementation, as well as funding of DRR in the education sector; risk assessments and early warning systems; DRR training for school staff, etc. An integrated approach is necessary and the E-HFA tasks can help cover the various important issues that need to be addressed.

9.8 Conclusion

It is expected that the new school construction will focus on the future disaster risk reduction issues, both in terms of school as shelter as well as school as educational facility. It is often considered to be contradicted that if school becomes a shelter, how to continue the education. School as the core vital community infrastructure becomes the shelter for natural reasons. However, how to continue education during the post disaster is also important. Thus, the education in emergency becomes of utmost importance, and the school manual should be modified to reflect these lessons. The location of the school, its structure, function, layout all becomes very important for countries like Japan, which has long coast belt and relatively less flat areas. Thus, it becomes very important that the E-HFA is properly realized and practiced in local governments. Still, even after the 2 years of the disaster, the total understanding of the resilience of education sector is relatively low in Japan, as well as other countries sin developing Asia. It is required to raise proper awareness of the local governments and other stakeholders in this regard.

The relationship of DRR education has different target groups, school education for students, teacher's training, regional education and education for parents. Common topics for education among all groups are: understanding the region, clarification of roles and responsibilities during disaster, operational methods of evacuation centers, rules on student handover and DRR drills. Good governance and coordination among schools and region, students and teachers, parents and community residents builds expectation leading to trust building, appropriate actions for evacuation, trouble free life in evacuation centers and quick recovery of school facilities. Coordination cannot be built solely under DRR activates, but also should be done through regional festivals and other events. Thus, educational governance becomes of utmost importance.

Acknowledgements This study is funded by CWS and MERCY Malaysia support to Kyoto University, which is highly acknowledged. The authors also acknowledge the support from the local government education board and teachers who cooperated with the survey.

References

Fernandez G, Shaw R, Takeuchi Y (2012) School damages in Asian countries, and its implication to Tohoku recovery. In: Shaw R, Takeuchi Y (eds) East Japan earthquake and tsunami. Research Publishers, Singapore

GAR (2011) UNISDR (United Nations International Strategy for Disaster Reduction). Global assessment report on disaster risk reduction. Geneva, UNISDR. http://www.preventionweb.net/english/hyogo/gar/2011/en/home/ index.html. Accessed 11 March 2013

Gwee QR, Shaw R, Takeuchi Y (2011) Disaster education policy: current and future. In: Shaw R, Shiwaku K, Takeuchi Y (eds) Disaster education. Emerald Group Publishing, Bingley

IEDM (International Environment and Disaster Management Laboratory) (2012) School recovery: lessons from Asia. Kyoto University, Kyoto

Matsuoka Y, Joerin J, Shaw R, Takeuchi Y (2012) Partnership between city government and community-based disaster prevention organizations in Kobe, Japan in community based disaster risk reduction. Emerald Publisher, Bingley, pp 151–184

Ministry of Education, Culture, Sports, Science and Technology (MEXT) (2011) Report of damage information by Tohoku earthquake and tsunami, no.176, p 26 (Japanese)

Nakagawa Y, Shaw R (2004) Social capital: a missing link to disaster recovery. Int J Mass Emerg Disaster 22(1):5–34

Shaw R (2012a) Community based disaster risk reduction. Emerald Publisher, Bingley, 402 pp

Shaw R (2012b) Overview of community based disaster risk reduction, in community based disaster risk reduction. Emerald Publisher, Bingley, pp 3–17

Shaw R, Goda K (2004) From disaster to sustainable civil society: the Kobe experience. Disasters 28(1):16–40

Takeuchi Y, Shaw R (2009) Gender dimensions in risk communication. A perspective from a sediment disaster in Hiroshima, Japan, Regional Development Dialogue (RDD). United Nations Center for Regional Development 30(1):63–75

Takeuchi Y, Shaw R (2012) Damage to education sector. In: Shaw R, Takeuchi Y (eds) East Japan earthquake and tsunami. Research Publishers, Singapore

Takeuchi Y, Xu W, Kajitani Y, Okada N (2012) Investigating risk communication process for community's disaster reduction with a framework of "Communicative Survey Method". J Nat Disaster Sci 33(1):49–58

UN/ISDR (2005) Thematic cluster/platform on knowledge and education. http://www.unisdr.org/eng/task%20force/working%20groups/knowledge-education/knowledge-education.htm. Accessed 10 November 2009

UN/ISDR (2007) Regional analysis on disaster risk reduction education in the Asia-Pacific region – in the context of Hyogo framework for action priority 3 implementation. In: 2nd Asian ministerial conference on disaster risk reduction, New Delhi, 7–8 November 2007. http://nidm.gov.in/amcdrr/acdrrbro.pdf. Accessed 29 November 2010

UN/ISDR (2008) Regional analysis on disaster risk reduction education in the Asia-Pacific region – in the context of Hyogo framework for action priority 3 implementation. In: 3rd Asian ministerial conference on disaster risk reduction, Malaysia, 2–4 December. www.preventionweb.net/.../3089_3089. FirstAnnouncementAsianMinisterialConference.pdf. Accessed 29 November 2010

UNEP (1992) Agenda 21/Chapter 36: promoting education, public awareness and training. http://habitat.igc.org/agenda21/a21-36.htm. Accessed 12 January 2010

UNESCO (2005) Education for sustainable development. Strategy. http://www.unesco.org/en/esd/strategy/. 9 December 2010

Wisner B (2006) Education for disaster risk reduction & school protection. http://www.unisdr.org/unisdr/eng/task%20force/working% 20groups/knowledge-education/Summary-findings-desk-study-review-knowledge-education-DRR.doc

Yoshida Y, Takeuchi Y, Shaw R (2009) Town watching as the useful tool of urban risk reduction: case of Saijo, Japan. In: Shaw R, Srinivas H, Sharma A (eds) Urban risk: an Asian perspective. Emerald Publication, Bingley, pp 189–206

Chapter 10
School Based Community Recovery in Kamaishi, Japan

Shohei Matsuura and Rajib Shaw

Abstract The East Japan Earthquake and Tsunami (EJET) had devastating effects on schools and communities. Not only did it interrupt educational activities, but also weakened community ties due to displacement of affected people, including students of affected schools. Under this situation, the Ministry of Education, Culture, Sports, Science and Technology (MEXT) of Japan has proposed the concept, "School Centered Community Building," which is composed of four main pillars, which are: (1) Securing safety of schools and relocating them to safe areas as necessary, (2) Retrofitting schools to improve their functions as evacuation center and hub for disaster management, (3) Making school building eco-friendly and (4) Making schools multi-functional public facilities to become the center of communities. Kamaishi is one of the disaster-affected cities in which in its city recovery plan has incorporated this concept. This chapter presents the initial findings of the school recovery process in Toni District, Kamaishi City, Iwate Prefecture as a study case to look into the actual implementation of School Centered Recovery and Community Building.

Keywords Community building • Disaster risk reduction education • School recovery • School–community linkage

---

S. Matsuura • R. Shaw (✉)
International Environment and Disaster Management, Graduate School of Global Environmental Studies, Kyoto University, Kyoto, Japan
e-mail: shaw.rajib.5u@kyoto-u.ac.jp

R. Shaw (ed.), *Disaster Recovery: Used or Misused Development Opportunity*, Disaster Risk Reduction, DOI 10.1007/978-4-431-54255-1_10,

10.1 Introduction

The East Japan Earthquake and Tsunami (EJET), which occurred at 2:46 pm on 11 March 2011 with magnitude of 9.0, left 15,879 dead and 2,712 missing as of 26 December 2012 (Reconstruction 2013a). In the most affected prefectures in the Northeast or Tohoku Region, Iwate, Miyagi and Fukushima, about 248,000 people are still displaced from their original homes with the majority taking shelter in temporary housing facilities after close to 2 years after EJET (Reconstruction 2013b). The education sector in the most affected prefectures was no exception to this catastrophe with reports of 653 deaths, 91 injuries and 75 missing of students and teachers (including public and private kindergartens, elementary, junior high and high schools, special needs school, vocational schools and universities). Twenty five thousand five hundred and sixteen students in which their schools have been damaged have transferred to other schools with over 14,000 of them to schools outside of their former residential prefectures. Over 2,400 education related facilities are still in need for rehabilitation at different levels and for this, majority of the students from the affected schools are now studying in temporary school buildings (MEXT 2012). This displacement of disaster affected people is the first case in Japan in which people became scattered all of the country providing serious physical, psychological, institutional and socio-economical consequences (Shaw and Takeuchi 2012). The conditions have greatly weakened community ties and have been one of the major bottlenecks in the recovery process.

As for the damages in Kamaishi, EJET had taken 888 lives and left 158 people missing, destroyed over 16,000 homes and numerous industrial and public facilities, including four kindergartens, nine elementary schools and four junior high schools (Kamaishi City 2011a). Out of these damaged schools, Toni Elementary School (ES) and Unosumai ES and Kamaishi Junior High School (JHS) received total damage from the tsunami and Toni JHS received damage by the earthquake and now planned to be demolished. Despite of the extensive damage to the schools, most of the students who were at the schools at the time of tsunami survived, being able to evacuate promptly with assistance of teachers and local residents (five students who were absent became causalities out of total two thousand nine hundred and twenty three students). Japanese media, including Sankei Newspaper on 17 March 2011, referred to this as the "Kamaishi Miracle" citing on the high awareness about disasters of the residents from traditional disaster risk reduction (DRR) culture, represented by *Tsunami Tendenko*, a teaching that has been carried on from generation to generation that states, "when you expect tsunami, you should escape without thinking about your parents and children; think only of yourself not to be sacrificed, and even if your loved ones die, no one can blame you," helped save many lives of school children and their families without being sacrificed trying to look for each other's safety in the disaster (Yamashita 2011) and DRR school education that is based on the 2010 Handbook for Tsunami Disaster Management Education of Kamaishi. In spite of this, Kamaishi, in its recovery plan, has pledged to build a safer and assuring society by further strengthening DRR measures to provide an environment to quickly proceed with the recovery process and build disaster resilient schools and communities for its future.

10.2 School Centered Community Recovery

On 11 October 2011, MEXT Minister Masaharu Nakagawa sent a notification to 15 prefectures and one city of EJET affected region to propose the concept of "School-Centered Community Building "for school recovery. The concept has four main pillars, (1) Securing safety of schools and relocating them to safe areas as necessary, (2) Retrofitting schools to improve their functions as evacuation center and hub for disaster management, (3) Making school building eco-friendly and (4) Making schools multi-functional public facilities to become the center of communities. As the concept addresses issues beyond the education sector, MEXT has teamed up with Ministry of Land, Infrastructure, Transport and Tourism (MLIT) and Ministry of Agriculture, Forestry and Fisheries (MAFF) in planning the concept.

Although the main purpose of this concept is to rehabilitate or newly construct safe schools, the significance of this concept lies in the new idea of rebuilding the new school to become a multi-functional facility by combining such facilities as children's daycare center, public library, community center and DRR facilities in which the whole community can utilize and benefit (Fig. 10.1). Placing the school at the center of recovery and community building is believed to be a feasible option for school recovery because schools commonly functions as a cultural and spiritual center for the community for different age groups, together with being a facility to provide education for children (Takeuchi et al. 2011). In Japan, elementary schools have functioned more to bring community members together compared with other public facilities, partly due to the fact that organizations and activities related to

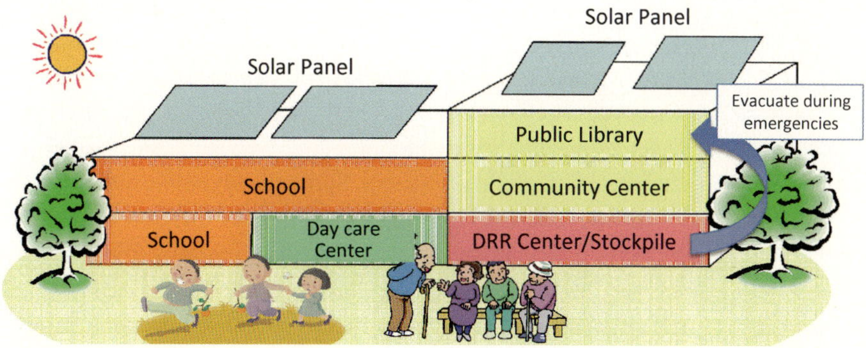

Fig. 10.1 Image of school as multi-functional public facility (adopted from image by MEXT)

social welfare has historically been established based on school districts (Sakagawa 2004). Therefore, if the concept is effectively implemented, school will not only be used for education, but a facility that will be able to serve the whole community, which could facilitate community integration among different age groups and strengthen community ties.

The concept also addresses important function of schools serving as evacuation centers during emergencies and as DRR hubs for related activities during normal times. At the time of EJET, 89.3 % or 30,513 public schools were designated as evacuation centers, however, it became clear that most of the schools were not properly prepared to function as evacuation centers when lifelines such as water supply and electricity were lost during EJET. National Institute for Educational Policy Research reported that schools equipped with proper storage room accounted for only 35.2 %, whilst water supply facility was 28.7 %, electric generators 18.0 % and telecommunications equipment 30.2 %, respectively (NIER 2011). As the case, conjugation of DRR functions with appropriate facilities and equipment is an effective way to strengthen the DRR capacity of schools and in making them DRR hubs for not only the school, but also for all of the community. Combining other public facilities also brings merit in regions where there is only limited elevated and safe land available. The school building itself may be constructed as a high-rise building so that people can evacuate during emergencies.

Actual implementation of the concept has been limited so far, because coordinated planning and action are required among local government, schools and community. Reaching consensus within various stakeholders for the new concept has also been a time consuming process as it involves multidisciplinary stakeholders. Despite of this, MEXT expects that the school centered recovery of public facilities and functions of public services will help accelerate the overall recovery of disaster affected communities (MEXT 2011). It is hoped that when public services and basic infrastructures, recover back to their normal status, displaced people will also be able to return to their respective communities.

Kamaishi is one of the disaster affected cities that has incorporated the MEXT's school centered recovery and community building concept in its recovery plan and for this, Kamaishi Board of Education (BoE) has been tasked to lead and facilitate in realizing the concept by starting off with the establishment of the School Construction Consultation Committee in December 2011, in which the members consist of leaders from the local government, school principles and community representatives. BoE has been responsible in explaining the benefits that the school centered recovery and in consulting with the Committee on what facilities and functions should be incorporated into the new school. Subsequently, while different public facilities and functions fall under various agencies and departments, BoE has been coordinating with these departments to materialize the requirements of the community, particularly for land use and peripheral infrastructures that will be necessary to operate the new school.

10.3 Case Study Area: Toni District, Kamaishi

Kamaishi is a historical coastal city in Iwate Prefecture known for its fishing industry and as one of the origins of the modern steel industry in Japan. Owning a rias coastline, the city has been historically prone to disasters, with the Meiji Sanriku Earthquake and Tsunami occurring in 1896, Showa Sanriku Earthquake and Tsunami in 1933 and Great Chilean Earthquake in 1960 (Fig. 10.2). Because of the city's history with disasters, the government has invested heavily in structural measures, as seen in the 60 m deep seawalls that took three decades to build amounting up to 1,200 billion yen (Kamaishi Port Office).

In the case study area, Toni District, the Meiji Sanriku Earthquake washed away 332 houses and took lives of 1,585 residents and the Showa Sanriku Earthquake took away 259 houses and 359 lives, both accounting for around 50 % of the population (Iwate Community Building Network 2012). The EJET destroyed 390 out of 956 houses and the causalities totaled 21, including two missing out of 2,106 people (Kamaishi City 2011a). Damages to public infrastructures were extensive, which included facilities for fishery, water supply and cultural heritage structures. As noted above, Toni ES and JHS are also included in the public facilities that were totally destroyed by EJET.

Toni ES was originally located in the higher ground together with the Toni JHS until 1982 (Fig. 10.3), then it was relocated to Katagishi in an area only about 100 m from the bay coast, surrounded by the 11.8-m dyke competed in Toni in 1981. According to a teacher who was working at the school during EJET, the sea could only be seen from the upper floors of the school. The tsunami, estimated

★Other EQ over M8: 1952 –Tokachi EQ, 1978 –Miyagi, 1994 –Sanriku Haruka EQ 2003–Miyagi Coastal and Northern EQ, 2008 –Iwate Northern EQ

Fig. 10.2 History of major disasters in Kamaishi at a glance (photos other than EJET are from Kamaishi city website)

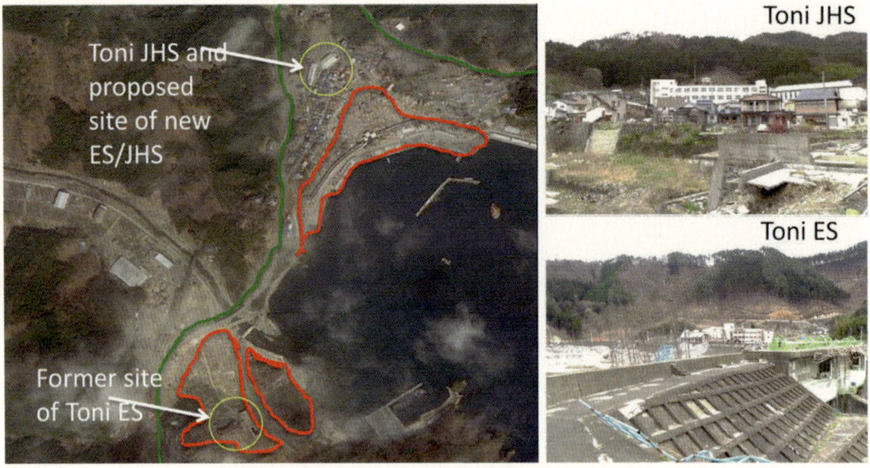

Fig. 10.3 Location of Toni ES and JHS and current situation in Kojirahama, Toni District (adopted from map provided by Kamaishi city government, photos by author) on *left* figure: *red lines* shows damaged areas, *green line* shows national road 45

around 10 m in height, breached the dyke and flooded the school up to the third floor. The ES building received devastating damage, leaving only the foundation and frame of the building remaining, thus difficult to rehabilitate the building for future use (Takeuchi and Shaw 2012). Plans to build back the new ES to its original location (before 1982), together with the JHS is now under discussion. Because safe highlands are limited in Kojirahama, the central commercial and administrative area in Toni, efficient use of land is required when considering new buildings. Hence, building the new ES together with JHS is seen as the most feasible option, as according to the discussions held in the School Construction Consultation Committee in Toni.

10.3.1 Recovery Plan of Kamaishi

Basic Plan for Recovery and Community Building of Kamaishi—"Scrum Kamaishi Recovery Plan," which was announced on 22 December 2011 places great importance on supporting school children and child raising for parents as shown in Table 10.1. The concept is planned to be implemented for schools in Unosumai and Toni Districts in which two elementary and two junior high schools received total damage. The plan annotates on both short-medium term goals of rebuilding the disaster affected schools into multi-functional facilities and longer term issues like support for child raising for working parents, enhancing DRR functions to be utilized as evacuation centers during emergencies and DRR hub for DRR education and drills during normal times as well as support for daily lives in social welfare and community education for the whole community (Kamaishi City 2011a).

Table 10.1 "Scrum Kamaishi Recovery Plan" Related sections from on school centered recovery

- Basic policy 2: Building community that places importance in community ties and support (2) Ensuring conditions for better child raising
  Build temporary facilities for affected Child Raising Club (child day care centers) to enrich after school programs, provide support for working parents and build new facilities in conjunction with Unosumai and Toni Elementary schools (…).
- Principle goal 6: Building community that nurtures children to survive
  Elementary (ES) and junior high schools (JHS) of the city were reconfirmed to carry an important function as evacuation centers after disasters. The affected ES (2) and JHS (2) need to be moved to safer places and rebuilt so that they will have new functions as hubs for supporting everyday lives and disaster management.
- Scrum 11*: Supporting the region by providing new functions to schools
  New ES and JHS schools will be built on safe, common land to provide educational environment for better coordination and for strengthening DRR functions. In addition, provisions to make the school a hub of community building by combining functions of social welfare and community education facilities and community centers should be considered.

*"Scrum" is a terminology used in rugby, but in the Recovery Plan refers to the community action plans that are aimed to achieve the goals.

The Recovery Plan expresses strong concerns on the population drainage, which is caused by both the immediate effects of EJET and due to chronic factors of low birth rate and aging society. Figures in the recovery plan states that Kamaishi has already been facing annual population loss of around 600–700 and in just 6 months after EJET, the city has lost more than 5 % or 2,000 of its population, in which the majority belongs to the juvenile (0–14 years old) and working (15–64 years old) age groups. For this, Issue of Concern for Recovery 2 in the Recovery Plan intends to provide better conditions for child raising to support working parents by building child day care centers together with the new elementary schools. In this way, preschoolers and ES school children will be able to join the after-school activities until their parents finish work. This support is much needed for households under difficult economic conditions after the disaster.

Principle Goal 6 refers in further enhancing the DRR functions of the schools by first relocating the four disaster affected schools to safer high rise areas and providing them with proper facilities, such as water supply facility, electric generators and portable toilets, and equipment including emergency goods (blankets, water, etc.) and communication system. Because schools will become community hubs in which various community members will utilize on daily basis, schools must be known to the whole community as a place to seek safety during emergencies. This goal also calls to further improve past DRR and environmental management education by incorporating the lessons learned from EJET, so that children will have the knowledge and the skills to survive disasters using their own judgment.

Scrum 11 under the Principle Goal refers on making affected schools into multifunctional facilities, as well as newly making joint ES and JHS schools. Following the MEXT concept, the initial proposal is to combine the schools with other public facilities, such as social welfare facility, usually used by the elderly age group, and community center, which is used widely by community members of all age groups in conducting administrative procedures and community culture/educational events. Joint ES/JHS schools are becoming more common in Japan as both urban and non-urban communities are losing their students, thus school operation has become difficult with less students. Joint schools do not only allow for efficient school operation, but also beneficial in allowing education programs to be better streamlined throughout ES and JHS.

With these plans and goals regarding school based recovery, the School Construction Consultation Committee has had active discussion on the ways to implement and realize the recovery plans. However, issues such as budget availability, finalization of land use planning, consensus building of the community and the sequence of implementation have been delaying the implementation process. As recovery and community building issues go beyond the schools and education, which are conventional responsibilities of BoE, related local government departments must place more efforts in coordinating and at the same time, conduct surveys and analysis to investigate the social situation and needs of the communities.

10.4 Outline of Interview Survey in Toni District

Following the MEXT concept and the discussions held with various stakeholders so far, series of interviews with time ranging from 45 to 120 min were conducted to 25 community leaders in Toni from July 11 to 15, 2012. To ensure that the questions are fully understood and answered in the same context, background information on MEXT's concept of Building School Centered Community and its relationship with Basic Plan for Recovery and Community Building of Kamaishi was explained before the interview. The interviewees can be categorized into three groups as below:

1. Local government (BoE and Toni Community Center),
2. Schools (Toni Elementary School (ES) and Junior High School (JHS), PTA representatives and preschool principle and
3. Community leaders of Toni Regional Council, town associations, temporary housing associations, Seigan temple, fishermen's association and volunteer fire fighting squad.

A report that conducted prior interview and questionnaire survey in October 2011 showed that even though 76 % of Toni households do not have children, close to 80 % of them have some sort of connection with the schools (Suda et al. 2012). As the case, there is relevance in targeting different groups of stakeholders for the interview. The same report was referred to in order to compare the changes in the community perceptions between the first 6 months and 1 year after EJET. The contents of the interview that was conducted can be sorted into three main issues below:

1. Profile of the stakeholder,
2. Roles in emergency response, recovery and DRR (preparedness),
3. Connections with schools and views on school centered recovery and
4. Comments on the current recovery process and next steps.

Table 10.2 shows the outline of the interview results that was shared with BoE and the School Construction Consultation Committee. Likewise, Table 10.3 shows the results of questions concerning connections with schools and perspectives on School Centered Recovery and Community Building.

10.4.1 Local Government

10.4.1.1 Board of Education (BoE)

BoE is responsible for the management of public ES and JHS and related facilities. During emergency response, BoE is responsible for confirming the status of schools and students from the school principle. After disasters, it is responsible for rehabilitating damaged school buildings and ensures that educational activities are quickly resumed and continued until situations are back to normal. As for DRR measures, BoE has been instructing schools to implement DRR education and activities based on the Handbook on Tsunami Disaster Management Education.

BoE has been organizing the School Construction Consultation Committee meetings to coordinate and consult with relevant city government department, school principles and PTA and other community leaders for the recovery of damaged schools and to implement the School Centered Recovery concept. For next steps, BoE plans to indicate a concrete schedule in building the new school and at the same time, apply for budget to the central government through the prefectural BoE. It will also look into revising the Handbook on Tsunami Disaster Management Education incorporating the lessons learned from EJET with experts from universities and research institutions.

10.4.1.2 Toni Community Center (CC)

CC functions as both branch office of the city government, center for community's lifelong education programs/events and other community service, including healthcare consultations. Equipped with wireless disaster radio, CC can be used as a temporary evacuation place during emergencies, but there are no facility or stockpiling to accommodate evacuees for long period. CC does not organize any DRR activities itself, although it has previously assisted community based groups as Women's Association in organizing "soup-run" demonstrations.

Currently, CC has no joint activities with the schools, but is indirectly connected with the school through students' participation in the annual CC Festival and Toni Day Festival. As a public facility engaged with the community on daily basis, CC is positive of sharing the new school building. In addition, since CC wishes to organize community events with the schools to encourage interactions between schoolchildren and elderly people. CC becoming an integrated facility with the school is a good option that is in line with the school centered community building concept.

Table 10.2 Outline of interview results (roles and status in response—recovery—DRR)

		Emergency response	DRR	Changes after 3.11	Roles and issues
Local government	Board of Education	• Status confirmation • Confirmation of whereabouts of students and teachers	• Making school related facilities safe (earthquake proofing) • Promoting tsunami disaster education handbook, providing advice to schools	• Reviewing schools' function as evacuation centers and DRR points • Renewing measures of schools for blackouts, suspension of water supply during emergencies • Revision of tsunami disaster edu. handbook	• Building new school building • Continuing with earthquake proofing of schools
	Toni Community Center (CC)	• Able to use center as temporary evacuation space (have communication equipment and warning alarm speakers	• Participation to March 3 City-wide evacuation drill • Previously, women's group conducted emergency cookouts and AED demonstrations	• Considering stockpiling for emergencies (even though CC is not a designated evacuation center)	• Considering planning more events that will facilitate community linkages

School related	Toni ES/JHS	• Confirming safety of students (primary responsibility) • Operation of evacuation center is responsibility of city government, but may have to support evacuees during emergencies • Use operation manuals of city and prefectural BoE, but the details are decided by the school	• Conducts evac drills six times/year • Considers three stages of DRR (at school, on way to school and at home) • Referring to advice provided by Iwate and Gunma University, implemented according to context of each schools by teachers	• Necessary revisions under review, but not major changes expected	• Psychiatric care for students who need it is priority • Counselors for students and parents visit school periodically
	ES/JHS PTA	• Support school in ensuring safety of students if possible (however, during 3.11, many parents working in city center could not enter Toni for several days)	• PTA members do not participate in school evacuation drills (but some members are community fire fighting members)	• No major changes in DRR activities expected	• Will be discussed as necessary
	Pre-school, child raising support club	• Act accordingly with ES	• Conduct evacuation drills three to four times/year, sometimes with support from surrounding residents • Show DRR videos depending manager of each facility	• New evacuation routes under consideration	• As many students need psychiatric care, the city government dispatches counselors • Old facilities and lack of staff are big issues • Relationship with surrounding community is good because operation is supported by them

(continued)

Table 10.2 (continued)

		Emergency response	DRR	Changes after 3.11	Roles and issues
Regional/town associations	Regional committee/ town associations	• Confirming safety of residents and disaster situation of town, information collection and dissemination to residents • Rehabilitation of lifeline • Procurement and distribution of relief goods	• March 3 city-wide evacuation drill is implemented in all towns, but the details are decided by each town • Some towns have well established DRR organizational setup and knows where the vulnerable people reside • Some towns have indigenous knowledge of observing the sea to predict tsunami • Self help DRR group have been established from request of city gov. since 2–3 years ago	• New designation of evac centers under request to city gov. Stockpiling is still not done (Kojirahama) • New evacuation center is planned to be built in the new public housing area (near foot of mountain) • Privacy of information became an issue when confirming safety of missing residents—need to consider taking measure for next event	• Improvements of access road to/ from town to other cities and smaller community road needed • Purchase of new residence cannot be decided because price of original land is not decided by city gov. • Difficult and slow to realize recovery plan suggested by town associations • Hoping that recovery funds can be utilized with more flexibility to accelerate recovery process
	Association of temporary housing facility	• Ensuring safety of residents	• Safety management of residents • Plan and conduct stretch classes, singing sessions for residents to connect with each other (hoping that sports experts and/or NPOs will be able to assist in this)	• For mega events as 3.11, response need to be sometimes improvised because DRR plans is only made under assumption of past disasters	• Many residents are not aware of the overall recovery process • There are opinions that the budget for rehabilitating seawalls should be used for community recovery • Schedule to acquire and build new public housing is not clear so many residents are worried when they will be able to move to new house

Local industry	Fishermen's association	• When EQ occurs, fishing boats are recommended to harbor to the nearest shore and evacuate to high lands	• Members join the March 3 drill individually as it is not possible to organize a joint drill with fishermen • Boats are recommended to carry communication radios and insurance	• Strong recommendation to fishing boats to carry communication radios and insurance, but further measures are difficult	• Operations are gradually commencing as most of the fish farming facilities have been recovered
Temple	Seiganji temple	• Have been used as temporary evacuation area in the past	• Have been talking to visitors on Showa Sanriku EQ disaster • Cooperated in emergency cooking drills of women's association		• Currently under rehabilitation from donations • Need to consider location of new public housing for maintain market price of apartments in the future
Syobodan	Community firefighting squad	• During emergencies, operate warning alarm, close water gates and assist evacuees when possible • There is manual, but measures were improvised during 3.11	• Assist March 3 evacuation drills • Inspection of fire extinguishers • Take part in DRR activities of students (at school, on the way and at home) • Syobodan members are aware of the community profile and locations of where most vulnerable reside	• During mega disasters, measures must sometimes be improvised, therefore, depending too much on DRR plan is not suitable for all cases	• Planning community events

Table 10.3 Outline of interview results (relationship with schools and perspectives on school centered recovery and community building)

		Linkages with school	Comments on School Centered Community Building	Next steps
Local Government	Board of Education	• Toni ES/JHS	• Prioritize recovering fundamental functions of schools • Need to discuss the details on the management of multi-functional facility • Further discussion needed for ES/JHS joint school	• Coordinating meeting with relevant divisions of city government • Firming up of schedule to measure, design and start new school • Revising Tsunami Disaster Education Handbook
	Toni Community Center (CC)	• Does not have direct linkages with schools (although indirectly connected through community events)	• Positive on joining school (but both admin/community service functions?) • May need to consider improving access road to and from school	• Since community events have not fully recovered would like to consider planning more community events for interactions)
School related	Toni ES/JHS	• ES has connections with pre-school (Gakudo Ikusei Club) for after school activities	• Prioritize recovering basic functions of school • Decisions on making school a multi-functional facility is up to the government • Entrance of school and community center should be separated for safety • New school should have simple design so that it can be rearranged in the future • Further discussion needed for ES/JHS joint school	• Taking care of students is first priority

School related	ES/JHS PTA	• Conduct BBQ events and garbage collection, grass cutting, window cleaning activities (ES) • Aside from school events, interact through community events	• Recovering basic function of the schools should be prioritized • Need to secure safety of school if it becomes multi-functional facility • Need to ensure School Centered Community Building does not become too much of a burden for the schools • Joint facility or building pre-school near ES is understandable	• Prompt recovery of school is needed
	Pre-school, Child Raising Support Club	• Good connections with ES because many students utilize the facility for after school activities	• There is no concrete plan to rehabilitate pre-schools, therefore joint facility with ES is desirable	• Waiting for recovery plan/ budget from city government
Regional/Town Associations	Regional Committee/ Town Associations	• Seeking to further strengthen community—school ties	• Need to build new school as soon as possible by prioritizing fundamental function of schools • ES/JHS joint school need to be decided quickly as it affects design of new school • Need to improve access road to school • Expecting school—community ties will be strengthened • Building school that shows the uniqueness of Toni is important for community recovery	• Economical effect and better accessibility to Toni ES/JHS are expected when Sanriku Coastal Highway is built • Improvement of community road within towns is expected • New school need to be built quickly
	Association of Temporary Housing Facility	• Residents with students	• Idea of ES/JHS existed before 3.11 (at former ES site) • New community was built in Katagishi area when Toni ES, pre-school and Toni Station were built in the area • Building good school in Toni is important to avoid further population drain in the area	• Quick implementation of Recovery Plan is expected to city government (especially building new public housing)

(continued)

Table 10.3 (continued)

		Linkages with school	Comments on School Centered Community Building	Next steps
Local Industry	Fishermen's Association	• The association sell school PE uniforms • Experience Classes (Wakame making, Salmon incubation, release and processing) are conducted in cooperation with schools (gradually commencing since end of 2011)	• Ideal to designate a facility that different groups of residents will use on a daily basis	• Upon request from school, would like to continue more experience classes for students
Temple	Seiganji Temple	• Experience Classes (meditation, grass cutting and Terakoya)	• Making school multi-functional has been considered before 3.11 • Land area in JHS is limited and access road needs improvements, so design should be carefully considered	• Upon request from school, would like to continue more experience classes for students
Syobodan	Community Firefighting Squad	• Upon request from school, assist in safety and disaster management activities	• School should be a simple place where students study and meet their friends and teachers • Worried that School Centered Community Building will become burden on the schools	• Local residents are too busy to rebuild their lives (secure jobs, housing) so thinking too much about the future is difficult at this stage

10.4.2 Schools and PTA

10.4.2.1 Toni Elementary School (ES) and Junior High School (JHS)

Toni Elementary School (ES) and Junior High School (JHS) have been heavily affected by EJET and in need for reconstruction. Both schools are planned to be rebuilt in the former location of the JHS, where the temporary school buildings are currently standing. As school ground on JHS is one of the evacuation areas designated by the city government, the new school building may be utilized as an evacuation center for future disasters, in which the school will be required to take initial response. During emergencies, the school principle is responsible for securing safety of students and reporting status to BoE, but may also be required to support other evacuees from the community. For DRR measures, both schools have conducted six evacuation drills since EJET, changing the evacuation scenario each time. DRR education, based on the Handbook on Tsunami Disaster Management Education, has also been conducted. The content of DRR education is decided by each school and teachers who implement them.

Regarding the School Centered Recovery and Community Building, the principles and teachers of both ES and JHS feel that the responsibilities are too large for schools. They are more concerned on immediate issues, such as psychiatric care of students and recovery of school building. As both principles of ES and JHS as well as numbers of teachers had been newly appointed in April 2012, it is assumed that the apprehensions to the concept come from their unfamiliarity with the community. On the other hand, the schools have had engagements with the community through joint education programs with local industries such as Fishermen's Association and Seigan Temple before EJET. Although only part of these activities has resumed, restarting these activities may encourage schools to regain their confidence in actively engaging with community.

10.4.2.2 PTA Members of Toni ES and JHS

The PTA functions to work in bridging students, parents and school in addition to providing requests to the school management regarding issues on school operation. Since EJET, the PTAs have been less active, mostly due to members being busy in recovering their lives and securing jobs. During EJET, as many of the parents work in city center, they could not return to Toni for several days. For the same reason, most PTA members have not been able to participate in school's DRR activities because they are out of Toni to work during school hours. In despite of this, most trusted and were confident that the school will safeguard their children from the disaster.

As for school recovery, the PTAs have been expressing their wishes to the city government for the prompt recovery of Toni ES/JHS, but most are not aware of the school centered recovery concept. A PTA leader noted that in case school is joined with other public facilities, such as CC, the school should have a separate entrance to ensure safety of the school.

10.4.2.3 Pre School (PS)

The preschools in Kamaishi are operated by Kamaishi Council of Social Welfare under the contract of Child Division of the city government. In Toni, there are two facilities, one for preschoolers (3–5 years old), which was washed away by the tsunami, and the other for elementary to high school students called the Child Raising Support Club, now operating in temporary building next to Toni JHS. PS have been conducting evacuation drills three to four times per year based on DRR manual. Videos on DRR were also occasionally shown to the children. Since EJET, new evacuation routes have been considered through town watching carried out by PS staffs.

PS is well coordinated with ES because many of its users are ES students who come there after schools. PS and its after school program has become an essential part of child raising for double income households, because the parents are not able to pick up and take care of their children during their working hours. As of now, as immediate funding for damaged PS recovery cannot be expected, joint facility with the new ES would be beneficial for both PS and users. PS may also help enhance the school–community linkage because PS usually possesses good linkage with its neighboring residents, as they are understaffed and requires support from the communities for operation. This will provide opportunities for community members, particularly the elderly people, to interact with schools and students.

10.4.3 Town Associations

10.4.3.1 Regional Council, Town Associations, Temporary Housing Associations

Toni Regional Council is a coordinating body that connects the town associations and communities with the local government, but does not plan or implement specific programs itself, as they are left for each town associations to decide.

There are seven townships in Toni, all with their own town associations. During emergencies, association members are tasked to collect information on the disaster situation and assist any residents who are in need for food and shelter. Response to EJET varied from townships and depended on the community profile. Towns like Kojirahama have many residents working in the city center at the time of disaster, while others like Kerobe have more elderly people whom needed help to evacuate. Some of the association members are also members of the community fire fighting squad (or *Syobo-dan*), hence, these members enabled town associations to work closely together with *Syobo-dan* during emergency response. Similarly, participation to the citywide DRR drills, held annually on 3 March, has varied from towns, according to their population profile. According to association leaders, DRR awareness in Toni is high that can be seen from the example in which some communities have indigenous knowledge of predicting the magnitude of tsunamis by observing the sea, but consciousness in taking actual actions may be low, possibly because of their confidence in the colossal seawalls and dykes protecting the bay areas. In the recovery process, town associations are tasked to compile the demand of the residents and report to the city government. Some town associations have already submitted their recovery plan, but most of them have are still under review by the local government.

Although town associations and schools have no mandate to work with each other, association members are connected with the schools through other channels, such as school committees, including School Construction Consultative Committee, PTA or involvement in community events. For this reason, town association members are positive with the idea of School Centered Recovery and Community Building because they understand the importance of school–community linkage that would facilitate community ties and allows all of the community to take roles in child-raising. In order to maintain the young population in the city, town association members are eager to see the new schools be built in a speedily manner. They also expect that the new interchange of the Sanriku Coastal Highway will provide better access to Toni, thus attracting students outside of Toni.

Currently, there are also Temporary Housing Associations for each of the five facilities in Toni, each belonging to the town associations of their respective locations. Their main function is the operational management of the housing facilities, including safety management, and organizing small events for the residents. In the prolonged lives in the temporary housing, each facility has become their own communities, already building ties among the residents.

10.4.3.2 Fishermen's Association

There are currently 439 members with age ranging from 20 to 60 years old. Ninety percent of the facilities that included cultivation facilities, refrigerated warehouses and food processing plants were destroyed by EJET. Unified disaster response is difficult as the fishermen are usually dispersed out in the sea, but association members are required to buy proper insurance and carry wireless radios on their boats. They are also instructed to reach the nearest shore, abandon ship and escape to high ground when tsunami is expected. As for recovery status, most of the cultivation facilities have been recovered and production of seaweed and scallop has resumed since end of 2011.

Fishermen's Association has been active in assisting Toni schools before EJET by cooperating in the school's experience education programs, including seaweed making and salmon hatchery, releasing and food processing. Experience education programs had provided invaluable experience for the students to understand how the economy of their communities is being supported. They expect that some of the students themselves will be engaged in the fishery industry in the future. Although part of experience classes have resumed since end of 2011, the association is keen in supporting School Centered Recovery and Community Building concept to build closer relationship with the schools and continue assisting in these kinds of regional education.

10.4.3.3 Seigan Temple

Seigan Temple owns a long history of 400 years and is one of the symbols of Toni. The head priest has been active in educating the local history, culture and DRR through its *Terakoya* or "temple school" activities. The first ES in Toni was established at Seigan Temple during the Meiji Era. Although premise of the temple was once one of the evacuation areas in Toni, the tsunami caused by EJET reached beyond the main road in which the temple is located, damaging the main building. Donations from current and former residents were received immediately after EJET to repair part of the damages, which depicts how the temple is an important spiritual symbol of the local people. Seigan Temple hopes that the rehabilitation of the temple will encourage Toni residents to overcome the difficulties in the recovery process.

Before EJET, the temple had supported Toni ES and JHS to conduct experience education programs in history and ethics. The head priest, who is also one of the members of the School Construction Consultative Committee, is also looking forward to resuming educational activities to support the schools.

10.4.3.4 Community Fire Fighting Squad (Syobo-dan)

Syobo-dan is a community based fire fighting squad in which the members usually have other jobs and voluntarily participate for about two to three years (per term) in firefighting activities (Ishiwatari 2012). In Kamaishi, each community would have around 15–20 members taking part in safety and disaster management activities in their respective communities, working closely with the town associations and receiving advice from the Disaster Management Division of the city government. When disasters occur, they warn the local residents by sirens and loud speakers and when tsunami is expected, they close the water gates and support residents to evacuate to high grounds. Although *Syobo-dan* has its operation manual, the magnitude of EJET forced them to make quick decisions and improvised actions to save lives. The *Syobo-dan* has its advantages in coping these situation because they are well aware of the community profile, thus is able to efficiently prioritize in reaching and assisting the most vulnerable community members for evacuation. During normal times, they conduct regular checkups of fire extinguishers/alarms in the schools. Occasionally, they assist schools to conduct DRR drills, in three different situations, at school, on the way and from the school and home.

10.5 Analysis of the Interview Survey (Roles and Responsibilities)

With the results of the interview survey, a SWOT analysis was conducted to evaluate the internal factors (*S*trengths and *W*eaknesses) and external factors (*O*pportunities and *T*hreats) of Toni to identify the benefits and challenges that Toni

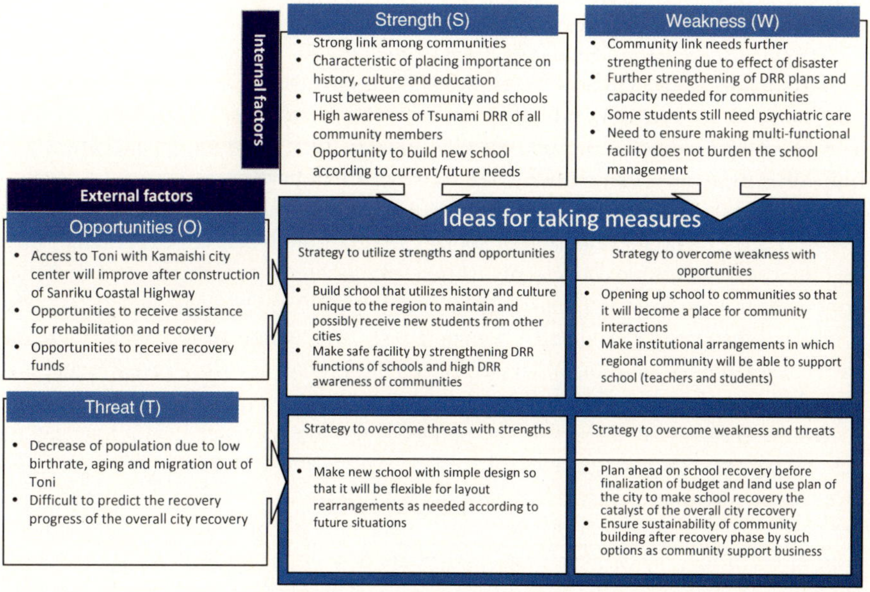

Fig. 10.4 SWOT analysis of interview survey and suggested measures

District possesses. This is followed by suggestive measures that can be taken for implementing the concept of Building School Centered Disaster Resilient Community (Figure 10.4).

10.5.1 Internal Factors (Strength/Weakness)

The strength of Toni primarily lies in its strong community linkage. Historically, Toni residents have a strong identity of belonging to the Date Clan (*Date-han*), a feudal clan that existed in Miyagi and southern Iwate prefectures in the nineteenth century, while other districts of Kamaishi belonged to the Nabu Clan (*Nanbu-han*). The two clans often had border disputes and for this, guard towers of both clans existed in Toni (Seiganji website). A former BoE official also noted on the fact that because majority of Toni residents were born and have spent most of their lives in Toni, they have strong pride of their hometown. Therefore, community ties and trust have been very strong, even among other Kamaishi districts. Toni is also known for its dedication to education, history and culture and has owned Kamaishi's first elementary school (former Toni ES which was located at Seigan Temple in Meiji Era). This dedication can be also seen from the fact that Toni residents refused the city government's request to build temporary housing facility on Toni JHS's school ground to avoid delay in school recovery. DRR awareness has also been high due to the lessons from past disasters passed on from generation to generation. In addition, as parents know that DRR activities have been extensively been conducted at the schools, they trust that the schools will ensure their child's safety during disasters. The combined elements of community's mutual trust, devotion to education and high DRR awareness that existed in Toni before EJET have been seen as the preconditions of making *Tsunami Tendenko* culture feasible during emergency response (Yamori 2012). The reevaluation of social capital through disaster experience, such as this, provides an ideal cycle of further strengthening community ties to better cope with future disasters (Matsui 2011). As for DRR functions of the schools, the school themselves have already conducted six evacuation drills, some with the assistance from Syobo-dan, and have revised the Emergency Response Manual that includes provision to assist evacuees outside of the schools.

On the contrary, as marked in the results of the interviews, EJET did weakened school–community linkage and for this, measures must be taken to rebuild and strengthen community ties of Toni residents to move forward in the recovery process. As noted by the school principles, there is also anxiety for the psychiatric conditions of their students, parents and teachers from the traumas of EJET experience. While posttraumatic stress disorder (PTSD) is more visible in adults, the conditions could vary among youths causing difficulties in identifying students in need of assistance, and for this, schools can be a viable place where larger scale screening and interventions are possible (Ronan and Johnston 2005). Therefore, in the school centered recovery plan, management of the new school facility must be carefully considered as not to place too much expectation on the schools. As for school's function as DRR hubs, there is still room for enhancing DRR functions of schools with better facilities and equipment and by exploring the possibilities of joint school–community DRR activities to enhance the DRR capacity of the whole community.

10.5.2 External Factors (Opportunities/Threats)

In Kamaishi, 90.88 billion Japanese Yen (JPY) or about USD 1 billion (1USD=90JPY) of recovery budget has been approved in four tranches, initially to be expended mostly for infrastructure rehabilitation projects (Kamaishi City 2012). Toni residents have high expectations for the new interchange of the Sanriku Coastal Highway that will improve access to Kamaishi city center, cutting the current travel time by half (10 min). The interchange will not only bring economical benefits, but may also attract new residents and students to Toni if living conditions, especially new public facilities, would be appealing to them. There are also opportunities to receive support from community based organizations, including NGO/NPOs, as well as funds with special preference that are available for recovery projects that can materialize some of the community building plans. Building network to receive such outside assistance can be effective in the initial stages of the recovery process, especially for aging communities that experienced mega disasters because they are not able to cope by themselves (Matsui 2011).

As for possible threats, as mentioned in the previous section, Kamaishi has been experiencing population drainage due to low birthrate and aging of communities and the situation is likely to worsen due to forced displacement of communities from the effects of EJET. With fewer children in the community, schools may have to be temporarily or permanently closed or merged with other schools in the city. Closing of schools may cause students and teachers become disconnected from their communities because learning opportunities that utilize local resources are lost and time to interact with the local community has lessened because of longer travelling time to the new school (Sakagawa 2004). Availability of good education will be a major factor that will influence families to decide to either remain or move out of Toni for better opportunities elsewhere, leaving the elderly residents behind. This could become a threat to the survival of not only schools, but also the community itself. Another threat is the unpredictability of the overall recovery process, which has been causing both worries and frustration among Toni residents. Delays in recovery, such as land use planning and especially building new public housing, may indeed hinder school recovery and hamper community building on the long run.

10.5.3 Suggested Measures for Implementation

The following measures are suggested to maintain strengths and opportunities and to conquer weakness and threats in implementing the School Centered Disaster Resilient Community concept.

10.5.3.1 Measures to Utilize Strengths and Opportunities

The new Toni ES and JHS must aim to become attractive schools that incorporate the unique history and culture of Toni in order to maintain current students number and to possibly receive new students from other cities. An example from Takaki ES in Saijo, Ehime Prefecture has shown that by actively promoting the school and the charms of its locality through its website and BoE allowing this particular school to receive students residing outside of the school district, the school was able to withstand the threat of being permanently closed due to low student number, by successfully attracting new students (Shaw 2012). Therefore, it is important that the school–community linkage is maintained and activities such as experience classes with local industries are resumed to keep the schools unique to the region. Another important element for the new schools is to preserve the DRR awareness of students and teachers by enhancing DRR education program. School's DRR functions as evacuation center should also be strengthened by equipping sufficient facilities to better cope with initial disaster response. It is important to ensure that the community is aware that they are able to seek safety at the schools when faced with disasters.

10.5.3.2 Strategy to Overcome Weakness with Opportunities

In realizing school centered disaster resilient community in Toni, the new school must be recovered on prerequisite that it will be opened to and used by the community, especially at the initial step of designing the layout and functionality of the school. On the non-structural aspect, the interviews revealed that Toni is rich in local human resources to help the schools in providing regional and professional education. There are limitations for school teachers to provide such education, and therefore, by taking advantage of "local teaching materials," students will be able to acquire knowledge that will enable them to be integrated and become contributors to their communities in the future (Sakagawa 2004). This will be the key element to maintain the uniqueness of Toni ES and JHS that will encourage students to stay and possibly attract new students to enroll. However, opening up the school to the community has been responded by a certain level of apprehension by the schools due to concerns on the burden that it might bring upon them. As the case, BoE is tasked to further consult with both schools and community for them to understand the school centered concept and the mutual benefits it will bring for Toni. Moreover, BoE will need to develop an institutional support system that will facilitate integration of these stakeholders. For example, the schools in Toni may consider in establishing Community School in which the voice of parents and community, by legal basis, is reflected in school management through their participation in the School Management Committee – a well-known school management system seen in Mitaka Dai-ni ES and JHA and other cities such as Okayama and Fukushima in Japan (Kainose 2010).

10.5.3.3 Strategy to Overcome Threats with Strengths

Considering construction of school building is a huge investment for local communities, there is pressure for local governments to efficiently utilize the classrooms, because they are commonly used only for few hours during weekdays and underused during weekends, evenings and school holidays (OECD 1998). In addition, Toni schools may not be able to resolve the declining student number despite of its efforts in maintaining student number due to the low-birth, aging phenomena. In this case, the school building adopts a simple design so that it will be flexible for layout rearrangements that will allow utilization of the facility for other purposes as according to future situations and needs. Whereas some school buildings nowadays adopt a modern design, veteran schoolteachers have remarked that schools that are simple and traditional in design are more functional, easier to maintain and commonly more resilient to disasters.

10.5.3.4 Strategy to Overcome Weakness and Threats

In the uncertain and dynamic changing recovery process in Kamaishi, school recovery should be planned ahead before the finalization of budget provisions and land use plan so that recovery projects can immediately start when decisions are made. In the first year after EJET, significant portion of the recovery related budget, amounting up to 5.8 out of 15 trillion yen, was left unused due to delays in examination of budget and consensus building of communities, which depicted the need to accelerate the recovery process (Japan Press Network 2012). These kinds missed opportunities should be avoided at all cost. Another dimension to the strategy should be on the sustainability of these community building projects that will continue after recovery phase. One option that can be explored is establishment of community support business that can be implemented by local community organizations, such as NGO/NPOs. An example from a network of welfare NPOs called "Nishi Suma Danran" from Kobe that started various community business based on the needs of the community during the recovery phase, such as cleaning, nursing and other support for elderly people (UNCRD 2003).

10.5.4 Image of School Centered Disaster Recovery and Community Building in Toni District

Figure 10.5 shows the image of the new Toni ES/JHS as a new multi-functional facility that is based on the analysis of discussions held so far in the School Construction Consultation Committee and the questionnaire/interview surveys conducted. The key stakeholders shown by their roles and functions in the community to support the schools and the actual activities they may jointly engage on. The Toni Community Center and the Toni Preschool/Child Raising Support Club are suggested to either share their functions or become joint facilities with the schools. Structural issues in building the new school will not be mentioned here, but may influence what can be actually built together with the school, given the limited high land available in Kojirahama area, around the former Toni JHS. The fact that the land use planning has not been finalized is another factor that may decide the locations of facilities other than the school will ultimately be built. Lastly, the more stories the building will have, the higher the cost. Hence, budget availability is also a key factor in the same regard.

10.5.4.1 Toni Community Center (CC)

As a branch office of the city government that provides administrative service and as one of the first model Life Support Center in Kamaishi that provides various community services, such as health consultations, in addition to planning and organizing community events, Toni CC is utilized by Toni citizens of different age groups on a daily basis. Because of its function of connecting the community with the local government, CC can be classified as having the administrative—institutional role in the School Centered Recovery and Community Building scenario. CC also serves as window for town associations, community clubs and external

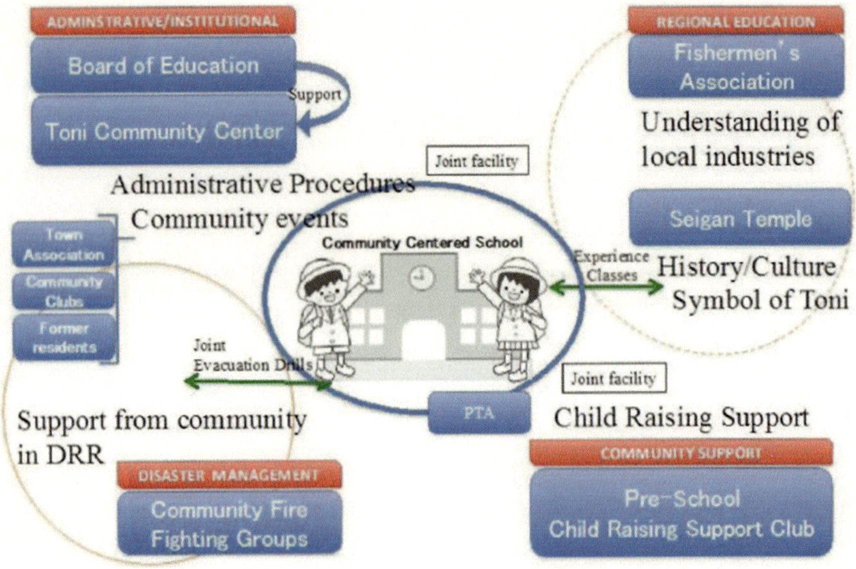

Fig. 10.5 Image of new Toni ES/JHS as multi functional facility

stakeholders, including former Toni residents and NGO/NPOs from outside the city, and for this reason, joining facilities with the school will help school to be more accessible to both internal and external communities. A good example of joint facility can be seen in Shiki ES in Saitama Prefecture, a school-CC-public library joint facility built in 2001, in which the facility is not a mere joint building, but a community hub for co-learning and interaction among different age groups of the community (Institute of City Structure and Life 2011).

10.5.4.2 Preschool/Child Raising Support Club (PS)

Toni residents, especially the double income households, rely on the PS to look after their children during working hours. For this, PS can be classified as one of the community support mechanism that is vital for its recovery. In the unstable conditions of the region after EJET, it is essential that these facilities be quickly restored so that parents can concentrate in recovering their normal lives. Furthermore, the support for children and child raising parents for the purpose to relieve some of the pressures of population decline from the city. In this context, Kamaishi has been continuing to provide child raising support through its "Action Plans for Assisting Strategies in Raising the Next Generation" since 2005, which aims to improve child raising conditions, enhancing educational contents and providing more opportunities for entering such facilities to ease both economical and time constraints, especially for working parents (Kamaishi City 2011b). Because PS are commonly well coordinated with ES in Kamaishi, integration of the damaged PS with the new Toni ES/JHS will enhance child and child raising support in the district. BoE must coordinate with the Child's Division of the Health and Social Welfare Department as PS is under its jurisdiction.

10.5.4.3 Local Industries and Religious/Cultural Organizations

As a spiritual symbol of Toni residents, Seigan-ji temple plays an important role in rebuilding community ties. Chamlee-Wright (2010) notes on how religiosity afforded a valuable set of socially embedded resources for bringing community people back and for them to contribute to community rebuilding after disasters. With regards to local industries, the benefits of making use of local human resources and the importance of school children to understand how their local economically is supported were mentioned above. Particularly for school education, involvement of local industries and religious/cultural organizations of the community plays an essential role in providing students with understanding, pride and love for the communities they live in. Through such activities as experience class programs, students can receive "real life" education in which schoolteachers are not able to provide. Therefore, local industries and religious-cultural organizations in Toni can be classified as having a role in providing regional education. It is worth noting that these industries and organizations also receive benefits from their involvement with the schools because they will be making a social contribution to nurture their children who are the future of their own communities.

10.5.4.4 Community Fire Fighting Corps (*Syobo-dan*)

Community Fire Fighting Corps or *syobo-dan* plays an important role in maintaining safety disaster management of the community as observed during EJET. In the case of Toni, according to one ES teacher, a *syobo-dan* member had worked effectively instructing schoolteachers and students to evacuate to the high grounds while they were still waiting on the school grounds, not being able to see the tsunami coming because of the 11.8 m sea dyke. Coincidently, this *syobo-dan* member was the father of a student who attended this ES and also the member of school PTA, so the instructions were trusted and followed immediately, coming from a familiar face. As this story illustrates how *syobo-dan's* activities become more effective when there is more familiarity with its community, close relationship and cooperation should be always maintained among school, *syobo-dan* and the community through interactions, not necessarily limited in DRR activities. In a small community like Toni, with less than 1,000 households, many of the *syobo-dan* members are family members of students, which will make them proud of their family if DRR activities are conducted by them at the school. This will be very effective in DRR awareness raising method. *Syobo-dan*, on the other hand, can also benefit from engaging with schools and Toni residents in DRR activities because the it will have the chance to better grasp the community profile that will be helpful in assisting the community during emergencies.

10.6 Conclusion

In the current recovery process in Kamaishi, even though many coordinating meetings within the city government and consultations continue with the communities, the first step of finalizing the land use planning of the city has not been done. Therefore, the location of the new ES/JHS joint school in Toni District has not been officially confirmed. Toni residents feel great anxiety for this, in that the more the school recovery delays, the more chance of young families leaving Toni. One of the reasons for this recovery pace is because the community and the local government are currently in the process on prioritizing among different recovery issues including ranges from housing, education, public infrastructures, hospital and welfare facilities (Takeuchi and Shaw 2012). This is not a simple task that involves different sets of government departments and stakeholders. From the interviews, the Toni residents place top priority on new housing, because they are not able to make long term commitments to their communities unless they know the location of their new permanent residence. However, for families with children, they would most likely avoid living in a community where there is no school or child raising support facility, hence, recovery of the school is equally important to them. Indeed, the experience from the 2005 Hurricane Katrina has shown that schools have played a central role in the recovery process, both for need for childcare, but also parents' strong desires to bring normal lives back to their children. This provided powerful reason for affected residents to return to their communities, even if this meant significant hardship (Chamlee-Wright 2010).

As suggested in the analysis of the conducted surveys in Toni, issues such as managerial aspect of the multi-functional facility, systemized methods and support for integrating schools and community, exploring possibilities of community business for community recovery should be carefully considered. Opportunities for building new schools that employ the local uniqueness to maintain and possibly attract new students should not be missed. Moreover, it is important that the DRR features of the school be further strengthened and be known to the community. Roles and responsibilities of school, community and local government for emergency response and in the recovery process need to be clarified and coordinated, incorporating the lessons learned from EJET to better respond to future disasters. These are all important elements for the sustainable utilization of the new school facility and for the continuity in building Toni a disaster resilient community.

Under the direction of the Kamaishi city mayor, the city government is now in consultation with each regional district to confirm the consensus built within communities and agree in implementing projects that include school recovery.

The recovery fund for the 2012 Japanese Fiscal Year (JFY) amounted up to 2,043.3 billion Japanese Yen (JPY) and for 2013 JFY, the Reconstruction Agency has requested for 2,823 billion (JPY) in which around half is planned to be used to rehabilitating public facilities for community recovery (Reconstruction Agency 2012). The local city governments, including Kamaishi, are tasked not to miss this opportunity and to promptly and efficiently utilize these funds to avoid delays in expenditures and implementation as experienced in the previous year.

Acknowledgements The authors would like to express their appreciation to Church World Service (CWS) Asia-Pacific and MERCY Malaysia for funding this research. In addition, extensive support was provided by Kamaishi Board of Education, members of Toni Regional Council and Toni ES and JHS.

References

Chamlee-Wright E (2010) In cultural and political economy of recovery: social learning in a post-disaster environment. Routledge, London, UK, pp 123–124

Institute of City Structure and Life (2011) City@life magazine, no.99. Archive Co Ltd, Ontario, Canada

Ishiwatari M (2012) Community-based disaster risk management. In: Shaw R (ed) Government roles in community-based disaster risk management. Bingley, Emerald, pp 21–27

Iwate Community Building Network website (2012) http://townplan.m27.coreserver.jp/wordpress/?page_id=92. Accessed 27 January 2013

Kainose S (2010) Elementary and junior high school joint community school – challenges. Director of Board of Education, Mitaka city

Kamaishi City Government (2011) Basic plan for recovery and community building of Kamaishi. http://www.city.kamaishi.iwate.jp/index.cfm/12,18690,78,447,html (in Japanese)

Kamaishi City Government (2011) Plan to promote child education in Kamaishi. http://www.city.kamaishi.iwate.jp/index.cfm/10,21138,78,html. Accessed 18 December 2012

Kamaishi City Government (2012) Recovery budget project plan. http://www.city.kamaishi.iwate.jp/index.cfm/10,19261,78,447,html. Accessed 29 January 2013

Kamaishi Port Office (2012) http://www.pa.thr.mlit.go.jp/kamaishi/port/km04.html. Accessed 14 October 2012

Japan Press Network (2012) Article on 19 August 2012. http://www.47news.jp/CN/201208/CN2012080901002539.html

Matsui K (2011) The sociology of earthquake disaster and recovery. Liberta Publishing, pp 134–139

Ministry of Education, Culture, Sports, Science and Technology (2011) Notification on school centered community building and notification on rehabilitation and reconstruction of public school facilities damaged by the East Japan earthquake and tsunami, October 2011

Ministry of Education, Culture, Sports, Science and Technology (2012) Information on damages from the East Japan Earthquake (Report #207)

National Institute for Educational Research (NIER) (2011) Result of factual survey on DRR functions of school facilities. http://www.nier.go.jp/shisetsu/pdf/bousaikinou2011.pdf

Organisation for Economic Co-operation and Development (1998) Under one roof: the integration of schools and community services in OECD countries, pp 13–14

Reconstruction Agency (2012) Overview of recovery budget for 2012 and overview of recovery budget request for 2013 JFY. http://www.reconstruction.go.jp/topics/120131%E3%80%90set%E7%89%88%E3%80%91111124%E6%A6%82%E7%AE%97%E6%B1%BA%E5%AE%9

A%E6%A6%82%E8%A6%81%20.pdf and http://www.reconstruction.go.jp/topics/20120907_H25gaisanyoukyuu.pdf

Reconstruction Agency (2013) Current situation of recovery and efforts (in Japanese). http://www.reconstruction.go.jp/topics/20130110_sanko03.pdf. Accessed 26 January 2013

Reconstruction Agency (2013) Number of evacuated people (in Japanese). http://www.reconstruction.go.jp/topics/20130125_hinansyasuutyousa.pdf. Accessed 26 January 2013

Ronan K, Johnston D (2005) Promoting resilience: recovery. In: Promoting community resilience in disasters: the role for schools, youth, and families. Springer, New York, p 15 and p 126

Sakagawa S (2004) School as a center of the local community (in Japanese), Kokon Shoinpp, pp 121–122 and pp 273–278

Seigan-ji (2012) From Seigan-ji website. http://seigan.or.jp/310.htm. Accessed 17 October 2012

Shaw R (2012) Knowledge note from workshop on enhancing disaster resilience of education sector and communities. IEDM, Kyoto University, Kyoto

Shaw R, Takeuchi Y (2012) Towards long term recovery and the mega disaster. In: Shaw R, Takeuchi Y (eds) East Japan earthquake and tsunami. Research Publishing, Singapore, pp 258–259

Suda Y (2012) Linking Human Behavior, Participatory Mapping Community Infrastructure for Disaster Risk Reduction: Approaches in Shisho and Kamaishi, Master's Thesis for Kyoto University's Graduate School of Global Environmental Studies

Takeuchi Y, Mulyasari F, Shaw R (2011) Roles of family and community in disaster education. In: Shaw R, Shiwaku K, Takeuchi Y (eds) Disaster education. Bingley, Emerald, pp 77–94

Takeuchi Y, Shaw R (2012) Damage of education sector and its recovery. In: Shaw R, Takeuchi Y (eds) East Japan earthquake and tsunami. Research Publishing, Singapore, pp 143–164 and pp 155–156

UNCRD (2003) In: Tsunehiro Y, Goda K, Shaw R (eds) From disaster to community development: the Kobe experience. UNCRD, Kobe, Japan

Yamashita F (2011) Tsunami Tendenko – history of tsunami in modern Japan. Shin Nihon Shuppansha, Tokyo, pp 52–53

Yamori K (2012) Revisiting the concept of "tsunami tendenko" Journal of Natural Disaster Science, J. JSNDS 31–1 35–46

Part III
Technology and Innovations

Chapter 11
Post-Disaster Housing Reconstruction in Indonesia: Review and Lessons from Aceh, Yogyakarta, West Java and West Sumatera Earthquakes

Krishna S. Pribadi, Dyah Kusumastuti, Saut A.H. Sagala, and Ramanditya Wimbardana

Abstract The post-disaster situation offers opportunities to rebuild liveable environment for achieving safer communities in the future, and housing reconstruction plays a crucial role in rebuilding the communities. In the past decade, Indonesia has experienced several major destructive earthquakes causing severe damages to infrastructures and human settlements. An ex-post review of the past experiences and challenges in post-disaster housing reconstruction after earthquakes in Aceh (2004), Yogyakarta (2006), West Java (2009) and West Sumatera (2009) reveals some strategic issues in implementing safer housing reconstruction that have to be addressed in the future for achieving "build back better" post-disaster reconstruction programs. Past experiences showed that training and capacity building of construction personnel and home-owners, through the dissemination of guidelines and manuals as well as building codes and standards for anti-seismic design, which have been implemented in the housing reconstruction programs by the national and local governments, NGOs and aid agencies have improved the practices by builders and masons in the areas that had experienced major earthquakes, but in general the attitudes of the building industry as well as local government building administrators in ensuring the housing earthquake safety still need to improve. Nevertheless, some good practices have been observed in several post-disaster housing reconstruction programs that shed the light to the development of better strategies for achieving earthquake safer housing, through the introduction of various supporting policies such as better project delivery systems, better mechanism for providing supervision and technical advices, more down-to-earth training and capacity building mechanism

K.S. Pribadi (✉) • D. Kusumastuti
Faculty of Civil and Environmental Engineering, Institute of Technology Bandung, Bandung, Indonesia
e-mail: ksppribadi@bdg.centrin.net.id; ksuryanto@si.itb.ac.id

S.A.H. Sagala • R. Wimbardana
School of Architecture, Planning and Policy Development, Institute of Technology Bandung, Bandung, Indonesia

R. Shaw (ed.), *Disaster Recovery: Used or Misused Development Opportunity*, Disaster Risk Reduction, DOI 10.1007/978-4-431-54255-1_11,

as well as smarter financing and incentives system, supported by appropriate technology approaches. Awareness building program is an important part of the mechanism and should be taken out seriously as it affect significantly the risk perception of the stakeholders, which is one of the key elements in the decision making process in investing for safer housing.

Keywords Anti seismic design • Awareness raising and training • Build back better • Building administration • Housing recovery

11.1 Introduction

Indonesia, the world's largest archipelago, is prone to earthquakes due to its location in the Pacific Ring of Fire. As four tectonic plates interact within the boundary of the country with approximately 240 million inhabitants, earthquakes of M 5.0 or larger can be recorded almost daily (Pribadi et al. 2008). In the last decade, several major earthquakes have caused massive destructions, such as the earthquakes in 2004 in Aceh (which has caused the killer Great Indian Ocean Tsunami), 2005 in Nias, 2006 in Yogyakarta, 2007 in Bengkulu, 2009 in West Java, and 2010 in West Sumatera, and claimed hundreds of thousands of lives and damaging half a million structures in total, mostly from the housing sector.

Experience shows that most of the casualties and economic losses in the earthquakes were due to damages in the housing sector. In many cases, houses in Indonesia are built as non-engineered structures, based on traditional practices of the building trades. Despite the existence of manual and guidelines for construction of earthquake resistant housing, many of them performed poorly during the earthquakes due to inadequate design and construction practices, which include poor materials and workmanship (Pribadi et al. 2008). Nevertheless, previous disaster events provide some opportunities to learn from previous mistakes and to rebuild a safer community against earthquake hazard. Considering the seismicity of Indonesia as one of the most earthquake prone area in the world, it is very important to ensure that all buildings, existing and newly reconstructed, perform well under earthquake loading. Improved house construction is needed to reduce vulnerability and to avoid worse impact in the next occurrence of earthquake.

Recovery process has become one of crucial steps that must be implemented after a disaster. Housing reconstruction is a critical factor in the recovery processes, whether one is addressing the phenomenon at the household or community level (Peacock et al. 2007). It is important because without establishing home, the ability of a household to carry out normal activities and to re-establish routine becomes limited and hampered. Delay in housing reconstruction and rehabilitation could delay other recovery effort, such as psychosocial, social and economic recovery (Barakat 2003; Lindell and Prater 2003). In addition Barakat (2003) noted that housing reconstruction and related activities can enhance communities' capacity by strengthening their physical, emotional and practical abilities to survive the disaster

Table 11.1 Recent earthquake disasters in Indonesia

No	Earthquake event	Magnitude	Loss of life	Homeless people	Damaged houses (units)	Destroyed houses (units)
1	Aceh earthquake and tsunami, December 26, 2004	M 9.4	110,000	700,000	57,137	69,932
2	Nias earthquake, March 28, 2005	M 8.6	850	40,000	71,891	12,010
3	Yogyakarta earthquake, May 27, 2006	M 6.8	5,700	100,000	260,000	154,000
4	Bengkulu earthquake, September 12, 2007	M 8.5	35	–	390,825	19,375
5	West Java earthquake, September 2, 2009	M 7.4	81	178,490	216,424	46,697
6	West Sumatera earthquake, September 30, 2009	M 7.6	1,117	–	249,833	114,797

Source: Bappenas 2007; Bappenas 2009; Bappenas 2010; Bappenas 2006; BRR and International Partners 2005

and facilitate reconciliation; improving institutional resources and informal social relations; increasing pride and self-esteem through participatory and stakeholder programming; and enabling disaster-affected people to look forward and invest in the future.

On the other side, housing reconstruction has been challenged not only by the limited lack of capacity in the technical or engineering aspect of the construction, but also by issues such as: (1) poor and chaotic planning and coordination between actors; (2) scarcity of resources and supplies in the time where projects and numerous aids are being launched simultaneously; (3) lack of knowledge and experience in providing and building settlements after disaster; (4) social-cultural consideration; and (5) appropriate beneficiaries' expectation and need (Barakat 2003; Davidson et al. 2007).

The recent post-earthquake recovery experiences in Indonesia, i.e. Aceh (2004), Yogyakarta (2006), West Java (2009) and West Sumatera (2009) (see Table 11.1) provide important lessons learned which are useful for developing better understanding and better guidelines on the housing reconstruction process. A review of the past experiences and challenges in the post-disaster housing reconstruction programs is conducted to identify the problems and strategic issues to be addressed in the future post-disaster reconstruction programs. The main question is: to what extent the experiences of post-disaster housing reconstruction provide effective lessons learned to the stakeholders?

The following sections outline the theoretical background that becomes a basis for analysis in this paper. This is followed by an explanation of the approach and methodology. Findings will be discussed with the reference to the theoretical background and the case study of four provinces in Indonesia affected by earthquake disasters. Finally, the paper suggests some conclusion and recommendation in developing sustainable housing reconstruction process.

11.2 Theoretical Background

Disasters leaves physical environment destruction, social disruption, and economic stagnation which have critical impact to human lives. Avoiding greater human, physical, and financial losses in the future, policy makers and praticioners are challenged to recover from those major impact. Despite the challenges attendant to the recovery process, the recovery phase offers important development opportunity to restore, rebuild, and reshape the affected area that it could reduce future vulnerabilites and affect sustainable development outcomes (Berke et al. 1993; Smith and Wenger 2007; Olshansky and Chang 2009). The influx of aid and assistance in the affected area can be used to build and support the recovery effort.

Appropriate reconstruction approach could provide a monumental window of opportunity to rebuild damaged structures stronger than before the event, reshape the existing social and economic system, and enhance disaster resilience. Many strategies can be opted, such as the repair and improvement of damaged buildings and infrastructure, stimulating local economic, enhancing public capacity and awaraness toof hazard, etc. The failure in the approach to establish the recovery goals can lead to poor reconstruction quality, a loss of jobs, a reduction in affordable housing stock, missed opportunities to incorporate mitigation into the rebuilding process, and an inability to assist the neediest recover (Smith and Wenger 2007). Oliver-Smith (1990), in his recovery study in Peru, found that sustainable recovery objectives, such as addressing issues of social inequality and the adoption of hazard mitigation practices during recovery, can be achieved when the strategies meet local needs, local capacity are considered by the donors and agencies, and the community understand programmatic assistance requirements (Smith and Wenger 2007).

Housing is fundamental for most societies because it relates to their well being (Barakat 2003; Peacock et al. 2007). As an asset, housing plays role in promoting family lifelines, such as health, education, economic, security, and social. It also generates social interaction, prides and cultural identity, and also political and economic resources (Barakat 2003). On the other hand, housing is also vulnerable asset when it is exposed to threats from natural hazards (earthquakes, landslides, floods). Approximately 97.7 % of the world's disaster homelessness occurs in developing countries (Gilbret 2001). After a disaster strikes, community needs a place to restart the business and proceed again for sustaining the economic activities.

Without adequate housing, individuals will have difficulties to start the economy and reopen businesses (Peacock et al. 2007).

After disaster strikes, housing usually get the main attention to restore back the community sustainability (Gilbret 2001). Depending on the scale of disaster, housing rehabilitation starts a few weeks until few months after the emergency phase and continues until the permanent houses are completed. In this stage, the housing need is solved by temporary shelters while taking time to confer with stakeholders and plan the reconstruction properly (Jha and Duyne 2010; Peacock et al. 2007). Government proposes post-disaster recovery planning based on the result of Post Disaster Need Assessment (PDNA) because if the reconstruction process is not well planned and implemented, further vulnerabilities might increase (Chang et al. 2010). Thus, five domains of strategy in disaster recovery is needed: (1) Institution, (2) Finance, (3) Community Participation, (4) Reconstruction Approach, and (5) Risk Management (Barakat 2003; Jha and Duyne 2010). This is followed by the common term of "Building back better" after a disaster (Monday 2002). Hence, a disaster recovery process can be seen as an opportunity to empower and promote community capacity in housing reconstruction.

At the outset, in the process of housing recovery, it is crucial to recognize all the involved stakeholders, the coordination structure and who will implement (Barakat 2003; Jha and Duyne 2010). In developing countries, various stakeholders may participate in the process, such as the community, local government, private sector, non-governmental organization (NGO), the United Nation (UN) agencies, etc (Barakat 2003). Jha and Duyne (2010) suggest that it is important to have a clear understanding of stakeholders' intentions and involvement, and how they can contribute to the reconstruction process. The involved stakeholders need to develop partnerships and coordination which is normally led by government officials. If necessary, it is important to enact laws, responsibility, regulations, and institutional arrangements which are used as guidance for coordination. The institutional strategy must also include monitoring and evaluation of the capabilities of the organizations involved and decide how their activities will be coordinated (Jha and Duyne 2010).

Barakat (2003) proposes several financing schemes in housing reconstruction process. These include outright gift, partial contribution combined with community self help, and loan. Outright gift is provided for very vulnerable communities by granting money to the beneficiaries as full recipient, without any requirement to return the fund in the future. In contrast, partial contribution provides the beneficiaries with only some limited funds, i.e. for provision of building materials and technical guidance, and supported by the community's source and, if needed, some long term special loans, with or without interest. Loan may also be provided normally through small amount of interest rate to help the community to carry out the construction. Jha and Duyne (2010) noted that the biggest challenge in housing reconstruction is to manage and control the financial spending. This is a complex set of decisions that have social, economic, and logistical implications.

Many stakeholders take community participation as their approach where the community is involved in contributing to project design, influencing public choices, and holding public institutions accountability (Davidson et al. 2007; Jha and Duyne 2010). Jha and Duyne (2010) said that participation can be seen as the direct engagement of the affected population in the project cycle—assessment, design, implementation, monitoring, and evaluation—in a variety of forms. On the other hand, others consider participation as an operating philosophy that puts affected populations at the heart of humanitarian and development activities as social actors with insights, competencies, energy, and ideas of their own (Jha and Duyne 2010). Community-based approach needs a different programming flow, one that begins not with assessment, but with mobilization of social groups and communities, which is then followed by a community based assessment (Jha and Duyne 2010). Davidson et al. (2007) found that some research showed that the beneficiaries of a housing project—depending on how it is organized—can, with proper, disinterested guidance, intervene at the levels of design decisions, material selection and preparation, construction, management and even financing. NGOs, local government, and national government can mobilize the beneficiaries and communication plays important role in a successful participatory process.

Reconstruction approach addresses how physical reconstruction will be taken out at the community level (Jha and Duyne 2010). Building back better becomes the goal which improves the safety and reduces the vulnerability to future hazards for the reconstructed housing. However, it depends on the role of households and reconstruction agencies and the forms of support, whether it is finance, training, or community facilitation (Jha and Duyne 2010). There are many ways to reach this goal, such as developing standards design and construction of housing, developing land use plan, relocating housing location, etc.

Housing construction and design become one of the main dilemmas in post-disaster housing reconstruction process. The common specific obstacle is related to the quality or quantity of material and construction specifications, which is also primarily caused by funding limitations, the people knowledge as well as the phenomenon of unplanned housing (Steinberg 2007). In terms of construction, Olshansky et al. (2006) also explain that in some cases and conditions of reconstruction process, the partial housing construction (or improvement) type is considered to be more efficient compared to the total housing construction (or rebuilding) type. Further during its development, housing construction, in the context of post-disaster recovery, is not only always identified by the technical approach, but also by the perception approach. This is supported in research conducted by Green (2008) in Istanbul, which states that externally, the confidence level to the various stakeholders such as governments, NGOs, and even engineers could ultimately affect the perception of impacted people in determining their own construction planning.

The housing location is also one of post-disaster recovery instrument that can also provide opportunities for the future disaster risk reduction effort (Usamah and Haynes 2012). On the other hand, impacted people certainly have their own discretion to determine the location of their damaged housing rebuilding, based on factors ranging from financial ability to even the possession of social capital. In some cases and post-disaster conditions, housing site selection may become a full responsibility of the government. It means that the government has the authority to choose the location of the rebuilding of the houses and also to relocate if necessary, or better known as re-settlement. However, despite the role of government in this regard, there are still many cases related to the rejection of relocation, with several contributing factors such as incompatibility with livelihood, limited access, or inappropriate condition to conduct economic activity (Dikmen 2006). In order to support the successful relocation scheme, as proposed by Ozden (2006), public involvement should be considered as a key factor in its planning and development. Jha and Duyne (2010) also pointed out risk management that need to be considered by policy makers in the housing reconstruction process. Poor management of the risks could result in unexpected outcomes, delays, and loss of credibility for the actors involved. Establishing a culture of risk management in reconstruction is crucial to mitigate future risk.

11.3 Method

The research is based on the review of the post-earthquake and tsunami housing reconstruction process in four cases in Indonesia that took place between the years of 2005–2012, namely Aceh, Yogyakarta, West Java and West Sumatera Provinces (Fig. 11.1). Documents for each of the cases were obtained from government officials, UN agencies, and previous research studies. In addition to the documents as the main information sources, the authors have had the opportunity to conduct field surveys and observation in the recovery areas, conducted interviews and some

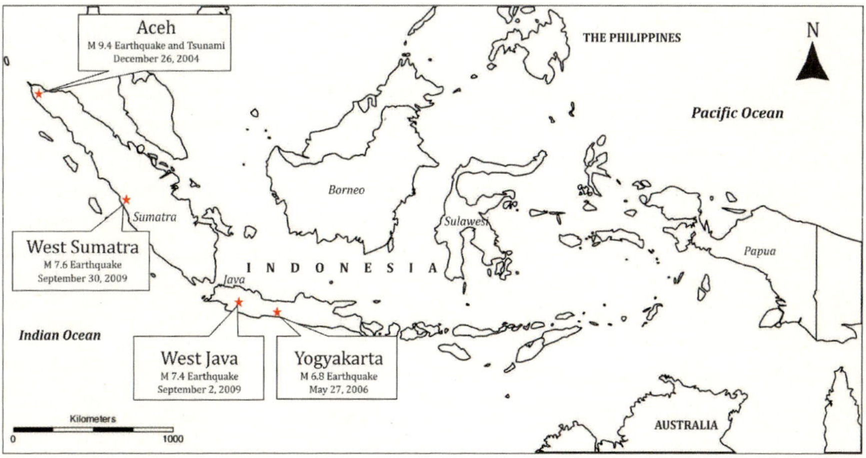

Fig. 11.1 Location of study area

focussed group discussions. The analytical method in this research is mainly qualitative, carried out through the discussion of the main issues and problems experienced in the cases, supported by some field as well as secondary data, followed by synthesizing the lessons learned, in line with the theoretical framework discussed in the Sect. 11.2. However, the depth of discussions in each case study varies due to limited information available in some issues in the case studies.

11.4 Housing Reconstruction in Four Affected Areas

The following discussion portrays the housing reconstruction processes and depicts the issues and problems experienced in the four affected areas in Indonesia, each of the case studies provides discussion on institution and finance, community participation and technical issues.

11.4.1 Aceh

On December 26, 2004, an M 9.4 earthquake struck the northern part of Sumatera and triggered a tsunami which swept the coastal areas of Aceh and North Sumatera Provinces in Indonesia. The tsunami wave affected also southern Thailand, Sri Lanka, Maldives, and some parts of the coast of Somalia in Africa (Steinberg 2007). The widespread destruction in Indonesia has been massive and larger than in other countries. The tsunami affected about 500 km of coastal areas along north part of Sumatera and swept away all physical objects along Aceh's western and north coastal line, flattening houses, infrastructures, and many other facilities (Ochiai and Shaw 2009). The districts of Aceh Jaya, Aceh Barat, Aceh Besar and Banda Aceh were identified as the most suffering area (Jayasuriya et al. 2010). The earthquake and tsunami caused more than 110,000 loss of life, 12,000 missing and 500,000 people became homeless (BRR and International Partners 2005). Total damage and losses were estimated at US$ 4.45 billion, the same amount of the total Gross Domestic Product (GDP) of Aceh province in 1 year. It was estimated that 127,000 houses destroyed (BRR 2005) and the housing sector was the most affected, suffering up to US$ 1.4 billion losses (BRR and International Partners 2005). In Banda Aceh, the capital city of Aceh, the local and regional governments were paralyzed and lost their function.

11.4.1.1 Institution and Finance

The Government of Indonesia, through its sector ministries, took immediate effort with financial and practical assistance when the mega-disaster occurred. An earlier housing policy was taken by Bappenas[1] and KPU.[2] However, since local government administration collapsed and central government had limited capacity in becoming operational on the ground, NGOs had become the "real" actor who drove emergency aid, rehabilitation and reconstruction (Steinberg 2007). There were 120 NGOs contributing to housing reconstruction. Further, many agencies and donors worked without coordination so that it led to a chaotic reconstruction environment (BRR and International Partners 2005; Telford and Cosgrave 2007).

Later, the Government of Indonesia appointed Bappenas to develop a master plan (Jayasuriya et al. 2010). It aims to coordinate, synchronize and integrate plans of various sectors, the business community and the community (stakeholders) in an action plan based on timeframes, locations, funding sources and the parties in charge (Bappenas 2005). However, many NGOs and Aceh local government refused the master plan's idea. They denied establishing a new agency to coordinate recovery activities. They insisted that this new agency would lead to centralized approach. In other words, it could add new layer of bureaucracy to the problems of working in Aceh and all of the construction work would be tendered in Jakarta, capital city of Indonesia, without proper local needs and involvement (Jayasuriya et al. 2010).

Hence, those who refused the master plan prepared their own reconstruction programs although it was not clear how they could be compatible with it (Jayasuriya et al. 2010). Many NGOs had been engaged in housing reconstruction work, but some of them were also inexperienced in this field (Steinberg 2007). According to Steinberg's (2007) field observation, they were still involved in housing reconstruction because they had available money from grant or their government fund and saw it as a good opportunity for them to exist. However, they worked without proper coordination between them, and there were competitions among them, such as: (1) competition to having beneficiaries; (2) competition in the completeness of the services; and (3) competition due to conflicting programs (BRR and International Partners 2005; Chang et al. 2011; Soelaksono 2009). Multilateral development banks (Asian Development Bank and World Bank) also delayed their housing assistance because of budgeting problem and procurement procedures (Steinberg 2007).

The central government realized that it needs more decentralized coordination and they still kept their mind to establish a new agency to solve this problem (Jayasuriya et al. 2010). In May 2005, BRR[3] was established by cabinet decision to coordinate agencies and donors in reconstruction work (Steinberg 2007). Further, BRR itself did not fully refer to the master plan when conducting its activities because it was felt that the plan did not ensure local involvement in many aspects (Jayasuriya et al. 2010). BRR changed its policy and completed a new housing policy in 2006. The approach gave opportunity to the community to participate in decision-making about where, how, and by whom houses were reconstructed (Jayasuriya et al. 2010; Steinberg 2007). However, the new policy still left debates

[1] Bappenas: *Badan Perencanaan Pembangunan Nasional* (Ministry of National Development Planning)

[2] KPU: *Kementerian Pekerjaan Umum* (Ministry of Public Works)

[3] BRR: *Badan Rekonstruksi dan Rehabilitasi* (Rehabilitation and Reconstruction Agency)

between housing reconstruction actors, such as site location criteria for resettlement and procedures of repair assistance (Steinberg 2007). After the establishment of BRR, Steinberg (2007) observed that the chaotic reconstruction environment still existed and BRR was overloaded with responsibilities including coordinating 120 NGOs contributing to housing reconstruction. BRR became less focussed in coordinating the large amount of actors when they were appointed by the Government of Indonesia to be involved in building houses (Jayasuriya et al. 2010). The recovery funding were allocated by domestic source of the Government of Indonesia, private sector funding, and foreign government aid, totalling about US$ 9.1 billion for 2005–2009 (Jayasuriya et al. 2010). Approximately, US$ 760 million were provided through different channels and accounts to reconstruct houses for 66,700 displaced families. Two hundred thousand houses were to be built or rehabilitated (BRR and International Partners 2005). The Government of Indonesia and donors allocated US$ 976 million assuming a 36 m^2 house with the cost estimated at US$ 3,000 per house (BRR and International Partners 2005; Jayasuriya et al. 2010). However, the cost of construction work was escalating in Aceh during the recovery phase. Lack of sustainable and legal building materials, increasing labour and building material cost, escalating transportation cost, and lack of access to affected area influenced the situation and slowed the progress of housing reconstruction work in Aceh (BRR and International Partners 2005; Chang et al. 2011). The situation was worsened by the fact that local manufacturing and supply facilities in Aceh were widely damaged, many labour died, and many affected areas were located remotely. Building material and labour were shipped from outside Aceh. Thus, by early 2006, BRR increased their estimation that a 36 m^2 would cost US$ 4,000 (BRR and International Partners 2005). In 2008, BRR estimated that 110,000 houses had been built and 70,000 houses had been rehabilitated (BRR 2008).

11.4.1.2 Community Participation

As noted before, BRR introduced in the master plan the new approach that local communities were to participate in housing reconstruction. Many agencies and donors offered to rebuild several houses in some villages, started discussing with local leader, and discussed the design of houses with the beneficiaries. Even though the Acehnese share the Indonesian tradition of having strong association between community members and community-related activities, it was not easy to drive this policy. Many agencies and donors had difficulties to drive the community in participatory planning because the mega-disaster has also affected the community structure and cohesion, as many local community leaders died and many communities were separated into barracks and tents which contributed to the weakening of the community's cohesion (Steinberg 2007). As BRR and International Partners (2005) observed, building communication between community and external parties took times and necessitates facilitators working with the communities. Trauma from personal losses and the losses of their family members and their personal belongings became obstacles in the participatory process (Steinberg 2007). Thus, some of them showed frustration and no interest in further community consultations. These factors hampered the facilitating and empowering the community in the reconstruction process and slowed down the housing reconstruction progress.

Chang et al. (2011) observed that community participation driven by donors was limited and failed in some particular affected areas. In the community participation based construction, the community or beneficiaries were asked to participate in the design and to partly manage the construction of their own houses (Boen 2006a). Lack of understanding, experience and knowledge of the way in which the community-based reconstruction can be organized has influenced the outcome (Dercon and Kusumawijaya 2007). In some cases, the community voiced discontent and were confused because many NGOs came to them and proposed to deliver similar assistance with community participation concept (BRR and International Partners 2005). Some NGOs scaled back the pace of their housing programs through community contracting mechanisms, including training local people in construction and supervising construction by themselves (Steinberg 2007). However, the initial response showed fast construction progress, but lacks of supervision and technical support produced low quality construction by the agencies (Boen 2006a; Steinberg 2007). The condition in Aceh was far from ideal for the implementation of community-based processes and the successful community participation for the local affected depends on the readiness of both aid agencies and the local communities (Chang et al. 2011; Steinberg 2007).

11.4.1.3 Technical Issues

Boen (2006a) observed that the collapsed buildings in Aceh by the 2004 Indian Ocean earthquake and tsunami are mostly non-engineered buildings. Non-engineered buildings are buildings that are built traditionally with very little or no assistance from qualified engineers (Okazaki et al. 2010). There were two types of non-engineered house buildings that collapsed due to the earthquake and tsunami: one or two stories confined masonry buildings and timber construction (Boen 2006a). These structures often do not follow minimum requirements for a good confined masonry building, and many of them use locally available materials to give a "masonry-like" feature, which are in fact very vulnerable to ground shaking (Boen and Pribadi 2007). Masonry buildings built with good quality materials and good workmanship will have better chance tosurvive in an earthquake, although they may not survive the tsunami generated by the earthquake. Okazaki et al. (2010) found that although only few existed prior to the earthquake, traditional timber construction demonstrated good performance under seismic loading. However, it could not survive the tsunami wave.

All of involved institution in the housing reconstruction process learned from the previous mistakes and rebuilt safer community for the future earthquake. Unfortunately, they did not have a clear housing reconstruction policy that could be referred to. In January 2005, Bappenas and KPU announced that all of the earthquake and tsunami victim households would receive housing reconstruction and

Fig. 11.2 A post-tsunami housing reconstruction project (*source*: Field Observation in Banda Aceh 2006)

rehabilitation, including a free 36 m^2 house for eligible households or house repair for partially destroyed houses (see Fig. 11.2) (Steinberg 2007). Soon, there was a debate about specifications and they did not reach any agreement. Many agencies and donors did not take serious responsibility at the housing design and planning process. They were successful in delivering housing reconstruction assistance, but as some of them were inexperienced in housing reconstruction , they had failed to deliver quality housing construction in term of good and permanent construction materials, earthquake-resistance building, and supporting infrastructures (Boen 2006a; Okazaki et al. 2010; Steinberg 2007). Many donors and agencies had also done poor and inadequate site engineering and feasibility study of the site area, which led to lack of supporting infrastructures (such as road, public transportation, electricity network, water supply, and sanitation) and lack of community livelihood consideration (Boen 2006a; Matsumaru et al. 2012; Okazaki et al. 2010; Steinberg 2007). Many relocated beneficiaries experienced this situation in the relocation site and they also faced a totally different environment from their original sites. Those units soon were refused, uninhabited, or abandoned by the beneficiaries (Okazaki et al. 2010; Steinberg 2007).

On the building safety aspect, many of those houses were constructed with poor quality materials and poor workmanship (inappropriate mixes of mortar and concrete, poor brick laying, poor detailing of reinforcement etc.) (Boen 2006b; Okazaki et al. 2010). The structures were also lacking in structural integrity, due to insufficient connections between each structural component (Boen 2008; Okazaki et al. 2010). Many projects demonstrated that the workers were not adequately skilled for the job (Boen 2008; Okazaki et al. 2010). For example, aggregates for concrete making were not sieved, stirrups were made without seismic hooks, concrete curing were not done, bricks were not soaked prior to laying, too much water in concrete mixture, etc. They just simply did what they thought as the easiest way for building construction with no concern about the quality because they were not equipped with knowledge of proper construction methods nor basic concepts of quality in structures.

11.4.2 Yogyakarta

An M 6.8 earthquake struck in the south part of Java Island on May 27, 2006. The earthquake directly affected the provinces of Yogyakarta and Central Java. The two most severely affected areas are Bantul District in Yogyakarta Province and Klaten District in Central Java Province. The affected areas were located in relatively poor rural and urban-fringe areas south, east and north-east of the city of Yogyakarta. As the earthquake struck in the early morning hours, many people were trapped in their homes. The earthquake had taken over 5,700 human lives. Injury estimates range from 37,000 to 50,000, and hundreds of thousands had been rendered homeless (Bappenas 2006). The damage was very heavily concentrated on housing and private sector buildings. Private homes were the hardest hit, accounting for more than half of the total damage and losses (US$ 1.5 billion). An estimated 154,000 houses were completely destroyed and 260,000 houses suffered some damage (Bappenas 2006). More houses will have to be replaced and repaired than in Aceh and Nias at a total cost of about 15 % higher than the damage and loss estimate of the tsunami. The high level of damage is mainly due to the high density of the population (1,600 persons/km^2) and the almost complete lack of seismic design provisions (Elnashai et al. 2006).

11.4.2.1 Institution and Finance

The recovery activity started 1 month after the disaster impact. In July 2006, after the Preliminary Damage and Loss Assessment was produced, Bappenas had released the Action Plan for Rehabilitation and Reconstruction for Post-disaster in Central Java. Three objectives were prioritized in the action plan: rehabilitation of housing and residential areas, rehabilitation of public facilities, and reactivation of the economy. Minister of Finance also invited multiple donors from various countries to mobilize support through a multi-donor trust fund. The aim of this strategy was to build on the ability to rapidly develop, finance and implement projects; coordinate international resources around common objectives; avoid duplication of effort; create synergies and reduce transaction costs for both donors and the recipient (World Bank 2012).

In October 2006, Java Reconstruction Fund (JRF) was established and contributed about US$ 94 million from seven donor countries. The JRF's work was coordinated by Bappenas, coordinator of the reconstruction, and adopted a phased approach to reconstruction in line with the action plan. On the other side, the central government set up approximately US$ 1,613 was allocated for each household by the central government (Resosudarmo et al. 2012). The earthquake survivor households were not given new houses constructed, but they were expected to reconstruct or renovate their collapsed houses using the funds provided. However, the lightly damaged houses were not given any housing support (Resosudarmo et al. 2012). In June 2008, the government had spent US$ 57 million on housing, and the JRF had spent US$ 60 million on various activities (mostly housing) (Resosudarmo et al. 2012). The JRF and the government had built more than 215,000 earthquake resistant houses in Java in less than 2 year. They claimed it as the largest and the fastest housing reconstruction after disaster in the world (World Bank 2012).

The Governor of Yogyakarta said that the beneficiaries should not only receive the government housing aid as it is only enough for the main structure of the houses—foundation, columns and beams, and roof main structure—which should be earthquake resistant, but also should receive all outside supports and funding sources as far as they are not loan (Ikaputra 2012). As Ikaputra (2012) observed in Bantul District, the beneficiaries also used fund from other sources, such as their own money, relatives, donors, NGOs or even from bank loan. They became alternatives because of the slow process of fund disbursement by the government (Raharjo 2007).

Some international NGOs started to collaborate with local NGOs (Raharjo 2007). Many of these international NGOs were also actors who were involved in housing reconstruction in Aceh (MacRae and Hodgkin 2011). To avoid the past experience in Aceh, they employed skilled and experienced local staff and also an increase in the proportion of international staff with in-country experience. They also recruited local university students and faculties, especially civil engineering and architecture, to conduct statistical surveys on the damage and number of victims and also as facilitator (MacRae and Hodgkin 2011; Raharjo 2007). Later, they introduced the concept of T-Shelter (Temporary Shelter) to be occupied by the community in the transitional phase of recovery while the government plan for preparing permanent shelter progresses. However, slow progress of funding disbursement and unclear formal coordination between the government and NGOs have caused uncoordinated and sporadic T-Shelter distribution in the affected areas (MacRae and Hodgkin 2011; Raharjo 2007). Further, as Ikaputra (2012) observed in Bantul, the beneficiaries had not only the reconstructed house, but also T-Shelter in their yard (Fig. 11.3).

11.4.2.2 Community Participation

The housing reconstruction policy adopted community-driven approach. In early August 2006, the Government's Decree instructed that reconstruction implementation must be organized and be implemented by the local community group—*Kelompok Masyarakat* (POKMAS) which made up of 10–15 families (Jha and Duyne 2010). They took the decision on how to distribute funds and chose which members should receive housing first (i.e. poor household, elderly household, large family household, etc). Training was provided to POKMAS members and local workers to ensure earthquake-resistant construction (Jha and Duyne 2010). The leader or other members from each group would attend trainings or workshops to improve their knowledge about earthquake-resistant housing and building materials. Later, they

Fig. 11.3 Dome houses in Ngeplen village, Sleman, Yogyakarta (courtesy of Prof. Sarwidi, UII Yogyakarta)

would convey what they had learned to the rest of their group. They worked together and helped each other in reconstructing houses.

To control their task, POKMAS must be provided technical assistance through facilitators (Raharjo 2007). *Lurah* (the head of the village) had the important role as technical coordinator for the distribution implementation on the ground. Facilitators are expected to assist *Lurah* in managing the process of funds utilization by POKMAS (Raharjo 2007). Facilitators were recruited and villages elected boards of trustees who organized community meetings and supervising implementation. Their tasks included (1) identifying beneficiaries and prioritizing the most vulnerable; (2) establishing POKMAS, who chose their leaders and a treasurer; (3) developing detailed plans to use the construction grants for each group; (4) opening group bank accounts for disbursed fund; and (5) obtaining approval of plans, disbursement in tranches, and group procurement, construction, and bookkeeping (Jha and Duyne 2010). Hence, the facilitators took important role because they had to ensure effective communication and adaptability of the program to local situations as well as compliance with program principles.

11.4.2.3 Technical Issues

In the affected areas, almost all of the collapsed buildings were non-engineered housing structures (Jha and Duyne 2010). They are masonry structures consisting of adobe, brick masonry, stone masonry, and RC frames with masonry infill (Meguro 2008). Many of them did not fulfil the earthquake resistant building requirements, used low quality building materials, lacked wall confinement elements, and lacked proper joints or have improper detailing (Pribadi et al. 2008; Satyarno 2009).

Nevertheless, Narafu et al. (2008) observed that the impact of the 2006 earthquake had increased the seismic safety in the affected areas due to provided technical guidelines and technical supervision by JRF and local university (See Fig. 11.4). Almost all the reconstructed houses implemented confined masonry.

Fig. 11.4 A reconstructed house was built according to earthquake resistant house manual by a local university in Yogyakarta (*Source*: Pribadi et al. 2008)

Regarding mortar for brick laying, using cement mortar is rare before the earthquake and it becomes to be almost 100 % after the earthquake. There is significant improvement in construction works. These prove that appropriate technical intervention could be accepted by people for enhancing seismic safety. However the technical interventions still need improvement especially from the view point of efficiency/ease of construction work (Narafu et al. 2008).

However, Narafu et al. (2008) and Suarjana and Sengara (2008) still found the problems related to detail in construction, such as insufficient compaction of concrete using simple tools like steel bars (not as effective as vibrators), inadequate lateral support for walls, congested rebar within small section of RC elements, etc. Suarjana and Sengara (2008) observed that these incorrect details were similar to what have been found in Aceh. They argued that it can be influenced by: (1) Contractors tend to chose minimum material quality and simple design for fast construction time that earn maximum profit, (2) Lack of tight supervision by engineer, and (3) Engineers and masons have minimum knowledge on earthquake resistant design.

Another emerging issue was found in dome housing in Ngeplen Village, Sleman Regency (Ikaputra 2008; Pandelaki and Shiozaki 2008). The technology was applied by an international NGO assisting the local people to rebuild their houses which were damaged by the earthquake triggered landslide. It was a new technology housing approach introduced as anti-seismic design house, providing an *iglo*-like concrete building casted as a single and integral structure which is supposed to be stronger, energy efficient, and cost effective (see Fig. 11.3) (Ikaputra 2008). In fact, the 38 m^2 circular houses which provide basic house need and the supporting infrastructure did not meet local need and local culture. Ikaputra (2008), Pandelaki and Shiozaki (2008), and Kondo and Maly (2012) found that the beneficiaries gave negative comments regarding the donated houses, such as the absence of ventilation, roof, terrace and eaves which are necessary in tropical house culture, the absence of living room, kitchen, garage, etc, and even it is difficult for them to construct a cattle shed in the site. Traditional habit, such as taking care of cow for livelihood, using wood for cooking in the traditional kitchen, welcoming guest and family in the living room, etc, was exchanged with modern family home (Ikaputra 2008). Using limited available material and construction capabilities, the beneficiaries had to adapt and expand some improvement on it, such building kitchen and extension room in the backyard of their house, add some eaves, canopy and veranda to be like their previous houses (Pandelaki and Shiozaki 2008). As the solution for their daily livelihood, later, the local government built cattle shed in the north part of the site location. It seems that the dome houses neglect the peoples' housing recovery needs to sustain their living.

11.4.3 West Java

West Java was hit by an M 7.4 earthquake on 2 September 2009, the epicenter was in the Indian Ocean. The earthquake caused damages and casualties in 15 districts/cities in West Java Province. Eighty one people loss their life while 1,917 people were injured and 50,964 families were displaced (194,719 inhabitants). Over 10,000 houses had been affected with various degree of damage (Yasaditama and Sagala 2012). The worst damage and loss due to the earthquake was suffered by the housing sector. From the total damage and losses estimated at Rp 7.9 trillion, the housing sector suffered Rp 6.9 trillion damage and losses.

11.4.3.1 Institutions and Finance

The West Java Earthquake was declared a national scale disaster event due to the large scale impact caused by the earthquake over seven districts and cities. However, not long after the earthquake, Government of Indonesia shifted the emergency phase into early recovery on 16 September 2009 to speed up the process of recovery (UNOCHA 2009). Bisri (2012) noted that provision of temporary shelter was the highest recorded number of post disaster activities while achievement on the housing reconstruction related activities located in Bandung District, Tasikmalaya District and Bandung City was limited. The small number of activities was particularly due to that all the budget for reconstruction process was solely dependent on government. The government allocated 15 million IDR/each heavily damaged house from annual national budget (APBN), 10 million IDR/each medium damaged house from annual provincial budget (APBD *Provinsi*) and 5 million IDR/each lightly damaged house from annual district budget (APBD *Kabupaten*) (1 USD equal approximately IDR 9,800 at the time of the writing of this article). The money in fact is not sufficient to reconstruct a decent house for each household. Thus, some people have to borrow money from banks or their relatives. Some people who have better savings were able to combine the money they obtained from the government with their own money to build earthquake resistant houses.

In term of coordination, the distribution of the money was conducted through community group (hereafter is called POKMAS). POKMAS is formed by the community and legalized by a formal letter from the Head of District. The funding was distributed through the POKMAS which consist of one chairperson, one treasurer and around 15–30 members representing the households that were affected by the earthquake.

11.4.3.2 Community Participation

Community participation observed in West Java post-earthquake recovery, particularly in hard to reach remote areas, where the communities were involved in the post disaster damage and loss assessment. In Pangalengan, a sub-district in Bandung District, West Java, where a lot of houses were damaged by the earthquake, the POKMAS participated in assessing the damage using the government provided criteria for the four damage levels: heavy, medium, light and no damage. However, in the reconstruction process, other than POKMAS activity, the community participation is still limited. Within the community, reconstruction works tend to be implemented individually by households. A small number of households with better economic saving were able to hire house builders to reconstruct their houses. However, most of the communities tend to rely on government grant to reconstruct their houses. Due to the limited amount of money provided by the government, the communities used some old materials to reconstruct their houses. In the sub-district, a large number of households are landless and most of them work at farms and large plantations as labors and thus have very limited economic capacity to recover from the disaster.

The Indonesian Red Cross (PMI) and the International Federation of Red Cross and Red Crescent Societies (IFRC) were involved in a lot of temporary shelter reconstruction, where the communities contributed in in-kind resources in the form of labors while PMI helped in the construction techniques, supervision and material provision.

11.4.3.3 Technical Issues

A survey conducted post 2009 West Java Earthquake showed that there have been some attitudinal changes observed in building houses in the affected areas. Observation in some examples show that some house owners would want to have better construction of houses that are more compliant to earthquake resistance house guidelines. Some households now have a general knowledge of the government directions about earthquake-resistant housing technical guidance. Response to questionnaire to some high and moderate damage level house owners indicated that they have followed the earthquake-resistant housing technical guidance in their own house rebuilding process despite their previous opinions which argue that the amount of government assistance was not enough to rebuild their houses. It was

Fig. 11.5 Housing Reconstruction in West Java (Left Picture – *source*: Field Observation in Tasikmalaya 2010 & Right Picture – *source*: Field Observation in Pangalengan 2006)

found that most of reconstructed houses have used shallow plate foundation or rock filled concrete pile foundation as well as columns and beams to confine the wall (see Fig. 11.5). Further the survey indicated that in the reconstruction process, people trusted the engineers who come from or recommended by the people in the village. On the other side, house builders suggested that until currently only a small number of people who follow the guidelines on earthquake-resistant house. Financial limitation as well as livelihood problems has constrained the people to implement cheaper housing construction techniques which are less resistant to earthquake. As estimated by Nazara and Resosudarmo (2007), the costs of construction materials went up due to limited availability of the materials. Therefore, to meet with the financial limitation, some strategies in reducing the construction cost through the adjustment of quality and quantity of materials, such as minimizing the reinforcing steel quantity or size or using an alternative material such as bamboo reinforcement, resizing of foundation (Yasaditama and Sagala 2012).

Alternative choice for replacing material actually had been highlighted in the Aceh Tsunami reconstruction (Steinberg 2007). In the West Java earthquake case, the people's financial condition was essential in the decision for alternative construction material selection. As the authority and responsibility for the use of alternative construction material were on the hands of the house owners, there were no difficulties in implementing the alternative solutions.

11.4.4 West Sumatera

On 30 September 2009, an earthquake of M 7.9 struck the coast of West Sumatera. The earthquake has caused a significant number of loss of lives and major setback in socio-economy aspects due to damage of structures and infrastructures. The official data reveals that in the city of Padang alone, the earthquake has damaged 9,635 buildings and houses and caused the death of 316 lives and injured 606 people, as well as damaging other infrastructures (Pribadi et al. 2011). Immediately after the quake, rapid emergency rescue teams arrived and aid organizations started assessing the damage and providing emergency relief.

11.4.4.1 Institution and Finance

Recovery in West Sumatera took place in November 2009, 2 months after the disaster (interviews with various Red Cross staffs). The UN Cluster approach was applied in the ground and some NGOs took on respective responsibilities for aid based on the coordination guidelines and the Sphere Standards. The funding for the West Sumatra Earthquake Recovery was received from four sources: Foreign Grant, National Budget, Provincial Budget and District/Municipality Budget (Bappenas 2009). The allocation of financing follows the Government Regulation number 22 Year 2008 on the financing and management of disaster aid. Similar to West Java, the government provided 15 million IDR/each heavily damaged house from annual national budget (APBN), 10 million IDR/each medium damaged house from annual provincial budget (APBD *Provinsi*) and 5 million IDR/each lightly damaged house from annual district budget (APBD *Kabupaten*).

11.4.4.2 Community Participation

There are several types of approaches in the housing recovery process found in West Sumatera. One common approach is to apply what happened in previous successful program as in other post-earthquake recovery, such as in Yogyakarta Earthquake (see previous sub section on Yogyakarta Earthquake). However, NGO workers admitted that these approaches did not work successfully as some of the T-Shelter houses were not used by the earthquake survivors. One argument on this was due to the lack of collective action (*gotong royong* in Bahasa Indonesia) among the communities (Vanhoebrouck and Sagala 2010). As Indonesia has a large number of cultures and ethnic groups, it should be understood that there are various ways to approach the community. In the study by Vanhoebrouck and Sagala (2010), community participation occured successfully in particular through sub-ethnic approaches, which mobilize the roles of formal, religious and cultural leaders. Some Minang people rely on self-help process to reconstruct their houses, which can be considered as valuable social capital. The self-help process is also supported financially by the diaspora of Minangkabau (West Sumatra ethnic) people who have migrated elsewhere in Indonesia or even abroad.

11.4.4.3 Technical Issues

There were three reconstruction practices in general implemented by the communities in the housing reconstruction, namely repair, retrofit, and rebuild (Pribadi et al. 2011). In repairing practices, the buildings were partially reconstructed and the

Fig. 11.6 Earthquake damaged houses in Padang Pariaman, West Sumatra. The house on the *left* was rehabilitated by the owner with own fund, while the one on the *right* waited for the disbursement of government aid (*source*: Field Observation in Padang Pariaman 2006)

reconstruction did not change the structural strength, thus the new structural strength is similar to the strength prior to damage. In retrofitting practices, the buildings were partially reconstructed, with the design and construction practice intended to increase the strength of the buildings, thus the reconstructed building should be stronger than before. In rebuilding practices, the buildings were totally demolished and reconstructed, hence for this practice there were two possible outcomes, i.e. strengthen the building or just maintaining the previous strength of the structure. The observation by Pribadi et al. (2011) suggested that 43 % of the surveyed respondents repaired their old houses, 49 % retrofitted while only 8 % implemented rebuilding. This is particularly due to some financial limitation in the community (See Fig. 11.6).

The recent program of "build back better" for post-disaster housing reconstruction in West Sumatera indicates that the role of building material supply stores played in this campaign was important. It clearly shows the critical role that these shops can play in ensuring that the general public has a better understanding about earthquake safe building standards. The previous evaluation by IDEP (2010) showed that not everyone knows exactly what materials they want before they go to material shops which were found at five different building supply stores in Solok District, Padang City, Padang Pariaman District, Pariaman City and Agam District.

There has been some changes on the attitude of house builders towards more earthquake resistant compliance. A (trained) house builder plays the role to explain the importance of building earthquake resistance house to the house owner which causes changes in the preference of construction types and use of materials (IDEP 2010). Therefore, house builders can be considered as agents of changes that communicate and influence the house owners to improve the housing quality. House builders explain that at least there are two reasons why people do not adhere to the earthquake resistant manuals. *First*, the cost of a normal house using cheaper materials and inadequate structural elements is cheaper. *Second*, some house owners are not aware of the method for preparing earthquake resistant structural design and details.

Study on non-engineered construction in Padang City (Pribadi et al. 2011) showed that most of house owners rely on the masons and carpenters for repair and rebuilding works of the damaged houses in Padang City. In most of the cases, the masons and carpenters were not trained properly to construct earthquake safer houses. They usually work using knowledge and skills obtained informally in the past experiences, and in many cases do not comply with the requirements for earthquake resistant construction. Providing the masons/carpenters with the appropriate skills and knowledge for safer houses becomes priority to improve the safety of houses in Padang city during the reconstruction process.

11.5 Discussion

This study attempts to examine processes and lessons learned from four case studies across provinces that have been affected by large earthquake and tsunami disasters in Indonesia. Limitation on the data availability in some of the case studies made differences in the depth of the analysis of each case.

In Aceh Province, which has sustained armed conflict for years before the Great Indian Ocean Tsunami struck, the local governments collapsed during the disaster. Therefore, the Agency of Rehabilitation and Reconstruction (BRR) had to develop not only physical reconstruction but also rebuild and strengthen the local governments in the post-disaster and post-conflict situation. The disaster has provided an opportunity to develop and strengthen the local governments and community as well as to restore peace in the province. In Yogyakarta Province, the post-earthquake reconstruction process has been used to implement better earthquake resistant standards for building houses, which before the earthquake were not known by the local population. In addition, as the local governance system was relatively intact during the disaster, the opportunity to involve the community in the recovery process resulting in better community participation in village development process has been well benefitted. On the other side, the example of the iglo-like construction which failed to include people's perception on housing seems to be problematic. In West Java, the earthquake has not been able to be used as an opportunity to change the behavior of the people towards better preparedness against earthquake risks. Many cases showed that people still built their houses in similar ways as they used to. In West Sumatra, the opportunity has been used to train local engineers and technicians as well as masons and carpenters to understand better earthquake resistant construction.

The experiences have provided a wealth of lessons learned for many actors in the housing reconstruction process on what have been working well and what have not worked, what to promote and what to avoid for better result. The review has showed that many factors have to be considered for an effective post-disaster housing recovery, which include not only technical or engineering factor, but also issues such as actors' planning, implementation, and coordination; aid and assistance mechanism; community participation; and risk management.

In the post-disaster housing recovery program, NGOs are essentials as operational actors who can drive reconstruction program straight to the community, in particular when the local authority effectiveness is hampered by the disaster impact while the central government has limited capacity in operational ground. However, a huge number of NGOs could bring a chaotic and uncoordinated reconstruction environment when there is no strong leadership. Experience in Aceh housing reconstruction showed that NGOs were working with different ways and different principles, even competing with each other to have beneficiaries and project location. The absence of a strong leading actor, strict regulation, and a reference plan contributed to this chaotic situation. However, the establishment of a new agency does not always promise better coordination. Limited communication, time, and involvement in planning process led to no clear policy and agreement on repair assistance.

The issues of the speed of physical delivery and financial management becomes another problem that most assistance agencies faced when providing aid. Many survivors need to settle so that they can carry out normal activities. In Aceh, the aid was provided through different channels and accounts of many agencies so that fund disbursement was very complicated. Learning from the Aceh experience, the government applied better funding mechanism in Yogyakarta case, where aid funding for physical work was provided through an appointed bank and distributed into facilitators' accounts. The facilitators have had the responsibility to manage and use the fund wisely with the consent of the beneficiaries. Nevertheless, slow and delayed aid disbursement by the government and NGOs in the various cases showed that it is important for affected community to have other resources, such as savings, relatives, donors or bank loan, which can be used to initiate the repair or the reconstruction of their houses. Increasing labour and building material cost, escalating transportation cost, and building materials avaliabilty are also issues that can delay the implementation of the housing recovery program. Using maximum local labour and local materials can be a strategy to minimize the spending and anticipate the increasing cost. This strategy has also the advantage of helping to move the local economy, needed for the economic recovery of the community.

11.6 Conclusion

Many actors have provided understanding on the factors influencing the outcome of the post-earthquake housing reconstruction program. Since the housing reconstruction work in Aceh, the community based approach has been introduced to the community. However, it met many challenges and has changed overtime since Aceh post-disaster reconstruction. Low interest in community participation, lack of supervision and technical support which led to low quality construction became the main challenges in Aceh. Many foreign experts and foreign agencies who had limited local experiences and local knowledge as well as little experience in housing reconstruction program have added to the complication of the process.

It is understood that many stakeholders have been involved in the training and capacity building of construction personnel and home-owners through the dissemination of guidelines, manuals, building codes and standards for anti-seismic design. The stakeholders include national and local governments, NGOs and aid agencies, local universities, and local communities. Tighter technical supervisions have been applied in Yogyakarta and West Java experience. Facilitators who are local expert, local university students, or local people supervised small groups of the beneficiaries in order to adapt the housing reconstruction program to local situation. They became the key actor to control the detail of construction work and shared their knowledge to the community. Nevertheless, the investigation shows varying progress in the change of the attitude toward construction practices among builders and masons, as well as among the local government building administrators in ensuring the earthquake safety in the housing reconstruction programs.

An important aspect that needs to be strengthened is the awareness and perception of the house builders and house owners as well as relevant government officials on earthquake safety issues. This paper suggests that it is important to influence the decision making process in investing for safer housing by carrying out disaster education for improving earthquake risk perception among those stakeholders so that mitigation efforts can be accepted at all levels. Inspections and quality controls should be conducted throughout the construction stage to guarantee that the structure will be built according to the design and earthquake safety guidelines. Combined efforts and coordination of government and research agencies, construction developers, construction personnel (contractors, technicians, masons and crafts) and the community are necessary in improving building performance towards future earthquakes.

Despite the problems found in housing reconstruction process in the four case studies, good practices have been observed in some of the programs which can contribute to earthquake safer housing. Crucial issues for improving housing reconstruction programs include better project delivery systems, better mechanism for providing supervision and technical advices, more down-to-earth training and capacity building mechanism as well as better financing and incentives system, supported by appropriate technology approaches. Stakeholders' coordination in some cases has proved to be difficult and lack of effectiveness due to the large number of stakeholders who were involved in housing reconstruction process. The recovery plan should build up the communication and agreement among the stakeholders that can provide specific or assigned types of assistance. The government can be a pioneer of post-disaster efforts and convene to engage in policy making. They can utilize all of available capital to establish a recovery policy framework.

The recovery policy should achieve sustainable development goals, integrated with disaster mitigation program. For example, it should consider risk mitigation in repairing the damaged houses by implementing anti-seismic house design and choosing lower risk locations for the rebuilt houses. It also need to facilitate disaster resilience, which implies an ability to "bounce back" more quickly following a disaster. Policy makers are often not aware of the need to build disaster resilience and its role in achieving sustainable development. The multi-stakeholders cooperation can bring massive capital to enhance communities in adopting disaster resilience approaches through risk communication and appropriate disaster risk reduction technology.

Acknowledgement The authors wish to acknowledge GRIPS and Building Research Institute of Japan for funding the investigations in Aceh (2006) and in Padang (2011), Oxfam GB in West Java and West Sumatera (2010) and International Federation of Red Cross (IFRC) in West Sumatera (2010). Specifically, the third and fourth authors wish to thank Osaka Gas Foundation and Institute of Technology Bandung for the grant on the research on Planning for Post Disaster Recovery, Case Study of Earthquake Disaster in Pangalengan, Bandung District, West Java (2012). A special thank goes also to Prof. Sarwidi from the Indonesian Islamic University (UII) in Yogyakarta for providing the nice pictures of the igloo-like houses in Sleman, Yogyakarta.

References

Bappenas (2005) Master plan for the rehabilitation and reconstruction of the regions and communities of the province of Nanggroe Aceh Darussalam and the Islands of Nias, Province of North Sumatera, Jakarta

Bappenas (2006) Preliminary Damage and Loss Assessment: Yogyakarta and Central Java natural disaster. In: The 15th meeting of the consultative group in Indonesia, Bappenas, Jakarta

Bappenas (2007) Laporan Penilaian Kerusakan dan Kerugian Bencana Gempa Bumi Bengkulu tahun 2007 (Preliminary damage and loss assessment: the 2007 Bengkulu earthquake), Bappenas, Jakarta

Bappenas (2009) Rencana Aksi Rehabilitasi dan Rekonstruksi Wilayah Pasca Bencana Gempa di Provinsi Sumatera Barat Tahun 2009–2011 (the post disaster action plan of rehabilitation and reconstruction in west Sumatera Province 2009–2011). Bappenas (ed), Bappenas, Jakarta

Bappenas (2010) Rencana Aksi Rehabilitasi dan Rekonstruksi Wilayah Pasca Bencana Gempa di Provinsi Jawa Barat Tahun 2009–2011 (the post disaster action plan of rehabilitation and reconstruction in west Java Province 2009–2011). Bappenas (ed), Jakarta

Barakat S (2003) Housing reconstruction after conflict and disaster. Humanitarian Policy Group, Network papers 43:1–40

Berke PR, Kartez J, Wenger D (1993) Recovery after disaster: achieving sustainable development, mitigation, and equity. Disasters 17(2):93–109

Bisri MBF (2012) Exploring intergovernmental cooperation in disaster management. Institute of Technology Bandung, Bandung

Boen T (2006a) Building safer Aceh reconstruction of houses, one year after the Dec. 26, 2004 tsunami. Presented during 40th anniversary of Trisakti University, "Answering the challenges in today's civil engineering", Jakarta, 26 January 2006

Boen T (2006b) Reconstruction of houses in Aceh, twenty months after the Dec. 26, 2004. In: 12th Japan association earthquake engineering (JAEE) symposium, Japan, November 2006

Boen T (2008) Lessons from the reconstruction of houses in Aceh after the December 26, 2004 tsunami. International symposium 2008 on earthquake safe housing, Tokyo, 28–29 November 2008

Boen T, Pribadi KS (2007) Engineering the non engineered houses for better earthquake resistance in Indonesia. The DRH contents meeting- EDM-NIED, Kobe, pp 202–208

BRR (2008) Synchronizing the recovery: semester report 2008. BRR, Banda Aceh

BRR and International Partners (2005) Aceh and Nias one year after the tsunami – the recovery effort and way forward. Badan Rekonstruksi dan Rehabilitasi, Jakarta

Chang Y, Wilkinson S, Potangaroa R, Seville E (2010) Resourcing challenges for post-disaster housing reconstruction: a comparative analysis. Building Res Inf 38(3):247–264

Chang Y, Wilkinson S, Potangaroa R, Seville E (2011) Donor-driven resource procurement for post-disaster reconstruction: constraints and actions. Habitat Int 35(2):199–205
Davidson CH, Johnson C, Lizarralde G, Dikmen N, Sliwinski A (2007) Truths and myths about community participation in post-disaster housing projects. Habitat Int 31(1):100–115
Dercon B, Kusumawijaya M (2007) Two years of settlement and recovery in Aceh and Nias: what should the planners have learned. In: 43rd ISOCARP congress 2007, Antwerp, 16–23 September
Dikmen N (2006) Relocation or rebuilding in the same area: an important factor for decision making for post disaster housing projects. In: Proceedings of the international conference and student competition on post-disaster reconstruction "meeting stakeholder interests", Florence, pp 17–19
Elnashai AM, Kim SJ, Yun GJ, Sidarta D (2006) The Yogyakarta earthquake of May 27, 2006, University of Illinois at Urbana-Champaign, Urbana
Gilbret R (2001) Doing more for those made homeless by natural disasters. World Bank, Washington
Green RA (2008) Unauthorised development and seismic hazard vulnerability: a study of squatters and engineers in Istanbul, Turkey. Disasters 32(3):358–376
IDEP (2010) Build back better: public awareness campaign, west Sumatera. IDEP, Bali
Ikaputra (2012) Synergy for house reconstruction of post-earthquake: a case study of Java post-earthquake 2006. J Appl Environ Biol Sci 2(1):28–34
Ikaputra (2008) People response to localize the imported culture study case: the dome house in the rural culture post Javanese earthquake 2006. The 14th world conference on earthquake engineering, Beijing, 12–17 October
Jayasuriya S, McCawley P, Resosudarmo BP, Weerakon D (2010) The Asian tsunami: aid and reconstruction after a disaster. Asian Development Bank Institute/Edward Elgar Publishing, Cheltenham/Massachusetts
Jha AK, Duyne JE (2010) Safer homes, stronger communities: a handbook for reconstructing after natural disasters. World Bank Publications, Washington
Kondo T, Maly E (2012) Housing recovery by type of resident involvement-providing housing vs. mobilizing residents. The first international conference for International Society of Habitat Engineering and Design (ISHED), Shanghai, 13–14 October 2012
Lindell MK, Prater CS (2003) Assessing community impacts of natural disasters. Nat Hazards Rev 4:176–185
MacRae G, Hodgkin D (2011) Half full or half empty? Shelter after the Jogjakarta earthquake. Disasters 35(1):243–267
Matsumaru R, Nagami K, Takeya K (2012) Reconstruction of the Aceh region following the 2004 Indian Ocean tsunami disaster: a transportation perspective. IATSS Res 36(1):11–19
Meguro K (2008) Technological and social approaches to achieve earthquake safer engineered houses. The 14th world conference on earthquake engineering, Beijing, 12–17 October 2008
Monday J (2002) Building back better: creating a sustainable community after disaster. University of Boulder, Natural Hazards Research and Applications Information Center, Boulder
Narafu T, Imai H, Matsuzaki S, Sakoda K, Matsumura F, Ishiyama Y, Tasaka A (2008) Basic study for bridge between engineering and construction practice of non-engineered houses. The 14th world conference on earthquake engineering, Beijing, 12–17 October
Nazara S, Resosudarmo BP (2007) Aceh-Nias reconstruction and rehabilitation: progress and challenges at the end of 2006. Asian Development Bank Institute, Tokyo
Ochiai C, Shaw R (2009) Post-disaster reconstruction in urban areas in Aceh
Okazaki K, Pribadi KS, Kusumastuti D, Boen T, Ando S (2010) Safety issues regarding reconstructed buildings in Aceh, Indonesia. UNESCO-IPRED-RIHS international workshop on surveys and activities on post-earthquake disaster, Padang, 6–8 July 2010
Oliver-Smith A (1990) Post disaster housing reconstruction and social inequality: a challenge to policy and practice. Disasters 14(1):7–19
Olshansky R, Chang S (2009) Planning for disaster recovery: emerging research needs and challenge. Progress Plan 72:200–209

Olshansky RB, Johnson LA, Topping KC (2006) Rebuilding communities following disaster: lessons from Kobe and Los Angeles. Built Environ 32(4):354–374

Ozden AT (2006) Developing a model for community involvement in post-disaster housing programmes. In: Proceedings of the international conference and student competition on post-disaster reconstruction "meeting stakeholder interests", Florence, 17–19 2006

Pandelaki EE, Shiozaki Y (2008) Social sustainability of new-Ngelepen dome housing as post-disaster housing reconstruction of Central Java-Yogyakarta earthquake 2006. 21st EAROPH world planning and human settlement congress & mayors' caucus, Japan, October 2008

Peacock WG, Dash N, Zhang Y (2007) Sheltering and housing recovery following disaster. In: Rodriguez H, Quarantelli EL, Dynes RR (eds) Handbook of disaster research. Springer, New York

Pribadi KS, Kusumastuti D, Febrin D, Fazan I, Okazaki K, Saito T (2011) Data collection on non-engineered construction in Padang city. Tokyo, Japan

Pribadi KS, Kusumastuti D, Rildova (2008) Learning from recent Indonesian: an overview to improve structural performance. The 14th world conference on earthquake engineering, Beijing, 12–17 October

Raharjo W (2007) Post-Disaster Housing in Yogyakarta, Indonesia. Shelter Urban Poor 165

Resosudarmo BP, Sugiyanto C, Kuncoro A (2012) Livelihood recovery after natural disasters and the role of aid: the case of the 2006 Yogyakarta earthquake. Asian Econ J 26(3):233–259

Satyarno I (2009) Socialization and training in earthquake-resistant housing construction for construction workers in Trimulyo village, Jetis sub-district, Bantul district, Yogyakarta, University of Gadjah Mada & International Recovery Platform

Smith GP, Wenger D (2007) Sustainable disaster recovery: operationalizing an existing agenda. In: Rodriguez H, Quarantelli EL, Dynes RR (eds) Handbook of disaster research. Springer, New York

Suarjana M, Sengara IM (2008) Lessons learned from Yogya and Aceh recovery program. The 14th world conference on earthquake engineering, Beijing, 12–17 October

Soelaksono A (2009) NGO and donor coordination to speeds up reconstruction and avoid NGO competition. In: 4th annual international workshop and expo on Sumatera tsunami disaster and recovery. Disaster Mitigation Research Center of Syiah Kuala University, Banda Aceh, pp 23–26

Steinberg F (2007) Housing reconstruction and rehabilitation in Aceh and Nias, Indonesia—rebuilding lives. Habitat Int 31(1):150–166

Telford J, Cosgrave J (2007) The international humanitarian system and the 2004 Indian Ocean earthquake and tsunamis. Disasters 31(1):1–28

World Bank (2012) REKOMPAK: rebuilding indonesia's communities after disasters. World Bank, Jakarta

UNOCHA (2009) West Java Earthquake: situation report number 5, UNOCHA, New York

Usamah M, Haynes K (2012) An examination of the resettlement program at Mayon Volcano: what can we learn for sustainable volcanic risk reduction? Bull Volcanol 1–21

Vanhoebrouck P, Sagala S (2010) Not so helpless victims: social capital roles in the 2009 West Sumatra earthquake recovery. Research report of International Federation of Red Cross and Red Crescent (IFRC), Indonesia

Yasaditama H, Sagala S (2012) Rebuilding settlements: learning from housing reconstruction process after West Java earthquake. The 2nd international conference on sustainable built environment, Islamic University of Indonesia

Chapter 12
Coastal Zone Management in Tamil Nadu, India: Challenges and Innovations

R.R. Krishnamurthy, K. Chandrasekar, and D. Shanmugam

Abstract More than two decades earlier the concept of integrated coastal zone management (ICZM), to involve the community in decision-making has been seriously debated and agreed upon for implementation in Tamil Nadu. About 1,000 km long Tamil Nadu coast has diversity of resources as well as issues due to various anthropogenic activities. Considerable efforts have been made during the last two decades, in making quantitative assessment on resource potential and alteration of coastal environment by using multispectral remote sensing data and spatial analysis tools such as GIS. Scientific database generated through these technology tools are used as a decision-support system by incorporating stakeholder's perception and field data, which helped considerably in effective coastal zone management in Tamil Nadu. Also efforts have been made in training and capacity building at different levels, which are mainly focused to implement ICZM in the field. This chapter is aimed to provide the genesis, different stages and present scenario of coastal zone management practices in Tamil Nadu coast and its direct and indirect impacts in helping/achieving disaster risk reduction in the aftermath of 2004 Indian ocean tsunami.

Keywords Coastal environment • Decision support system • GIS • Remote sensing • Stakeholder participation

---

R.R. Krishnamurthy (✉)
University of Madras Arts and Science College and University Constituent College, Kanchipuram, India
e-mail: rrkrishnamurthy@gmail.com

K. Chandrasekar
National Remote Sensing Centre, Indian Space Research Organization, Hyderabad, India

D. Shanmugam
Department of Applied Geology, University of Madras, Chennai, India

R. Shaw (ed.), *Disaster Recovery: Used or Misused Development Opportunity*, Disaster Risk Reduction, DOI 10.1007/978-4-431-54255-1_12,

12.1 Introduction

Coast is a boundary between land and sea; however it is not as discrete as it appears. It can be considered as a transition zone between land and sea due to the various interactions between these two entities along this zone. The transitional zone refers to the two main environments, terrestrial and marine, and their main influences to coastal zone. As a transitional region, the coastal zone is a unique system because it is subjected to dynamic influences from land and ocean ecosystems. Each ecosystem has its specific characteristic and the interaction between these remains ever changing with natural fluctuation in the biological, chemical and geological attributes. The complex and dynamic character of the coastal zone has the physical action on the area, as well as the interaction of three bio-ecological systems: land and sea, sea and air, and sea and sediments (Siry 2007). The uniqueness makes the coastal zone one of most productive ecosystems which abounds with natural resources, and is often considered highly scenic. This interface offers enormous resources that it attracts and supports more than half the world's population. It has been estimated that 23 % of the world's population lives both within 100 km distance of the coast and <100 m above sea level and population densities in coastal regions are about three times higher than the global average (Small and Nicholls 2003). Within the coastal population, 71 % live within 50 km of estuaries, and in tropical regions, settlements are concentrated near mangroves and coral reefs (Brown et al. 2006). The ever increasing population and migration into the coastal regions will increase the demand for the resources like the food, fuel and shelter. This in turn has resulted in urbanization, industrialization and pollution of coastal zones. The urban conquest of coastal zones goes back to the beginning of humankind, as a result of the consolidation of its fundamental activities, such as fishing and riparian agriculture, among others (Ducci 1995). Timmerman (1997) established that as of the year 2000, for the first time in history, more than 50 % of the world's population would be living in cities, and 50 % of that population would be in coastal cities, which has proven true. The closeness to the ports, easy access water and raw materials, disposal of industrial waste etc. makes coastal regions, very attractive to set up industries. Asian rivers account for nearly 50 % of the total sediment load (13.5 billion tones per year) transported by the world's rivers (UNEP 1992). Industrialization leading to settlement and economic activity has reduced the resilience and adaptability of coastal systems to climate variability and change, as well as to sea-level rise (Watson et al. 1997). The main source of pollution of the coastal ecosystem is the urbanization, industrialization and land use changes happening in the coastal zones. As much as 70 % of the waste effluent discharged into the Pacific Ocean has no prior treatment (Fuavo 1990). Over 40 % of marine pollution in the region is derived from land-based activities (via riverine discharge) and maritime transport contributes a further 12 % (Weber 1993). In India, for example, exceptionally high concentrations of lead (820 mg/L) and cadmium (336 mg/L) have been observed in Thane creeks on the Bombay Coast and the mercury concentration was 778 mg/L. Sediment along the creeks and off-shore stations was also reported to contain significant concentrations of lead (Government of India 1993). Therefore it is important to protect the coastal zone from the adverse impact due to population, urbanization, industrialization and pollution.

To protect the coastal zone a well conceived management is essential for the sustainable development. Integrated Coastal Zone Management (ICZM) is seen as the management of the coastal zone taken as a whole in relation to local, regional and international goals with a particular focus on the interactions between various human activities, resource demands both within and between activities from the coastal zone and from other regions. A part of the management was the integration of environmental protection goals into economic and technical decision making process (OECD 1993).

12.2 Tamil Nadu Coast

Tamil Nadu is the southern most state of India which has a coast line of more than 1,000 km in length (Fig. 12.1). Tamil Nadu coast is punctuated by a number of landforms from the brackish water Pulicat Lake in the north to the rocky coast of Kanyakumari in the south. Other landforms include are mud flats, sandy beaches, mangroves, marshes, delta, creeks, palace beach ridges etc are present along the coast. Geomorphologically Tamil Nadu coast is being considered as a submerging coast due to its topography and drainage pattern. The drainage pattern is controlled by major rivers such as Cauvery, Palar, Vaigai, Tamaraparani and there are several minor rivers. Though the intertidal flat is about 1 m there are considerable area

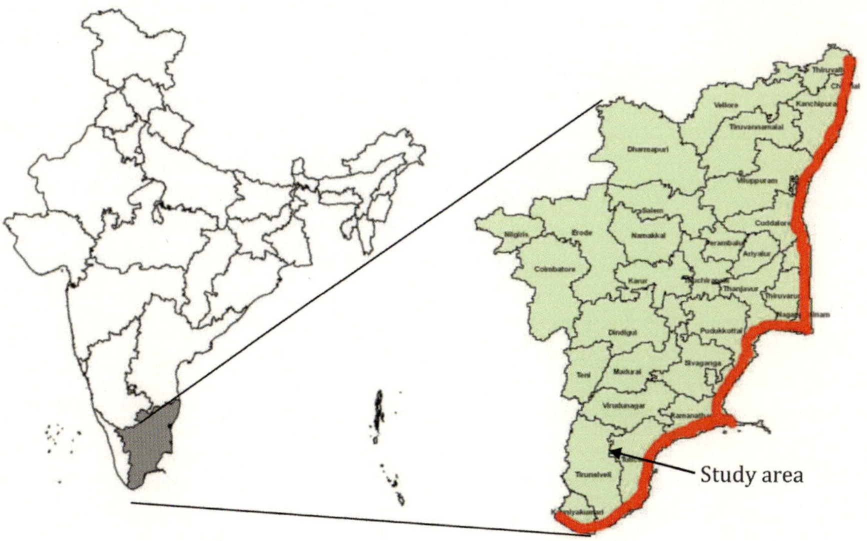

Fig. 12.1 Study area—coastal zone of the state of Tamil Nadu, India

under mangroves occur in Pichavaram and Muthupet. Due to the river discharge and availability of source rocks, few stretches of Tamil Nadu coast is enriched with placer minerals. Pulicat Lake and its wetlands located on the northern part is being considered as a Ramsar site and the Gulf of Mannar, which has about 21 islands/islets closer to Northern part of Sri Lanka is an important Marine Biosphere Reserve. The shore temple at Mahabalipuram is a UNESCO world heritage site, which receives considerable tourists every year. The Union Territory of Puduchery is located in the northern middle part of coastal Tamil Nadu, where the unique inter-state coastal issues can be witnessed.

Chennai, Puduchery, Cuddalore, Tuticorin and Kanyakumari are the important urban centers along the Tamil Nadu coast. Due to its proxy location one cannot exclude the coast of Puduchery, which is one of the Union Territories of India. Tamil Nadu coast has experienced the impact of a major catastrophic tsunami event on 26th December 2004 and subsequent cyclones in 2005 and 2011, causing deaths of millions and severe environmental issues due to heavy inundation of coastal land, destruction of coastal ecosystems and adjacent agricultural lands and infrastructure facilities. Thane cyclone, which had a landfall between Cuddalore and Puducherry on 30th December 2011 killed 46 people with severe damages to coastal ecosystems, coastal agriculture and infrastructure is the recent natural calamity to this coast.

12.3 Coastal Pollution

Coastal pollution is one of the important environmental hazards, which affect both the quality of life and livelihood security of coastal communities (Fig. 12.2). Chennai (earlier called as "Madras") is the capital of Tamil Nadu state, which has a population more than five million and the fourth largest urban centre in the country. There is a growing need for electricity, which is being obtained from various sources including thermal power plants located along the coast. The discharges from thermal power plants i.e. the fly ash mixed with water called slurry being discharged in to the sea. Additionally, the city is discharging about one million liters of sewage everyday through its water ways connected to the sea. This has resulted in coastal pollution which not only contaminated the coastal waters but also the aquatic life and the dependency of coastal communities on this resource for their livelihood. Several studies have been carried out on the quality of ground water in the city are reported that the quality of ground water is deteriorated beyond recovery. As city is growing rapidly particularly in its southern and western parts due to the boom of IT industries during the last decade there is a steady demand for ground water. Major parts of coastal aquifers are contaminated by saline water incursion due to over exploitation apart from pollution due to industries and domestic discharges. The pristine water ways such as Cooum and Adyar rivers which pass through the city are heavily polluted due to anthropogenic activities. The 400 km long Buckingham canal which is used for navigation during historical times between

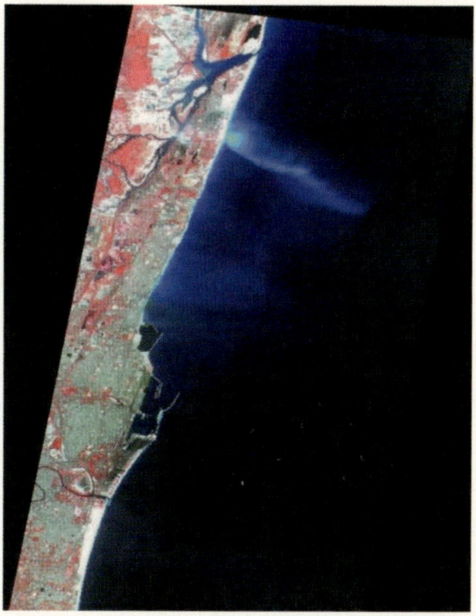

Fig. 12.2 Coastal pollution along Chennai

Tamil Nadu and Andhra Pradesh States has been severely damaged and contaminated due to heavy metal pollution. This canal is facing the problem due to siltation and is in the verge is complete extinction. Recent study by Jayaprakash et al. (2011) confirmed that rapid urbanization and industrialization especially in the in northern part of Chennai city, yields uncontrolled discharge of sewage, garbage, and industrial effluents into the Buckingham Canal and the elevated concentration of trace metals is not unexpected in view of this situation. The level of enrichment of trace metals has also increased by 30–70 % for most of the elements when compared with all other ecosystems in the world as well as nearby locations in the study area. Part of Chennai especially the Manali, which is an industrial complex located about 20 km north of Chennai is facing the problem due to air pollution from the industries. There are about 28 categories of industries like oil refineries, fertilizer plants, chemicals, fabric yarn and steel, etc. which have existed for more than two decades. A comparison of key air borne pollutants with primary health data shows that the coastal population in this area is severely affected by respiratory problems (Jayanthi and Krishnamoorthy 2006). Unplanned and uncontrolled way of urban development has resulted in severe environmental degradation, which poses threat to the future generation to use its resources effectively.

Figure 12.3 shows the IRS FCC imagery of the general configuration of Pulicat lake mouth, which is connected by the Buckingham canal from north of Chennai and the discharge of minor rivers from its landward surroundings. Pulicat Lake ecosystem is severely threatened by the release of chemical contaminants from the industries located in the north of Chennai. A study by Kamala Kannan et al. (2008) on the concentrations of three heavy metals chromium (Cr), cadmium (Cd) and lead

Fig. 12.3 Morphology of Pulicat Lake mouth

(Pb) present in water, sediment and green algae were examined. The samples were collected from six different stations at Pulicat Lake. This study has confirmed that the concentrations of these heavy metals are found to be high in sediment, whereas in green algae the concentration of Pb was higher than water and sediment samples. It was recorded that the concentration of these heavy metals are invariably high in the mouth section of this Lake, which is the point of discharge from industries through the Buckingham canal and also due to the solid waste from nearby fishermen hamlets. Mouth section of this lake connected to the Bay of Bengal through a narrow water way, which is often blocked by the growth of sand spit. The overall environmental quality of this important ecosystem is being mainly influenced by the geomorphological set up at the mouth section of this lake. Frequent closure of lake mouth due to sand accretion prevents the free flow of industrial discharge into the sea. Sand accretion and closure of lake mouth has become an annual phenomenon that the contamination of aquatic species within the lake as well as prevention of movement of fishing vessel to the sea through this lake mouth has become a problem.

Conflict among the primary stakeholders always exists in this region due to this pollution issue, which has direct impact on the livelihood of coastal community. Coastal zone management experts still struggle to bring together all the stakeholders, in sorting out this unresolved issue during the last two decades. An initial multi-stakeholders approach carried out in this region through ICZM model during 1998–2000 have helped to fully understand the complexity of management issues. Important recommendations of this practical exercise carried out by key decision-makers which include polluters pay policy, has resulted in curtailing the impact of industrial pollution.

12.4 Coastal Ecosystems

Tamil Nadu coast is endowed with important ecosystems such as mangroves and coral reefs, which has direct livelihood support to considerable coastal population. Pichavaram mangrove wetland ecosystem is of estuarine type located in between two estuaries called Vellar and Coleroon. Muthupet mangroves are bordered around the lagoon and along the beach, which directly faces the sea front. Few isolated patches of mangroves also occur in the Islands of Gulf of Mannar. The estuarine mangrove wetlands of Pichavaram are severely altered by anthropogenic activities and hence there are several attempts by the Government and other NGOs to implement Joint Mangrove Management (JMM) approach involving the local community. A study on the multi-disciplinary approach to manage this important coastal ecosystem during 1990s by the M. S. Swaminathan Research Foundation under the India–Canada Environment Facility programme paved way to thoroughly understand the complexity of biophysical factors responsible for the degradation of mangroves. The advancements in remote sensing technology using the multi date—multi sensor remote sensing data have helped to make quantitative assessments on a real extent of mangroves, species distribution, changes in coastal morphological changes, which have direct influence on water quality within this ecosystem. Satellite remote sensing technology greatly helped to understand this ecosystem through change-detection analysis. However, greater field skills are required to understand the causes for natural felling of mangroves due to changes in micro-topography, salinity changes, soil quality etc. Apart from tidal amplitude of sea and freshwater input through rivers keeps up an ecological balance for the sustenance of mangroves. Also the composition of mangroves is mainly determined by the tolerance of different species to substrate and saline conditions, which in turn determines the dominance of one or various species at a specific site. In general, the sustenance and sustainable management of mangrove ecosystem in the State is not only based on the control of local anthropogenic activities but also due to the impact of upland activities especially the restriction on freshwater inflow in to this ecosystem due to the use of freshwater for agriculture and aquaculture. Several drought years during 1990s have witnessed with the natural felling of mangroves due to changes in salinity and micro-topography and with natural regeneration once the degraded sites start receiving freshwater during high rainfall and flooding years. Palynostratigraphical and geochemical evidences from Pichavaram estuary show that during the past three millennium the climate has ameliorated from warm and humid to comparatively drier condition (Srivastava et al. 2012).

Gulf of Mannar Biosphere Reserve (GOMBR) is 130–275 km wide and has a length of about 160 km considered as very important “hot spot” in the Indo-Pacific region, which received global attention towards sustainable management of its resources during the last two decades. Government of India’s initiative on “Conservation and Sustainable Use of the Gulf of Mannar Biosphere Reserve’s Coastal Biodiversity” has received a major funding support from the Global Environment Facility (GEF). Vast area of coral reefs especially fringing reefs can be observed in majority of these islands, which supports large number of ornamental fishes and other important coastal resources. References show that there were about 117 coral species in this area, which was famous during Roman Empire in export of pearls from this region. It has a record of about 3,600 species of plants and animals. A series of shoals which connect the Rameswaram Island and Northwestern part of Sri Lanka is called Adam's bridge. Gulf of Mannar Biosphere Reserve Trust is a registered trust of Government of Tamil Nadu http://gombrt.org. Several studies were carried out to document the resources, preparation of management strategies towards conservation and sustainable utilizations of GOMBR during the last two decades. However, the recent initiative on Sethusamudaram Shipping Canal Project has threatened the ecology and environmental quality of this biosphere reserve. This project was initiated by the Government of India with a view that this will boost marine time trade by shortening the travel distance of about 780 km and a total travel time of about 30 h. Though this project is not yet completed for operation the scientific community is now focusing on the impact of this canal project on coastal ecosystems.

12.5 Coastal Vulnerability Studies

Tamil Nadu coast is vulnerable to both natural as well as man-made disasters. It faced the serious impact of 26th December 2004 Indian Ocean Tsunami for the first time in the recent memory. It is also exposed to the fury of climate disasters like cyclone and monsoonal floods almost every year since 2005. Before 2005, majority of the Tamil Nadu coastal zone was facing severe shortage of freshwater and ground water resources due to shortage of annual rainfall and the government was seriously advocating rain water harvesting and effective utilization of freshwater resources both in urban and rural areas. However in the last 8 years there has been an erratic change in rainfall pattern leading to above average rainfall and flooding in majority of the coastal districts of Tamil Nadu without any drought period in between. "Nilam" cyclone which crossed South Chennai coast on 31st October 2012 is the recent disaster in the State. More frequent crossing of cyclone has been recorded only in the recent past. Similarly "Thane" cyclone crossed the southeast coast making its landfall in Cuddalore, Puducherry and Nagapattinam districts on 30th December 2011. During November 2010 the coastal districts of Tamil Nadu has faced the impact of Cyclone "Jal" and Cyclonic storm "Ward" affected the southernmost part of coastal Tamil Nadu. Almost every northeast monsoon season i.e. October–December since 2005 the Chennai city has faced the problem of urban flooding. Hence the civic agencies especially the Corporation of Chennai have a major challenge to handle climate-related disasters more effectively.

12.5.1 Tsunami and Storm Surges Wave Modeling

Under the Ministry of Earth Sciences (MoES), Government of India, national programme on Establishment of National Early Warning System for Tsunami and Storm Surges, has been carried out for data processing through field survey and spatial analysis for modeling and mapping for tsunami and storm surges under the coordination of Integrated Coastal and Marine Area (ICMAM) Project Directorate. This study was aimed to analyze and forecast possible risks of an ocean originating inundation, such as tsunamis and storm surges, upon low-lying and densely populated coastal areas of India. Under this study tsunami and storm surges model has been constructed using numerical equations. It was observed that numerical model found to be an excellent tool for understanding past events and stimulation for future forecasts. Since the source parameters that triggered 26th December 2004 Indian Ocean tsunami are well known these parameters are used as first to set in order to capture the past events. The prediction using the model developed under this study is directly related to the quality of data used to create ocean bathymetry and topography of the coastal land. The salient outcomes of this programme are: (1) creation of extensive database on coastal landforms elevation and (2) near shore bathymetry. These two parameters are of immense use for the preparation of tsunami and storm surges model, which could be used for demarcating vulnerable sites.

The model results have been obtained using high resolution bathymetry and land elevation data through validation using extensive field observations. Other important data used for this study are seismic parameters, in conjunction with deep sea and shallow water bathymetry and land topography. Using the above data, grid has been generated for the entire Indian coastal region. Simulation for inundation model has been carried out by the two parameters, namely computation of initial deformation due to earthquake and propagation from source to inundation. Different scenarios of extreme inundation have been created using this model to study the extreme inundation and run-up by varying the source parameters that actually trigger the tsunami. This study outcome is found to be more suitable to assess and demarcate the vulnerable coastal areas along Indian coast.

12.5.2 Tsunami Impact Assessment

A fast tract project funded by the Department of Science and Technology, Government of India was carried out during December 2004 to May 2005 along Chennai coast starting from Pulicat Lake mouth to Kovalam to quantitatively measure the extent of tsunami water inundation along this coastal stretch. It was observed that inundation of areas is higher in the coastal stretch of south Chennai (Cooum Estuary—Kovalam Creek) when compared to the coastal areas of north Chennai. The inundation areas are invariably covered by fishermen's settlements, build-up areas including beach resorts/hotels and amusement parks, beaches, and sand dunes with and without vegetation.

Remarkable highlights of this study outcome are:

1. South Chennai coast, which has numerous waterways and protected by natural barriers such as sand dunes and coastal vegetation suffered more damages due to inundation of tsunami water. The maximum inundation of 538 m was recorded in Sholinganallur and its surroundings. The total area submerged by the tsunami waters is 1.31 km^2. The intensity of the tsunami impact, including changes in coastal geomorphology, has been strongly observed in this coastal stretch. Along the south Chennai coastal zone, it was observed that while inundation is higher in Sholinganallur and its surroundings, the coastal areas of Injambakkam and its surroundings suffered more damages on coastal settlements,
2. Middle part of Chennai coast, which has two important waterways such as Cooum and Adyar rivers allowed the entry of tsunami water towards urban areas, which houses more of coastal slums and fishermen hamlets. The tsunami water inundation recorded here was up to a depth of 278 m near Gandhi Statue on Kamarajar Salai. The tsunami waters submerged the entire Marina beach and extended up to 585 m in the Chepauk coastal area,
3. A situation similar to that in the Adyar and Cooum estuaries was observed in Ennore Creek and its surrounding areas. Extensive damage to mechanized boats and settlements was observed in Ennurekuppam and Nettukuppam. However, damages were relatively restricted in the Ennore industrial belt, and this might be attributed to factors such as shoreline erosion protection measures, near shore bathymetry, sea floor topography, and so on, and
4. Along the Pulicat lake mouth, the tsunami water inundation was up to a depth of 450 m and washed out an area of about 1.23 km^2. Along this coastal stretch, the fishing hamlets located very close to the sea such as Koraikuppam and Sattangkuppam recorded the maximum devastation. These two fishing hamlets mainly depend on sea fishing and thus lived on the coast paving the way for major damages compared to the damages suffered by other fishing hamlets in and around Pulicat Lake.

It was observed that the artificial protection measures such as harbor structures, groynes and sea walls constructed along north Chennai coast to control shore protection measures have helped considerably to reduce the impact of tsunami waves. Also the nature of near shore sea floor topography and the elevation of coastal landforms also play a vital role in controlling tsunami water inundation (Satheeshkumar et al. 2008). However, the situation is different in south Chennai since the majority of natural protection measures such as sand dunes and coastal vegetation are severely altered by human activities, which resulted in severe damage due to tsunami.

12.5.3 Climate-induced and Chronic Disasters

As a first step towards the assessment of coastal urban centers with reference to climate change induced disasters the Kyoto University in Japan jointly with the University of Madras and Corporation of Chennai (COC) has carried out a study in Chennai during 2010–2011. Ten administrative zones of COC were assessed with reference to five important parameters i.e. physical, social, economic, institutional and natural to understand its resilience in the event of climate disasters. Another important outcome of this study is an action-oriented resilience assessment (AoRA). It is proposed as a subsequent tool to examine the responsibilities of different stakeholders (local government, communities, academia, private organizations and NGOs) in implementing defined actions aimed at enhancing the resilience of Chennai to disasters (Joerin et al. 2012a, b). As a follow up, micro level study has been carried out in Chennai to understand the community's coping and adaptive capacity which is crucial in absorbing, managing and bouncing back from a disaster event (Joerin et al. 2012a, b).

Coastal process, interference of man-made structures and morphology of the shoreline are vital parameters towards coastal erosion as well as accretion, which is being aggravated by the changes in sea level. Approximately, 26 % of India's coastline is prone to coastal erosion and facing the environmental problems due to inundation by saline water as well as incursion of saline water in the coastal aquifers. Several parts of Tamil Nadu coast are also facing the fury of coastal erosion as well as accretion during the last three decades. It has been confirmed and accepted that the most important anthropogenic cause for shoreline erosion is the construction of harbor and other coastal engineering structures along the coast. There are about 15 fishing hamlets still facing the issue of severe shoreline erosion in north Chennai. Though there are shore protection measures including the construction of groins to prevent erosion, the severity of this issue is not being controlled in this area. Based on the analysis of shoreline boundary in different periods, it has been quantified that about 77 m^2 of land along the 3 km coastal stretch is being sacrificed to the sea every year, which resulted in the submergence of community well, temples and other properties during the last two decades (Krishnamoorthy et al. 2002). This has resulted in resettlement of communities towards the landward side which gets further complicated when other existing settlements or features like highways or industries does not allow landward resettlement. This situation leads to conflict among the stakeholders–especially the fishing community, the port trust, the industries and the highways. Experts are of the opinion that the shore protection measures like groins and sea walls should have been constructed before the construction of harbor structures. Field observations in this coastal area since 1995 have clearly confirmed that the present erosion protection measures are not sufficient enough to completely control this situation and again the safety of the community living on this coast is a question mark. Experts are also of the opinion that in order to prevent erosion, an appropriate offshore structure is essential to dissipate the wave energy before it approaches the coast. However, we are yet to arrive at concrete solutions to resolve this issue

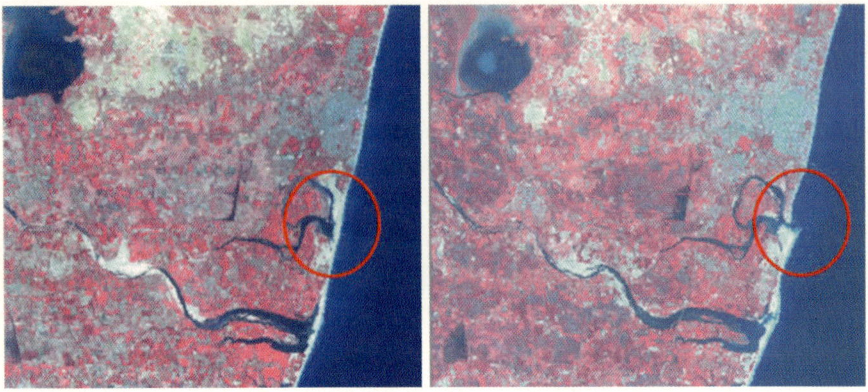

Fig. 12.4 Landsat TM imagery (www.landsat.org/ortho/) show coastal configuration in 1991 (*left*) and in 2000 (*right*)

permanently (Krishnamurthy 2010). Ultimately, the community's live in this place are forced to move out of their original place of origin as "climate refugees."

More important issue arising out of shore erosion is witnessed with the trans-boundary between Tamil Nadu and Puducherry coasts. Various field based studies have confirmed that the construction of harbor structure in 1989 at the southern tip of Puducherry coast altered the sediment movement, which resulted in erosion on its north and deposition at its south. Experts arguments highlight that two breakwaters constructed at the harbor prevented the littoral drift i.e. movement of sand from south to north. The prevailing coastal erosion in this area is similar to that of north Chennai coast, where the harbor breakwaters affected the littoral drift, which resulted in severe erosion on its north and deposition at its south where the marina beach is expanding every year. Similar to the situation in north Chennai the Puducherry government has spent considerable money to arrest erosion by putting sea walls for about 7 km length during 2002–2003, which also proved to be a failure in terms of controlling erosion. About 8–10 km length of coastline just on the north of Puducherry harbor facing the problem of erosion only after 1989, which has changed the configuration of the coast from a straight line in to a punctuated line as shown in Fig. 12.4. There is a considerable change in the configuration of the coast only after the construction of harbor breakwaters as shown in Fig. 12.5.

Based on the assessment on shoreline change by erosion and accretion the Ministry of Environment and Forests (MoES) has classified the India's shoreline in to (1) high accretion, (2) medium accretion, (3) low accretion, (4) stable coast, (5) low erosion, (6) medium erosion and (7) high erosion. Quantitative assessment on shoreline for the entire country is on progress, which is mainly based on the utility of multidate satellite imagery using onscreen digitization method. The MoES has taken up this initiative to make an assessment for a period of 38 years starting from 1972 to 2010. The outcomes of this national programme will provide a clear insight on the present scenario as well as future coast on the potential damages due to this chronic disaster.

After experiencing the devastation by 2004 Indian Ocean tsunami event the government came out with suitable warming system for tsunami and storm surges,

Fig. 12.5 Shoreline of 2005 (in *red*) is overlaid on 2011 image (www.earth.google.com)

which is operational now. Tsunami wave model based on the input of submarine earthquake parameters will indicate the tsunami wave height and the likelihood area of submergence for disseminating warning message. Similarly efforts are also taken in studying the resilience of coastal urban centers with reference to climate change induced disasters like cyclone, flooding etc. Both the combination of field based survey and spatial analysis have helped to make quantitative assessment on vulnerability.

12.6 Technological Intervention for ICZM

The availability and easy accessibility of multisensor data has helped the different end users to quantify the coastal resources and also the environmental hazards such as coastal pollution, shoreline change etc. As a part of ICZM capacity building end users like Port Trust, Department of Forest etc were trained in the use of remote sensing and GIS technology tools. Considerable efforts have been taken up to scientifically study the coastal resources using technology tools like remote sensing and GIS, which helped the government and other field based NGOs to take up restoration of degraded coastal ecosystems such as mangroves and coral reefs. For example, the Joint Mangrove Management (JMM) and the institution of Gulf of Mannar Marine Biosphere Reserve Trust by the government with international and national funding support are important initiatives, which resulted in considerable improvement in enriching the coastal resources in order to support the livelihood of local community. Understanding on the impact of biophysical factors on the coastal ecosystems has helped to devise appropriate implementation strategy such as man-made canals to bring more influx of fresh water in to the degraded mangrove sites. Figure 12.6 shows the "fish bone pattern" of manmade canals inside Pichavaram mangrove forests after the implementation of JMM. Change-detection of mangroves using optical

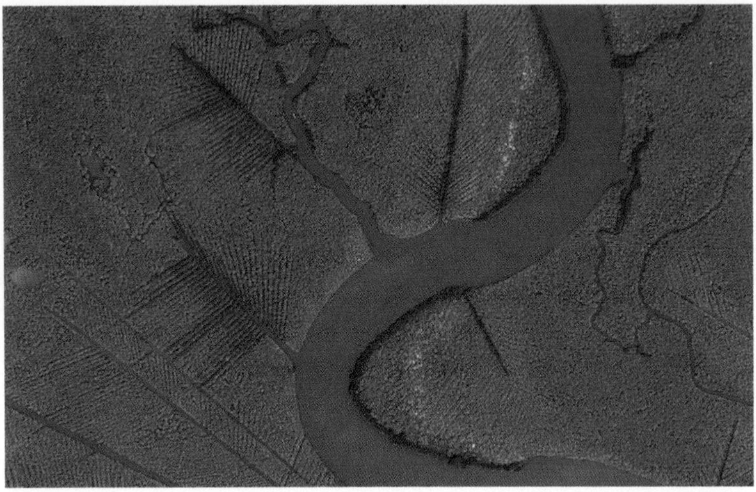

Fig. 12.6 "Fish bone pattern" of manmade canals inside Pichavaram mangrove forests (www.earth.google.com)

remote sensing data of pre and post JMM has proved that there is a considerable increase in mangrove area after the implementation of JMM (Selvam et al. 2003). Also it has been noted that the real advantage of these technology tools in making assessment, interpretation on underlying causes for ecosystem and environment degradation, to demarcate sites for taking up restoration and monitoring of changes after the implementation of field based projects were effectively carried out using these technology tools. More of science based people centered approach has been found to be successful in these initiatives as being done in several other developed countries.

Technological interventions have not only helped to restore the degraded coastal ecosystems but also to assess and forecast the vulnerability of the coast with reference to both episodic and chronic disasters. Many field based research results such as CDRI for Chennai city, tsunami vulnerability atlas for Chennai coast, etc were being used by the end users towards effective DRR measures.

12.7 Conclusions

Tamil Nadu coastal zone is facing multifaceted issues and problems, which are very complex in nature attributed by the following factors:

- High population density and unsustainable utilization of coastal resources still continues despite several awareness measures,
- Lack of multi-institutional cooperation mechanism towards development of ICZM plan and its implementation even though there is an establishment of national centre for coastal zone management,

- Lack of participation of local community in development and implementation of ICZM, and
- Transboundary issues such as sharing of freshwater resources, for example, discharge of river water from Karnataka to Tamil Nadu, affect the mangrove ecosystems, which in turn linked to the coastal fisheries and the use of this ecosystem as natural barrier.

This chapter throws light on these issues by discussing the coastal ecosystem and coastal pollution. Urbanization, industrialization, population pressure and poor coastal management results in coastal ecosystem degradation due to intense coastal pollution. Pollution control measures through appropriate legislation and its implementation in the field is need of this hour. It also discusses the vulnerability of the coastal zone to the disasters like tsunami, cyclones and associated storm surges. The damage caused by the sea water incursion, shore line erosion and sand deposition due to anthropogenic activities are also discussed. Detailed discussion on the tsunami impact and associated storm surge wave modelling helped in identifying the vulnerable coastal areas. The outcomes of these study results can be used effectively in the field by end users especially towards disaster preparedness.

This chapter emphasizes the need for multi-institutional cooperative mechanism and participation of coastal community in the ICZM. An effective and implementable ICZM will be the vital part in achieving disaster risk reduction with reference to coastal areas. The scientific community must advocate that ICZM is an important component of disaster management since this is an end product achieved through multi-institutional cooperation, community participation, etc and its integrates anthropocentric and eco-centric approaches.

Acknowledgement The author (RRK) is thankful to the Department of Science and Technology and Ministry of Earth Sciences including its ICMAM Project Directorate for the funding support and guidance to carry out ambitious projects and its results were interpreted and used for the preparation of this chapter. Also he is thankful to Dr. Rajib Shaw, Kyoto University, Japan, for his help and support under the bilateral cooperation programme carried out for Chennai city. The authors also thankful to the reviewers for their comments and feedback to revise the chapter.

References

Brown C, Corcoran E, Herkenrath P, Thonell J (2006) Marine and coastal ecosystems and human well being. Synthesis. UNEP-WCMC, Cambridge, 65 pp

Ducci ME (ed) (1995) Introducción al urbanismo. Conceptos básicos. Editorial Trillas, México, 94 pp

Fuavo V (1990) Areas of environmental concerns in the South Pacific region. Secretariat of the Pacific Regional Environment Programme, Samoa

Government of India (1993) Environmental action programme. Ministry of Environment and Forests, New Delhi

Joerin J, Shaw R, Takeuchi Y, Krishnamurthy RR (2012a) Actionoriented resilience assessment of communities in Chennai, India. Environmental Hazards 11(3):226–241

Joerin J, Shaw R, Takeuchi Y, Krishnamurthy RR (2012b) Assessing community resilience to climaterelated disasters in Chennai, India. International Journal of Disaster Risk Reduction 1:44–54

Jayanthi V, Krishnamoorthy R (2006) Key airborne pollutants – impact on human health in Manali, Chennai. Curr Sci 90:405–413

Jayaprakash M, Nagarajan R, Velmurugan PM, Sathiyamoorthy J, Krishnamurthy RR, Urban B (2011) Assessment of trace metal contamination in a historical freshwater canal (BuckinghamCanal), Chennai, India. Environ Monit Assess. doi:10.1007/s10661-011-2509-5

Joerin J, Shaw R, Takeuchi Y, Krishnamurthy RR (2012) Action-oriented resilience assessment of communities in Chennai, India. Environ Hazards, Taylor & Francis. http://dx.doi.org/10.1080/17477891.2012.689248 # 2012, ISSN: 1747–7891

Kamala Kannan S, Prabhu Dass Batvari B, Lee KJ, Kannan N, Krishnamoorthy R, Shanthi K, Jayaprakash M (2008) Assessment of heavy metals (Cd, Cr and Pb) in water, sediment and seaweed (*Ulva lactuca*) in the Pulicat Lake, South East India. Chemosphere 71:1233–1240

Krishnamoorthy R, Bharathi GS, Perikali P, Ramachandran S (2002) Coastal zone hazards in India: study based on remote sensing and GIS techniques. In: Begni G (ed) Observing our environment from space. A.A. Balkema Publishers, Rotterdam, pp 79–86

Krishnamurthy RR (2010) Coastal morphological changes in India and its relevance to coastal management. In: Shaw R, Krishnamurthy RR (eds) Communities and coastal zone management. Research Publishing, Singapore, pp 47–59

Organisation for Economic Cooperation and Development (OECD), (1993) Coastal Zone Management: Integrated policies, OECD, Paris

Satheeshkumar C, Arul Murugan P, Krishnamurthy RR, Prabhu Dass Batvari B, Ramanamurthy MV, Usha T, Pari Y (2008) Inundation mapping – a study based on December 2004 tsunami hazard along Chennai coast, Southeast India. Nat Hazards Earth Syst Sci 8:617–626

Selvam V, Ravichandran KK, Gnanappazham L, Navamuniyammal M (2003) Assessment of community-based restoration of Pichavaram mangrove wetland using remote sensing data. Curr Sci 85:794–798

Siry HY (2007) Making decentralized coastal zone management work for the Southeast Asian region: comparative perspectives, division for ocean affairs and the law of the sea. Office of legal affairs, the united nations, New York

Small C, Nicholls RJ (2003) A global analysis of human settlement in coastal zones. J Coastal Res 19:584–599

Srivastava J, Farooqui A, Hussain SM (2012) Vegetation history and salinity gradient during the last 3700 years in Pichavaram estuary, India. J Earth Syst Sci 121(5):1229–1237

Timmerman P (1997) Ciudades costeras: una nueva agenda. In: Simposio de Ciudades Costeras del Cono Sur. International Federation of Institutes for Advanced Studies (IFIAS), University of Toronto, Toronto

UNEP (1992) Marine pollution from land-based sources: facts and figures. UNEP Industry and Environment, Paris

Watson RT, Zinyowera MC, Moss RH, Dokken DJ (1997) Summary for policymakers, the regional impacts of climate change: an assessment of vulnerability. A special report of IPCC working group II. Published for the intergovernmental panel on climate change. Cambridge University Press, Cambridge, and New York, ISBN: 92-9169-110-0

Weber PC (1993) Abandoned seas. Worldwatch paper 116, Washington

Chapter 13
Post-Aila Community Recovery Innovations and Planning

Fuad Mallick and Aminul Islam

Abstract In the aftermath of super cyclone Aila in 2009, there was neither adequately or appropriately response planning and implementation from disaster management including relief and rehabilitation agencies, as was the case with the super cyclone Sidr which hit almost same South West Region of Bangladesh in 2007. Cyclone Aila damaged >50 % of houses, affecting ten million people and totally damaged shrimp business—a source of second largest export earnings for the country and collapse of the agro-ecosystem as well which was critical base for livelihood source of the inhabitants. Although government undertakes sizeable recovery operation (USD16.95 million for food, cash and shelter and international community and UN USD84.74 million) for public infrastructure after each disaster but restoration at the vulnerable populations' livelihood and shelter in particular faces deficient and needs remain unmet. At the same time there was the rethinking about the actual usefulness of post disaster early recovery including livelihood restoration and build back better shelters and possible risk reduction alternatives linked to sustainable development. Recovery resilient issue is emphasized to focus on investment on non-infrastructure resilience, livelihood and knowledge as well as governance to create enabling environment for early recovery. About 7.7 % of the coastal populations are covered by the existing cyclone shelters. The cyclone track seems to be extended from South East to South West. The state of habitat in the Aila affected areas, although deplorable provided the opportunity to try out innovations, one such being the concept of the Disaster Resilient Habitat (DRH). DRH called for an endeavor towards disaster/climate proof cluster settlement as the resilient habitat. This effort includes strengthening an entire village and its community together in a comprehensive manner with cyclone-resilient houses and infrastructure so that they are able to

F. Mallick (✉)
Postgraduate Programs in Disaster Management, BRAC University, Dhaka, Bangladesh
e-mail: fuad@bracu.ac.b

A. Islam
United Nations Development Prorgamme, Dhaka, Bangladesh

R. Shaw (ed.), *Disaster Recovery: Used or Misused Development Opportunity*, Disaster Risk Reduction, DOI 10.1007/978-4-431-54255-1_13,

withstand winds and storm surges to protect both lives and livelihood assets and even if there is destruction, getting back to life is easier. This chapter will describe how this innovative idea came into being and how it was implemented. The approach ensured that solutions emerge from the community rather than being top down with a combination of package including mobilization of social, physical, ecological and economic resource, and owner driven initiatives from planning to implementation having both software and hardware aspects managed by the community. The combination of technical expertise and local skill would emerge as creative solution. The DRH is considered a more structurally and ecologically sustainable solution than cyclone shelters. Perhaps in the future huge investments in top down cyclone shelters can be better utilized in this bottom up sustainable approach of comprehensive risk reduction.

Keywords Community participation • Cyclone Aila • Disaster resilience habitat • Ecosystem • Expert inputs

13.1 Introduction

Bangladesh coastline is more than 700 km long comprised of about 20 % of the country's land mass, and yet it is home to more than 30 million people. Risks to life, property, and the environment from coastal natural hazard events will increase with the climate change and its variability along with growth of coastal populations over the next several decades. Unprecedented tropical cyclones or hurricane and storm surges damage and personal despair during the super cyclone Sidr in 2007 and Aila in 2009 has tested the government at all levels, and has left many unanswered questions regarding the level of our preparedness to deal with and respond to disasters of this magnitude. Cyclones, tidal surge, and potential tsunamis are compounded by other trends, including coastal inundation (due to rising sea level and land subsidence). Increasing trend of storm intensity, uneven enforcement of building codes, lack of adequate zoning ordinances, poor planning and construction, continued development in high risk areas and the inexperience of coastal dwellers to prepare for and recover from these events need to be addressed. Despite significant progress in the application of science and technology to disaster risk reduction, communities remain challenged by disaster preparation, response, and recovery. These observations underscore the need for a dedicated national effort to provide science-based information towards the reduction of the social, economic, and environmental costs of natural hazards to our coastal communities.

The innovation of resilient recovery which starts with humanitarian assistance but ends up with sustainable development is a case analyzed in this chapter with "Disaster Resilient Habitat" (DRH) as a concept beyond Cyclone Shelter emerged in response to following three questions:

1. Contemporary loss and damage due to disaster reveals the fact that death casualty decreased significantly but damage in assets and properties are dramatically increasing. How to reduce loss of assets and properties of the poor community through converting existing settlement in vulnerable coastal zones into disaster-resilient housing and settlement? How early recovery can reduce humanitarian needs on a timely scale and facilitate affected people to recuperate quickly?
2. This chapter is an attempt to explore the possibility whether disaster resilient habitat is economically cost effective, socially acceptable and environmentally sound. Evidences needs to be established to justify the case that DRH is comparatively better choice than cyclone shelters?
3. How to ensure peoples' driven initiatives and their ownership with sustainability concerns appropriately taken into consideration? How to promote combination of technical expertise and indigenous resource and knowledge for building back better and restore heritage as well?

A range of stakeholders including national governments, the United Nations (UN) system, international financial institutions (IFIs), donors, international non-government organizations (INGOs), local NGOs and communities are engaged in post-disaster recovery operations. Their interventions in recovery are guided by various recovery needs assessment exercises by individual and or multiple agencies. The methodology, time line, processes and systems for such recovery needs assessments vary across stakeholders. Such variations in methods and related assessment outputs, challenges comparability across assessments and, often presents conflicting images of needs. These conflicts lead to coordination challenges, both within and across recovery agencies, with recovery donors and with other humanitarian and development actors. These experiences and related lessons call for development of a comprehensive and system-wide recovery needs assessment framework, which may be used by different recovery stakeholders to conduct joint and independent assessments at different points, during the disaster time line and share findings in a comparable way (Bollin and Khanna 2007).

13.1.1 Background: Cyclones and Bangladesh

Bangladesh's socio-economic condition coupled with geographical location, topography and dense population constitute high vulnerability that often lead to high loss of lives, shelters and economic damages due to disaster. About a third of its population—or some 50 million people—live in chronic poverty. Historically, deaths from single events, such as cyclones, reached into the hundreds of thousands in Bangladesh. But the trend is one of progress, with data suggesting that preparation for and response to disasters is improving over time. There have been massive reductions in the number of lives lost. An important element in this is a substantial improvement in early warning systems, and the establishment of cyclone shelters in particular. However, this is not enough and seems ineffective in the face of emerging trend of disaster and its magnitude, frequency, extent and severity. The Centre for Research on the Epidemiology of Disasters estimates that close to 229 million people have been directly affected by natural disasters between 1979 and 2008, with over 191,415 killed and economic damage in the order of US$5.6 billion.[1] At the same time, Bangladesh is one of the countries most at risk from the impacts of climate change. It is therefore likely that both acute hazards (such as flooding or cyclonic events) and chronic hazards (drought, sea level rise and saline intrusion, for example) will become more frequent and severe in the coming decades. The nature and scope of the hazards are well-documented in a wide range of existing literature.[2]

13.1.2 Post Aila Early Disaster Recovery (EDR) Challenges

In recent years, early recovery as a goal has evolved into an approach, termed as "build back better," which builds on vulnerability research and the theory that the post-disaster context offers a window of opportunity for disaster risk reduction and improved re-development. In this sense, the recovery period is seen as a tool for implementing policies and programs designed to remedy the weaknesses in developmental policies, infrastructure and institutional arrangements.[3] Evaluation findings on the newly initiated response programme together with other humanitarian and development principles, Bangladeshi concept of EDR could respond more efficiently provided following issues were given due considerations and three core components formed a basis for planning and implementation as well.[4]

- *Recovery comprehensive.* Early Recovery (ER) is a complex process which requires cross-cutting nature of multi-sectoral interventions towards building back better. This calls for integration and mainstreaming the principles of ER in all development and recovery interventions.
- *Recovery early.* That support should be provided to the affected households (HHs) and local economy to recover as early as possible. There should be

[1] EM-DAT: The OFDA/CRED International Disaster Database (2012); for further details see http://www.emdat.be/Database/CountryProfile/countryprofile.php

[2] Natural Disaster Hotspots Case Studies. World Bank (2005). Hotspots online mapping application (World Bank website) http://www.worldbank.org/ieg/naturaldisasters/maps/

Reducing Disaster Risk, A Challenge for Development UNDP (2004)

[3] Detailed theoretical analysis and discussion on Post-Disaster Recovery and Vulnerability by Erin Joakim is licensed under the Creative Commons Attribution-Non Commercial No Derivs 3.0 Unported License. To view a copy of this license, visit http://creativecommons.org/licenses/by-nc-nd/3.0/

[4] Khurshid Alam (2012): Lesson Learning Study on Early Recovery programme of UNDP Bangladesh: Potentials, Challenges and Policy Implications, Dhaka

organisational preparedness for assessment, planning and funding to facilitate such a process.

- *Recovery enabling*. The national and local context conditions ER planning may vary by nature and scale of disaster. But no doubt that affected people will have their own plan to recover their losses. The evidences clearly suggest that effectiveness of any recovery operation is higher when it is built on local capacity with affected people in driving force. Therefore, the key concept of any recovery to adopt is "recovery enabling" as oppose to recovery provisioning. The central idea of this thinking is a careful understanding of households' recovery process and ability and strengthening that. This requires thorough assessment, deeper understanding of location specific human-environment interaction for adaptation planning, higher quality engagement and nurturing support.
- *Recovery resilient*. Considering Bangladesh's disaster risk—the recovery operation should focus on sustainability, self reliance and resilience building as an opportunity of building back better.
- *Recovery confident*. This needs an understanding and address of disaster related traumas that people experience after a disaster. Experience from Sidr and Aila suggest that people's ability to re-start their livelihood has greatly conditioned their mental and physical well-being as well as building poor people's analytical ability to understand implication of climate change on their existing risk.

Therefore, any early recovery operation should consider three internally cohesive components to achieve desired impact:

- Enabling household to recover—enhancing confidence building through psychological support, ensure basic needs assistance, reduction of incidental cost, providing income opportunities and strengthening security.
- Creating enabling environment—such as market, infrastructure, social and economic opportunities including practice of right based approach for victims.
- Investing on governance to recover and enable to coordinate recovery planning and implementation.

Even after 2 years, most people living within prolonged inundation were not able to rebuild their houses and livelihood. But areas where water receded in 3 weeks time—people had their houses built and many started their livelihood. Broadly, the economy of Aila affected area is based on fish and rice—which got heavily affected and people were not been able to get back on feet again even after 2 years. Secondary impact of the disasters on households and local economic levels was significant due to inadequate past response to recovery. Most households visited by the external team were indebted. This was also due to limited crop yield and reduction in purchasing capacity that had an impact on local business. More women were engaged in fish fry collection today—an indicator of cash need of the HHs.

13.1.3 Post Aila Economic Recovery

- Market emerged as a powerful actor that recovered itself through formal and informal cooperation between business people; big business provided materials and cash to small retailers to re-open their business.
- Microfinance played an important role at both HH level and local economic levels; Some schemes introduced soft loan by partner NGOs of MFI, played a positive role for local business to get back on their feet again.
- External assistance was unpredictable—"something was provided suddenly"; recovery support was spread over a long time even though many of them were useful; if provided within the first 6 months people would have recovered much earlier;
- Development partners support in some cases was smaller/inadequate that what was required by the beneficiaries to recover their business meaningfully;
- Contribution to HH's liquidity need—incidental costs such as health, children's education limits HH's ability to produce surplus for investment in recovery

13.2 The Resilience Approach of Early Disaster Recovery (EDR)

In Bangladesh, UNDP first utilized EDR principles in its programming in 2007 for post Sidr and Aila disaster response. It launched Early Recovery (ER) working group, helped government in the assessment and formulated recovery plan while managing a large scale livelihood and shelter programme. Government and humanitarian actors accepted this concept as dependable contributor with new thoughts. The most significant contribution of UNDP at this stage was promotion of resilience principles in infrastructure development in shelter, resilient habitat and enterprise. Such an idea in practice helped UNDP to shape concept of "building back better." This forward looking principle had an influence over the Aila response as well where UNDP initiated resilient habitat and alternative livelihood in the context of climate change.

The resilience principle was applied consistently with some exception in Aila livelihood recovery. It was used more in the context of physical infrastructure than softer aspect of recovery operation. Within livelihood, it was felt necessary to pay more attention to asset protection and develop the concept of resilient livelihood based on the lesson from Aila. The notion of differential vulnerability caused by inequality should be systematically included in assessment, planning and implementation.

13.2.1 Rationales of Advocating for DRH

In coastal areas, concentration of people under poverty line is comparatively higher than rest of the country. Potential impact of natural disasters seems to be greater on the coastal communities as they are becoming more vulnerable because of increasing trend of occurrence of hazard and extreme events. Vulnerability is as important a cause of disasters as the physical events that trigger them. Poor people's vulnerability is often increased when development goes wrong. Thus, development is a contributing factor in the occurrence and scale of disasters. At the same time, disasters, when they happen, cause serious setbacks to development. To get out of the vicious circle, more attention will have to be paid to mitigation and tackling the causes of vulnerability. Formal approaches to mitigation, initiated mainly by the public sector, have often been inefficient and at times have left people more vulnerable. A successful alternative approach, community-based disaster mitigation, can reduce vulnerability by engaging popular approaches, local knowledge and social capital, whilst addressing their weaknesses. Some examples of community-based mitigation are derived as lessons: learn from the past, build relations with communities, encourage participation, involve local builders and artisans, build local capacity, document and share lessons, and influence formal education. Integrated Disaster Preparedness combined with capacity building is a key factor for poverty alleviation and for improved daily life.

13.2.2 Building Disaster Resilient Community[5]

It is becoming increasingly apparent that the future of disaster management is preparedness at the community and household level. Many of the 2,400 cyclone shelters in the southwest sit empty because builders did not consult residents, who then felt little stake in their upkeep or use, noted the local NGO, Bangladesh Disaster Preparedness Centre. "We know how to build strong houses; the challenge is building stronger communities," said the new disaster-resilient project's lead architect.

UNDP Bangladesh as a part of research and development supported with the policy concept and grant resources to the tune of US$130,000,—an amount close to the cost of a standard cyclone shelter to demonstrate resilient habitat in collaboration with the NGO, BRAC and Architecture Department of BRAC University in partnership with the community. Two additional disaster-resilient communities are planned in nearby villages. Planners expect these constructions to cost half as much as the first one. The new residents are encouraged to take part in identifying disaster threats, mapping escape routes and helping to build the homes. In the process of building these habitats, the community has grown stronger, allowing for a quick and coordinated response in the event of a disaster.

[5] Adapted version quoted from IRIN, 9 November 2011

13.2.2.1 New Settlement Concept to Meet the Emerging Climate Change Challenges

Traditional settlement pattern in the coastal areas are no more climate resilient. Poor and displaced households are compelled to live in the vulnerable zones out of the protection. Particularly location of shelter and home of the poor are scattered in the vulnerable zones. It is difficult to reach them with warning message and evacuation by the volunteers in critical moments. Their life and livelihood is not secured at all and fully depends on the mercy of nature. These poor isolated households in the coastal vulnerable areas are excluded in most cases from the public service facility and even their children hardly go to school. These isolated households are out of micro-credit coverage and health care services.

One way to increase the demand for sustainable buildings is through the description, assessment, and communication of their economic advantages and reduced risks in comparison with conventional buildings. These benefits can also be expressed through favorable lending and insurance conditions. However, a precondition is the integration of sustainability issues into the processes used by the financial and insurance industries for assessing property assets (e.g. risk assessment, rating, and valuation). Thus, the valuation and property rating process itself and those involved with it have a key role.

Spatial planning and land management provide various tools to reduce risks and prevent natural hazards while undertaking resilient habitat initiative. The prevention of catastrophes in general is a consideration of spatial planning and land management on the regional and local level. Therefore a more active role of planning and land management is necessary. They have to support a sustainable settlement development and a sustainable land use on consideration of the different public and private interests because of their important influences on environmental disasters.

Strategies of coastal regional planning is necessary, which attain a concentration of settlement within a poly-centric structure, an optimized building density and density of population, a variety and mix of coastal land use, private and public spaces with high ecological and social quality and a transport system compatible with the environment and the cluster settlements. This kind of settlement structure will support at the same time the prevention of environment and climatic hazards and mitigate the negative effects. Apart from the statuary framework, the regional and the communal level have a great legal capacity to prevent environmental disasters by the following tools and strategies:

- Strengthen the decentralized disaster resilient cluster habitat development with poly-centric structure;
- Assessment framework to choose suitable location and spaces of the future settlement development apart from tidal zones, river valleys, natural retention areas and unstable slopes;
- Space saving development of new residential areas with mixed use and increased density;
- Cooperation between urban and rural areas in the fields of settlement, infrastructure and protection of the environment

- Reduction of soil sealing
- Safeguarding and efficient realization of coastal cluster settlement concepts,

13.2.3 Learning from Cyclone Sidr 2007: Build Disaster Proof Settlement

- A good number of people in the vulnerable coastal area did not believe the cyclone Sidr signal as previous warning was not effective at the last minute.
- Since there is no provision of cattle/livestock shelter in and around the cyclone shelter, many of the hard core poor families did not go to cyclone shelter as they do not want to leave behind of their cattle being source of their livelihood.
- Poor women headed household, families having physically handicapped person, old and disabled living far away from the cyclone shelter were not interested to move for a safer shelter.
- Accommodation capacity of cyclone shelters are inadequate and not gender friendly.

13.3 Salient Features of the Disaster Resilient Eco-Habitat

Structural Safety:

- Cyclone resisting structural design
- 100 years tidal surge safety measures
- Saline proof structure
- Decision Support Systems; integrating stakeholders for assessing vulnerabilities for critical infrastructures

Adaptation Interventions:

- Renewable energy
- Rain water harvesting
- Growth Centre based shelters
- Air to water technology
- Bio-gas digester and
- Common grazing land

Social Interventions:

- Common recreational facility
- Children's school program
- Health Care
- Educational Programme (schools…)

Livelihood supportive Interventions:

- Common grazing land and mini-dairy farm
- Common pond for aquaculture
- Common production centre (handloom, tailoring, handicrafts, bakery etc.)
- Kitchen garden/backyard farm

Early warning systems are made up of, and rely upon, four main elements:

- Observation and recording
- Risk knowledge and recognition
- Warning and dissemination
- Appropriate response

Sustainable Land and Water management aspects:

- Land and water are treated as precious resources, so that the natural and beneficial functions of floodplains, wetlands, and coastal areas are protected
- There is a natural mitigation of flooding
- The market favors sustainable development, so that flood prone construction rarely occurs
- New development is designed and built to have no adverse impact on flood levels, sedimentation, erosion, riparian or coastal habitat, or other community designated values, and is responsive to climate change

Develop hazard mitigation strategies and technologies:

- Develop an understanding of the social, cultural, and economic factors that promote or inhibit adoption and enforcement of promising mitigation strategies or technologies;
- Develop outreach and training programs to enhance state and local government capacity to adopt improved mitigation strategies and policies;
- Develop strategies for mitigating negative impacts on coastal zone ecosystems;
- Develop improved and more accessible mitigation strategy models (e.g., HURREVAC computer software, risk and vulnerability tools, improved DEMs and maps) and other technical assistance to state and local governments that are adopting new mitigation strategies and policies;
- Develop a coastal inundation GIS system using information on historical and projected probabilities of various categories of sea level incursion to help identify socio-economic impacts of vulnerable regions/areas/populations.

Reduce the vulnerability of infrastructure:

- Model the impacts of events affecting the infrastructure, including the effects of disaster events once in a 100 years, waves, and coastal change (i.e., erosion, inlet formation);
- Examine the interaction between wind and inundation to determine the impact on building foundations and critical infrastructure;

- Focus research on new mitigation technologies for purpose of avoidance, resistance, rapid repair and restoration of critical infrastructure and other essential facilities;
- Model the impacts of changes in coastal zone ecosystems on infrastructure vulnerability;

13.3.1 Disaster Challenge as a Development Opportunity: A Case of Tourism Promotion

The communities in eco-habitat may explore the possibilities of tapping local tourist opportunities. The UNDP in partnership with Universities and local NGOs, through the Parjatan (Tourism) Corporation and District Development Committee, may initiate a "Tourism for Rural Poverty Alleviation in Coastal Vulnerable Zones" programme as part of which they are exploring the possibilities of ferrying tourists between the place of interests with sharing local wisdom and culture as well as local marketable resources such as natural mat, cap, handicrafts, herbal medicines, dry fish and so on..

13.4 Changing Policy and Practice

A major objective of the programme has been to demonstrate affordable, replicable and sustainable practices in disaster risk reduction which can be incorporated into government and NGOs policy and practice. To this effect UNDP will concentrate its efforts on influencing pro-poor disaster resilient settlement with focus on sustainable livelihood policy at horizontal and vertical level addressing community to Government and the practices of actors at the Disaster Management Committees at Union, Upazila, District and National level. It has also aimed to demonstrate the applicability and relevance of its technological solutions and ways of working both with and through communities.

13.4.1 Policy Options

- Disaster risk reduction in terms of saving life, property, assets and relocation.
- Learning by establishing an ecologically sound disaster resilient habitat with multiple advantages such as renewable energy facilities, safe water supply/rain water harvesting arrangement, health and recreational facilities are socially welcoming, economically feasible and environmentally acceptable venture with livelihood opportunities.

- Expansion of the strategic role of coastal vulnerable regions and "gateway places," giving particular attention to the development of peripheral regions.
- Improvement of the economic basis, environment and service infrastructure of cities, particularly in economically less favoured regions, in order to increase their attractiveness for mobile investment.
- Promotion of an economic diversification strategy in shelter sites which are too dependent on a single branch of economic activity, and support for the economic development of coastal areas which are less favoured regions.
- Promotion of integrated settlement development strategies sensitive to social and functional diversity. Particular attention should be given to fighting social exclusion and the recycling and/or restructuring of underused or derelict rural sites and areas.
- Promotion of a wise management of the coastal ecosystem.
- Promotion of better accessibility in neighbouring areas through an appropriate location policy and land use planning that will stimulate mixing of service functions and the use of public service including transport.
- Support for effective methods of reducing misuse of precious land; reduction of excessive settlement pressure, particularly in coastal regions.
- Policy should be based on an integrated approach to new developments and contribute to the creation or restoration of attractive landscapes.
- Better co-ordination of spatial development policy and land use planning with sustainable use of natural resources and telecommunications planning.
- Promoting the interconnection of inter-modal junctions for freight transport, in particular for transport.
- Coordinated and integrated infrastructure planning and management for avoiding inefficient investments and securing the most efficient use of existing infrastructure.
- Wide-ranging integration of knowledge-relevant policies, such as the promotion of innovation, education, vocational training and further training, research and technology development, into spatial development policies, especially in remote or densely populated areas.

13.5 A Self Sustained Resilient Habitat

This concept tries to put together all possible risk reduction, elements of resilient ecological landscape and socio-economic provisions to address disaster and climate change risks, create opportunities for access to wider natural and environmental resources, and poverty reduction in an integrated manner. One step would be to design and begin some work along these lines on a modest scale. While such an initiative would have some original features, we should not pretend that it is entirely novel. We are confident that the fields we have identified have not been addressed in an integrated way before. On the other hand, similar ideas of a more integrated approach have been proposed and tried before in other domains and other contexts. Perhaps a more systematic effort is needed to draw upon these experiences. Model innovation demonstrated based on the principles discussed are piloted in super cyclone Aila affected area under Early Recovery Facility (2011) and Comprehensive Disaster Management Projects supported by UNDP and its Development Partners in Khulna (Dacope) and Satkhira (Shyamnagar).

13.6 The Concept of the Disaster Resilient Habitat (DRH) in Practice

The core idea behind the DRH is the ability of the habitat and its occupants to spring back to normal life as soon as possible after the occurrence of a disaster, in this case cyclone and storm surge. While the concept of Multipurpose Cyclone Shelters remains, the DRH looks at each structure in the habitat as a cyclone shelter of sorts (Fig. 13.1). The structures and the infrastructure of the DRH should have the capacity to resist cyclonic winds and storm surges to a certain extent and that

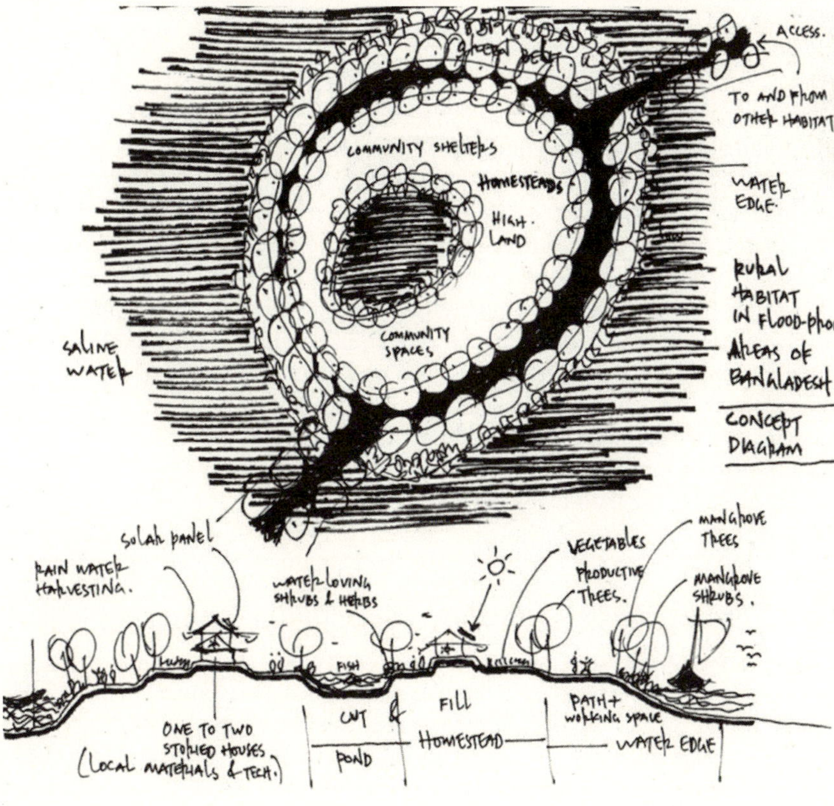

Fig. 13.1 Concept of the disaster resilient habitat (*source*: Hasibul Kabir)

Fig. 13.2 Post Aila landscape (photo: Tahmina Rahman)

there should not be total destruction. Apart from the structures themselves the roads and embankments should also not suffer total destruction and should be able to operational after a disaster.

To be able to make a habitat resilient to disasters, it is important to involve the community. Although it could be a top down processs as well, like the cyclone shelter, participation was considered important to allow a sense of belonging and take care of replication issues. Participation would also induce certain skills into the community, skills that would be useful when they add to the structures.

Engagement of the community is also to make them self sufficient in other helping other coastal communities to build DRH's of their own. Thus as the skills are transferred from community to community the vision for having all communities along the coast having their own DRH can be realized.

13.6.1 From Concept to Reality; The Satkhira Experience

After cyclone Aila some habitats in the coastal areas became completely inhabitable (Fig. 13.2). This was not so much for the destruction to the habitats Aila had caused but more because of effects of the embankments that had collapsed as a result. These embankments had been made weak because of the holes dug in them to let in sea water for shrimp farming. The weakened embankments collapsed and sea water came in and the habitats also became subject to the daily tides. Saline water had made agriculture impossible and the inundation made earning a livelihood nearly impossible.

UNDP identified in collaboration with the District and Local administration one village called "Cluster Village for Landless Families" as a potential location for constructing the DRH. This habitat was like an island in saline water and the high tide inundated its surroundings. In the UNDP survey there were 55 homesteads that needed to be rebuilt. The salinity all around made the site unfriendly for construction activity. Transportation was difficult and the soil unfriendly for construction.

The homeowners here had been given the land on which there homes stood as a grant for a resettlement project and they could not sell this land but had to live on it. The choice of this village was therefore good since developments on the homeowners lands could not be sold to someone else.

The people here were very poor and had little money to make meaningful improvements to their houses. The only resource that could be tapped was their labour i.e. participation in the construction. The situation automatically called for a participatory approach, something that the project was conceived on.

Building materials would have to be brought over difficult terrain and the participation by all was absolutely necessary. Whatever skills these people had, had to be shared. Some difficulties were also envisaged since part of the structure was to be in reinforced concrete, which required strict quality control measures.

13.6.2 Process vs. Product

Right from the beginning it was decided that a fixed design would not be imposed on the site. The DRH would evolve through a process that was to be initiated on site. Past experiences in housing reconstruction in cyclone Sidr affected areas where a preconceived design had been imposed had yielded unsatisfactory responses from the users.

The process of the design would evolve in to an acceptable product. It is only when there is ownership of the product that it will be useful. Being involved in the construction of one's own home would also ensure quality as would step by step supervision. Certain problems may evolve during the construction phase that had not been envisaged in the design phase. Solutions to such problems would also be site and experience based.

A combination of technical and indigenous knowledge would be used in the design process. Architects and engineers would provide the technical knowledge and do the necessary calculations. The local people would provide their knowledge in local building terms and a right combination of both would be used in optimizing the final design solution.

Fig. 13.3 Workshop on site (photograph: Hasibul Kabir)

Fig. 13.4 Workshop at the department of architecture, BRAC university (photo: H. Kabir)

13.6.3 Engaging the Community

A team of architects from the Department of Architecture, BRAC University made a reconnaissance visit to the site in August 2009. This was the initial contact with the community. A workshop was held in the site where the householders were asked to design their dream house with no parameters attached (Figs. 13.3 and 13.4). This would provide an insight into the aspirations of the community. A wide variety of

Fig. 13.5 Concept sketch of the house and possible development (Hasibul Kabir)

solutions were developed, some practical others visionary. It was seen that almost all the house solutions were raised above ground with space underneath.

In the next step representatives from the community were invited to the Department of Architecture in Dhaka and another more practical workshop was held over 3 days. Here the skills of the community were identified more precisely. It was discovered that they had good carpentry skills because quite few of them were boat makers by profession. A mason who had been engaged in another project in developing stronger houses for another rural community provided his inputs from his practical knowledge. The engineers developed a frame design in reinforced concrete on which the houses would be built. The community was introduced to the concepts in reinforced concrete construction. Thus the gaps in skills of both the professionals and the community were identified and also filled in with the expertise of both. A general idea about how the DHR would be constructed was now developed. However the design was not finalized. That would be done when faced with the practicalities of the site at the site itself.

13.6.4 Design Developments

The design solution developed after the workshop consisted of a reinforced concrete frame 10 ft by 13 ft and of a height of 8 ft. On this frame would be attached another timber frame which would hold the house. This called for careful detailing of how the timber frame would be fixed to concrete frame. The structure on top would essentially be the mini shelter while the homeowners would be free to build whatever they wanted to underneath and on the sides on their own (Fig. 13.5).

The reinforced concrete frame design had to incorporate into it considerations special to the site. Since salinity was a major issue and it has been seen in other reinforced concrete structures in the locality that the reinforcement bars inside the concrete would rust and expand, extra clear cover from the outer faces to the steel bars was allowed. Also a chemical mixture was to be introduced into the concrete for resistance against salinity. Casting concrete in such a saline location had its own problems but it was managed on the site. Timber was available in the site and this was from timber plantations and not from the forests. Extra strengthening details were incorporated in the timber frame, such as cross bracing and closer purlins at the edge of the roof. The roof material was not decided but several options weighed. It was decided that local solutions would be looked at in order to arrive at a final decision

13.6.5 Building the DRH

The basic design decisions being taken the team from Dhaka moved to the site to start construction. The part of the design that was finalized was the concrete frame. The timber structure on top was designed but was open to interpretation depending on conditions at site. Some parts of the construction needed some expertise such as cutting and binding of reinforcement bars, mixing of concrete etc. Specialized labor would be hired for this, but for the rest of the construction the people of the community would be utilized. The team of architects, engineers, researchers and students would be there to help and to make drawing and document the process. Accountants from BRAC the NGO would deal with money and accounts. As mentioned communications to the site was difficult and cement, sand, steel were all brought by boat during high tide times.

13.6.6 Problems and How They Were Overcome

As mentioned earlier the site is difficult to reach and materials could only be carried by boat at high tide. After locating the site for the test house, the first thing that was needed to be done was to dig the trenches for the foundations of the columns, which were done by laborers. The laying of bricks at the bottom over a layer of sand was entrusted to the house owner. The reinforcement bars had to be laid in place and this required skilled labor as far as cutting and bending the bars were concerned. Smaller tasks such as tying the bars could be handled by the locals (Fig. 13.6).

The main problem was concrete casting in saline conditions. Saline water cannot be used for casting and fresh water is required. After several meetings to decide how to handle the problem it was decided that it would be the owner's responsibility to ensure that fresh water was used for casting. Since the house was his or her own this decision ensured that quality would be maintained. Post casting curing i.e., keeping the cast concrete wet after the shuttering was removed also needed fresh water and this was again made the responsibility of the house owner (Fig. 13.7).

The roofing was a problem in the sense that there were very few options. There wasn't enough money to make concrete roofs although that would be a good option. The most preferred option by the house owners was corrugated iron sheet roof but the material is thermally inappropriate and in the event the roof is detached

Fig. 13.6 A student volunteer and local lady tying reinforcement bars (photo: Hasibul Kabir)

Fig. 13.7 The test house (photo: Hasibul Kabir)

due to very strong winds there is the danger of its sharpness in resulting of injuries. The local material most commonly used is "golpata" a kind of reed but was not preferred by the community. While researching the surroundings for a suitable option it was found that there was 150 year old structure with a tile roof that was unaffected by previous cyclones. One closer look it was seen that each tile had a small hole in the center and the entire roof was "stitched" together. It was decided to adopt this option but instead of "stitching each" tile was screwed on to the wooden frame below. This ensured that tiles wouldn't fly off in high winds.

13.7 The Concept Realized

The DRH now stands as a linear village with 43 houses standing on stilts. In one case two houses are attached where the two owners paid for the attachment. There is school building in two levels, with one side having raised earth up to the upper level. This is to bring in cattle if a cyclone warning is issued. The school is a replacement for the one that was completely destroyed by Aila (Fig. 13.8).

It is interesting to note the adaptations the owners have made to the houses. In most cases the owners assigned little use for the house on stilts. Instead they use the space underneath and have constructed rooms and have added structures around it. These changes are constantly happening and will continue to do so. The intention of the DRH is thus realized where each house is a mini cyclone shelter of sorts. The house on stilts can, not only accommodate the household it belongs to but also another small family from a neighboring village if needed (Fig. 13.9).

There isn't much difference in the way the DRH was conceptualized and the way it has taken shape. It was part of the concept that the homeowners would be actively involved in the design and construction of the habitat. They were not only involved in the construction but have now taken it on themselves to add to the structure to suit their needs. The DRH has bought about some changes in the social fabric of the community. There are no recorded observations but in conversation with people it was seen that they are more confident on their own abilities and have brought about developments in their livelihoods. The security of a home is one possible reason. On the down side the aid reliance attitude of the people has been affected. Since they are now house owners and apparently in good living conditions, they are denied aid by agencies. A part of the population was very aid dependent and they now consider their improved status to be detrimental to their previous ability to get aid.

The success of the DHR also depends on how well the houses are maintained. The lower parts of the concrete stilts have been protected well against the elements. The house itself is made of timber and is attached to the concrete frame. The joinery here needs maintenance and upkeep. This requires the people to do periodic maintenance work, which they are expected to do. It is till too early to comment on these issues. The community is now experienced in construction and management issues and it was envisaged that with this experience they will be able to contribute to future efforts in building DRHs. A few other such habitats have been built by other agencies but none of them applied the participatory methods used here in a comprehensive sense.

Fig. 13.8 The school (photograph: Fuad Mallick)

Fig. 13.9 A house modified by the owner (photo: Hasibul Kabir)

For coastal communities to build such a habitat on their own is difficult to comprehend since most don't have the economic means. Any future construction of similar nature will require an element of subsidy. The entire cost of the habitat was less than that of a 2,000 capacity cyclone shelter. Money from such cyclone shelter projects could be diverted to efforts in constructing DRHs to make the coastal area a safer place to live (Fig. 13.10).

Fig. 13.10 The disaster resilient habitat (photo: Hasibul Kabir)

13.8 Beyond the Benefit Innovation on Disaster Resilient Habitat

(a) Disaster resilient habitat is an integrated approach to address the poverty, disaster and climate risk reduction ;
(b) Manage environment, biodiversity and ecological aspects to enhance natural resilience as well of the human habitat;
(c) Renewable energy and rainwater harvest system will add value to living comfort with more productive facilities;
(d) Get full preparation with the secondary protection of the cluster village from the cyclone and storm surge from false security in case sea front protection fails;
(e) No need to relocate frequently while early warning served and household assets including livestock, poultry, duckery and crops could be protected in place.

13.9 Concluding Remark

The combination of technical expertise and local skill would emerge as creative solution to disaster recovery and risk reduction endeavor as well. The disaster resilient habitat is a combination of both hardware and software along with people's participation resulted a more structurally and ecologically sustainable solution than cyclone shelters. Perhaps in the future huge investments in top down cyclone shelters can be better utilized in this bottom up sustainable approach of comprehensive risk reduction. This innovative approach to implement disaster recovery requires further thought and debate from the perspective of geographical pattern of nature-human interaction, disaster type, frequency and magnitude and its trend over time and place. The nature of response to disaster suggest that current approach tend to emphasize on "provisioning approach" based on the idea that external support takes primacy over spontaneous effort of the community or the affected population. Hence, policy priority and management strategy should give attention to local risk reduction and recovery capacity building with focus more on mobilizing technical inputs from experts, learning and education, technology transfer and governance for creating enabling environment those produce some solid knowledge on resilience technologies. This study would like to highlight on the recovery enabling approach where government, academics, researchers and development partners as well as policymakers role would be much more as supporter to ensure the lead role of affected communities. A number of priorities can be considered: formulating advance policy/strategy for large scale disaster response, especially for urban areas with strong focus on governance. Disaster recovery needs coordinated effort with engagement of multi-sectoral and multi-agencies work closely linked to disaster management and response for greater synergy. It is also important to consider capacity of the local government on recovery planning so that national recovery planning can evolve from the local one after each disaster.

References

Alam K (2012) Study report on lesson learning study on early recovery programme of UNDP Bangladesh: potentials, challenges and policy implications, Dhaka

Bollin C, Khanna S (2007) Review of post disaster recovery needs assessment methodologies. UNDP

Early Recovery Facility (ERF) (2011) UNDP in Bangladesh. Available in www.undp.org.bd/projects/proj_detail.php?pid=84

EM-DAT: the OFDA/CRED international disaster database (2012). For further details see http://www.emdat.be/Database/CountryProfile/countryprofile.php

Reducing Disaster Risk, A Challenge for Development UNDP (2004) Bureau for crisis prevention and recovery. United Nations Development Programme, New York, USA

World Bank (2005) Natural disaster hotspots: a global risk analysis by Margaret Arnold, Maxx Dilley, Uwe Deichmann, Robert S. Chen, Arthur L. Lerner-Lam. http://www.worldbank.org/ieg/naturaldisasters/maps/

Chapter 14
Incorporating Traditional Knowledge in Post Disaster Recovery to Integrate Climate Change Adaptation and Disaster Risk Reduction

Anshu Sharma and Sahba Chauhan

Abstract The most visible and extreme impacts of climate change are being felt through hydro-meteorological disasters. These disasters can be avoided and mitigated through integrated climate change adaptation and disaster risk reduction approaches. Traditional knowledge is a rich source of time-tested solutions for early warning and locally appropriate coping mechanisms in times of disasters. At the same time traditional knowledge is also a valuable source of information about local climate systems, longstanding adaptation practices and adaptive capacities of vulnerable communities. Consequently, traditional knowledge can aid the design of highly customized and integrated climate change adaptation and disaster risk reduction solutions that are locally contextualized, appropriate and sustainable. In this chapter, we present two case studies of post hydro-meteorological disaster situations from India to show used and misused opportunities for incorporating traditional knowledge in post disaster recovery to integrate climate change adaptation and disaster risk reduction. Post disaster situations present a unique opportunity to facilitate integration of climate change adaptation and disaster risk reduction because of increased levels of sensitivity and willingness to engage in risk reduction from all major stakeholders. The case studies presented in this chapter, explore disaster recovery programmes of the 2006 flash floods in Barmer, Rajasthan and 2010 flash floods of Leh, Jammu and Kashmir. Both these districts lie in highly arid and climate sensitive agro-ecological zones of the country. The unprecedented flash floods in these usually drought prone districts and consequent disaster recovery efforts provided us a unique opportunity to study and demonstrate the role of traditional knowledge in integration of climate change adaptation and disaster risk reduction. We conclude that traditional knowledge can be used to understand and facilitate the integration of climate change adaptation and disaster risk reduction in

A. Sharma (✉) • S. Chauhan
SEEDS India, Delhi, India
e-mail: anshu@saferworld.in

R. Shaw (ed.), *Disaster Recovery: Used or Misused Development Opportunity*, Disaster Risk Reduction, DOI 10.1007/978-4-431-54255-1_14, © Springer Japan 2014

post disaster recovery and unless this is done it is not only a missed but misused development opportunity.

Keywords Arid zones • Climate change • Disasters • India • Risk reduction • Traditional knowledge

14.1 Understanding the Relationship Between Climate Change, Disasters and Traditional Knowledge (TK)

Extreme weather events are perhaps the most visible and destructive consequences of climate change. More and more people across the world are now experiencing "unknown" and "unimaginable" summers or winters, rainfall, droughts, floods and cyclones (Holland 1989; UN 1994; Munich Reinsurance Company 2001; Van Aalst 2001; Anderson and Bausch 2006; Webster et al. 2008). Increasingly these extreme weather events are turning into disasters with unprecedented economic damages due to a lack of preparedness or incapacity of communities to deal with such unprecedented conditions (Stern 2006; UNISDR 2011). Those already affected by and vulnerable to disasters because of poverty, poor health and other forms of social, economic, religious and political marginalization are also the most vulnerable to and least able to deal with the impacts of climate change (IPCC 2012). On the other hand, not all disasters are unprecedented or a result of climate change impacts or vice versa. Historically, human societies have coped and adapted in various ways to changes around them. Dealing with climate related disasters is therefore not entirely new. In particular, indigenous communities living in harsh local climate systems and highly disaster prone areas have accumulated a wealth of traditional wisdom and innovative practices that form a rich source of knowledge accrued over time to cope with disasters and adapt to environmental changes or constraints (Stitger et al. 2005). However, the existence of TK alone does not make communities immune to climate change impacts or disaster risks. This gives rise to various important questions,[1] for instance, why do communities with rich traditional wisdom and effective coping mechanisms remain vulnerable to climate change and disaster risks? And, why has TK not been formalized in designing response strategies to climate change and disasters? Also, given the obvious value of TK, how can it be used to reduce the risks of climate change and disasters in the future? The answers to these questions can also help explain further the relationship between climate change, disasters and TK. The first being that climate change much like disasters is the unforgiving outcome of unplanned and unsustainable practices that ignore ecological principles necessary for sustainable development, within which TK is understood to be beneficial for societies (Foster 2009). Secondly, social and political

[1] We acknowledge that there are many fundamental questions regarding TK some of which are out of the purview of the discussions in this chapter, however those pertinent in understanding the relationship between climate change, disasters and traditional knowledge are discussed above.

marginalization processes are the main reasons why people are made vulnerable and forced to live in high-risk zones and unfavorable environmental conditions in the first place. This marginalization process extends to the generation and use of knowledge. Often knowledge generated by the most vulnerable and marginalized sections of society remains marginalized and confined in use too, as it is of little value to those removed from the immediate risks.

Thirdly, climate change adds new dimensions and scales to disaster risks and to vulnerability from disasters in general and therefore requires immediate planned risk reduction and adaptation measures in response (Schipper 2007). These response measures must be developed using various forms of knowledge including scientific and TK. Traditional knowledge has the advantage of being a time tested and dynamic practice of knowledge accretion and resource building that a community inculcates over a period of time and is based on observations, experiences, cultural, religious and political beliefs and behaviors that help assess vulnerability and adaptive capacities. TK can also help to address the root causes of the climate change and disasters by offering localized perspectives on how communities' relationship with their immediate environments develops over time and by offering solutions that are relevant and sustainable.

The interaction of TK with more formal and dominant forms of knowledge such as modern scientific knowledge has existed for a long time. The resurgence of the importance of TK as a vital repository of information in addressing sustainable development issues such as natural resources management and biodiversity conservation has been well recognised and documented (ICSU 2002; Strigl 2003; Gernier 1998). Increasingly disaster risk reduction and climate change mitigation and adaptation studies have also recognised the value of TK as it not only provides insights and tools for risk reduction but also incentives for sustainable development. However, its importance and use in understanding vulnerabilities to climate change and disaster risks, inherent and acquired adaptive capacities of at- risk communities and use in designing planned risk reduction and adaptation strategies still remains limited (Nyong et al. 2007; Mitchell et al. 2010).

In this chapter, we present two case studies from post disaster recovery programmes in India, to understand how TK can be used to plan more appropriate and contextualized planned climate change adaptation and disaster risk reduction strategies. We argue that post disaster situations present a unique opportunity to facilitate integration of climate change adaptation and disaster risk reduction because of increased levels of sensitivity and willingness to engage in risk reduction from all major stakeholders. The case studies presented in this chapter, explore disaster recovery programmes of the 2006 flash floods in Barmer, Rajasthan and 2010 flash floods of Leh, Jammu and Kashmir. Both these districts lie in highly arid and climate sensitive agro-ecological zones of the country. The unprecedented flash floods in these usually drought prone districts and consequent disaster recovery efforts provided us a unique opportunity to study and demonstrate the role of TK in integration of climate change adaptation and disaster risk reduction. The interventions look largely at physical planning issues related to housing and its ecosystem, but extend to softer components of climate change impacts and adaptation actions as a long term strategy. We conclude that TK can be used to understand and facilitate the integration of climate change adaptation and disaster risk reduction in post disaster recovery and unless this is done it is not only a missed but misused development opportunity.

14.1.1 Opportunities for Integration of Climate Change Adaptation (CCA) and Disaster Risk Reduction (DRR) in Post Disaster Recovery Programmes Through Use of TK: Case Studies from Barmer and Leh, India

In India climate change impacts are being felt through a rise in extreme weather events and increasing weather related development stressors including water shortages, losses in crop yield and rise in vector borne diseases (Seath 2009). Those living in fragile ecosystems of the country are feeling the severest of these impacts. Leh district in the upper Himalayan region of the country is one of the largest and remotest districts of India. In recent years Leh, which normally has a cold desert climate witnessed some of the worst flash floods in its memory. Other impacts of climate change in Leh are being felt through water stress, loss of grassland and locust attacks. In another part of the country, Barmer a district in Rajasthan's Thar Desert area also witnessed unusual flash floods that caused severe damages. Climate studies have shown that Rajasthan falls within the areas of greatest climate sensitivity, maximum vulnerability and lowest adaptive capacity in the country. Communities living in Leh and Barmer and other similar climate sensitive and remote regions of India have limited capacity to deal with these "new" dimensions of disasters and climate change impacts. Consequently the need to address climate change concerns with the aim of reducing disaster risks and enhancing adaptive capacity is high in such climate sensitive regions as losses due to climate change are already negating hard-earned development gains. The need to address DRR and CCA in national development policy is high also given that climate change is adversely impacting the lives and livelihoods of about two thirds of the country's largely rural and resource dependent population (Ganguly and Panda 2010). While the Indian government professes its high vulnerability to climate change and is also party to a number of international treaties and development plans, including the HFA, UNFCCC and the Bali Action Plan that formally acknowledge that integrated DRR and CCA actions can help reduce vulnerability, its adaptation investments remain focused on technological reforms to enhance the efficiency of energy use, including efforts to improve the air quality in major cities and enhancing afforestation (Ministry of Environment and Forest India 2002). On the other hand, the National Disaster Management Policy professes mainstreaming of DRR, remains a fragmented domain. Significant work is yet to start on mainstreaming of climate change adaptation, and it's linking with DRR.

There is a need to study the effectiveness of local multi-stakeholder action as an enabling factor for mainstreaming DRR-CCA in post disaster programs and ultimately in state and national policies. The impact of consolidation of local change agents, their enablement through knowledge tools, and their strategic actions as an institution as enablers for Shifts towards linking DRR and CCA and mainstreaming DRR CCA actions into development processes are going to be critical to address the rising risks and vulnerabilities in the country. All of these will lead to a set of influencing factors that can enable local actors be more effective in influencing programs and ultimately policies towards mainstreaming DRR CCA. However, this conceptual understanding does not reflect in humanitarian and development policy actions, especially at the national and local level, where it is much needed.

Policy action that supports the integration of DRR and CCA can be generated by

- Being able to identify and leverage "windows of opportunity" for policy action through processes of agenda setting, coalition building and policy learning activities in the context of climate and disaster resilience.
- Measuring the outcomes of this process by conducting an in-depth evaluation of these efforts to understand how systematic approaches to policy change can be replicated elsewhere.

Another related aspect to generating policy action towards integration of DRR and CCA is to be able to know the "extent" to which the integration of DRR and CCA in development policies has taken place.

The need to address the integration of DRR and CCA at district level arises because of two main reasons. The first one being that bottom up approaches facilitate consideration of activities at the local level, which is where the impacts of climate change, climate variability and extreme events are witnessed and where risks ultimately need to be handled (Olhoff 2011). The second reason is that the district level administrative units in India are the lowest tier of policy development where government policies and programmes converge. As the policy and strategy level closest to the communities, the district level is the best suited to bring about upward filtered transformation.

14.1.2 Rationale

14.1.2.1 The Post Disaster Window of Opportunity

Post disaster relief and recovery work demonstrates the disconnect between environment centric vulnerability and technology centric post disaster action. At the same time, the post disaster situation opens windows in the form of suddenly increased levels of sensitivity and willingness to engage both from local governments and the civil society. These factors make the post disaster context the best suited to study, demonstrate and tap the relationships between local actions and strategic and policy environments.

14.1.2.2 The Cases of Barmer and Leh

Barmer receives an annual average of 277 mm or rainfall. In August 2006 it received 577 mm of rainfall in 3 days, leading to catastrophic floods of the scale and nature never ever recorded in the meteorological history of the region. The 2006 floods in Barmer, highlighted the need to take up concerted efforts on long term DRR and CCA. The Rajasthan State Climate Change Agenda (Rajasthan State Pollution Control Board 2009), which feeds into the state's environment policy, and the Rajasthan State Action Plan on Climate Change (Government of Rajasthan, TERI, GIZ 2011) are both steps in the right direction, but only begin to scratch the surface of the immense problem that the state faces. The position taken is largely mitigation centric, and though sustainable development is an agenda item, the integration of DRR and CCA is minimal. The case study of Barmer will thus provide a canvas to map the intense disaster impacts on the community and the specific points of disconnect in the state programs and policies.

Leh came into the DRR-CCA limelight after the 2006 and 2010 floods that were uncharacteristic for the region. Leh receives about 20 mm of rain, which has risen nearly ten times in the last 6 years (Behera and Vaswan 2008). Leh has made significant progress in integrating climate and disaster resilience in its policy actions to deal with the recent challenges faced by its communities. Innovations in Leh can provide examples to other parts of the country facing similar challenges and also for national and state level policy learning. A recent study by Ladakh Renewable Energy Development Agency (LREDA) has pointed out that Leh provides a good case study for exploring the ways in which adaptation to environmental change is taking place at community scale, both through the implementation of technologies and adaptation to suit extreme and changing conditions and through decision making by adaptive and resilient communities (Daultrey and Gergan 2011).

14.2 Barmer: Flashfloods in the Desert

Barmer, a district on the southwestern boarder of Rajasthan had particularly low literacy rates and extremely adverse sex ratios. The district of Barmer is the westernmost district of the state of Rajasthan, India. Located along the border of India and Pakistan, this district falls completely under the Thar Desert region. Heavy monsoon rains that started on 16 August 2006 in the district and its surrounding areas engulfed about a hundred villages. The local community is characterized by sparse and scattered living. There are four to five circular structures in one cluster bound by a low boundary wall, which forms a family's abode and is called a *Dhani* in the local language. Each structure is used for a different activity such as sleeping, storage, cooking and daily activities. A cluster of *Dhanis* constitutes a village. These communities are living in very harsh climatic conditions and making judicious use of the sparse resources available within their surroundings for their day-to-day requirement, and also for construction of houses. The population density of

Barmer district is among the lowest in India. Water is a major problem in this area. Village women walk long distances with headloads of pots to fetch drinking water, sometimes making more than one trip a day.

By midnight of 21 August 2006, Barmer had received 577 mm of rainfall in 3 days, 300 mm more than the annual average rainfall of 277 mm. Three villages of the district: Kavas, namely Malua, Bhadkha and Shiv, were the worst affected. The water level reached close to 30 ft above the ground level. Local communities and administrative system were not prepared for such an emergency situation and as result the floods took a toll of 139 lives while almost 50,000 people lost their houses. About 95 % of the families in the affected villages were rendered homeless. Even where parts of their houses were standing, they were rendered uninhabitable. Since the structures were mostly made of mud, many were badly damaged and destroyed by the flood. As the local people had never experienced this kind of disaster before, they were shocked and did not know how to handle the situation. Some local people thought that it was an act of displeased gods, while those linked to the scientific world pointed fingers at climate change. The impact of the damage was more because houses in this region are normally built in depressions in between sand dunes so as to protect them from sandstorms. This worked to their disadvantage as these low-lying pockets got flooded worst, and due to the impervious sub-soils, the water stagnated for weeks. The impact was felt more because the region is very sparsely populated and has very low level of infrastructure facilities, thus making access to services very difficult.

SEEDS carried out a damage assessment along with a study of the local natural and built environment immediately after the disasters. The team found that the traditional construction practices in the area were based on mud walls and thatch roofs, with circular shelter designs. This was very environment friendly as the materials created no ecological or carbon footprint and the houses were very conducive and thermally comfortable in the extreme weather conditions prevalent in the area. Moreover, the circular design protected the structures from strong winds and earthquakes and the construction processes were simple and suited to the local skill levels.

It was also realized that though the traditional practices of shelter construction were appropriate, they did have certain shortfalls in the water resistant capacity of the mud structures, due to which the houses had suffered severe damage during the floods. While traditional wisdom had provided a very high level of performance for generations, it needed the support of some technological intervention so as to help it face the challenges posed by unprecedented disasters linked to climate change, which will become more frequent in the future.

14.2.1 Indigenous Knowledge for Shelter and Sustainability

Communities living in rural Rajasthan are used to constructing houses with local materials and indigenous technology for many generations. For construction of their *dhani*, all the family members play a major role and have assigned responsibilities. While the men of the family collect soil of good quality from nearby places, the womenfolk gather cow dung, which they mix with the mud to prepare the basic construction material. The women of the family do the plasterwork for the new house, as well as for regular maintenance of the walls and floor. The roof is made by tying and weaving the dried stalk and by-product of the local *Jowar* crop. The house is oriented in such a way that the wind direction and sun path ensure good ventilation and thermal comfort, which is very critical since summer temperatures in this region reach about 50°C. Normally the size of the openings is very small as it reduces heat gain, and also gives less exposure to sand storms, which are a common local threat.

The people generally go for houses that are circular in plan and opt for lower heights. This is basically due to the location in the High Wind Velocity Zone due to which they face heavy winds especially during the summers. The circular plan helps to streamline the airflow with least resistance.

As this area also falls under moderate to high seismic zone based on the Earthquake Vulnerability Map of India,[2] the circular shape can also give good lateral resisting strength to house. During the 2001 earthquake in Kutch, Gujarat, which is very near to Barmer, very less damage was observed in houses with similar designs.[3]

14.2.2 *Survival and Propagation of Indigenous Construction Knowledge*

The indigenous technology for constructing shelters is widely used in the area and the community members themselves are the messengers for transferring this technology to the next generation. As all the members of the family are part of the construction activity, they have a sense of ownership of the shelter and an understanding of the materials and processes.

There are five main factors why this technology of shelter construction is still surviving in the remote desert areas and how it got disseminated to other communities in the larger region. These are shown in Fig. 14.1 and elaborated below.

14.2.2.1 Community Leaders Set an Example by Using this Technology

One of the typical traditions followed in any rural community in India is to see the way respected people in the community live and the rest follow their way of life.

---

[2] *Vulnerability Atlas of India*. Building Materials and Technology Promotion Council, Government of India, 2006.

[3] Local communities in parts of Kutch build circular houses with conical tiled roofs. These are called *Bhungas*.

Fig. 14.1 Factors for the survival and propagation of indigenous shelter technology

This is very common yet a very important aspect of community. In Barmer villages, most of the respected people in the community live in these kinds of *Dhanis*. Seeing this, other community members are encouraged to follow.

14.2.2.2 Community Involvement in Construction of Shelter

All the community and family members are involved in various activities of shelter construction. Involvement of family members as well as relatives eases the burden of construction and strengthens community spirit. This is also one of the reasons this technology is surviving in this still rural and tradition-centered area.

14.2.2.3 Extreme Climatic Conditions

Barmer is a witness to summer where temperature is as high as 50°C and winter where night temperature is near the freezing point. Concrete houses become ovens in the heat and chillers in the cold. There is no electricity and fuel is very scarce and unaffordable for thermal control. In order to survive in these extreme conditions, an appropriate house is required. Though some people have started opting for modern materials, they are not as comfortable in these modern houses as they are in traditional ones.

14.2.2.4 Availability of Local Materials at No Cost

Availability of local materials, which is free of cost and transportation, is a major attraction for a community already impoverished by inadequate livelihood options and a harsh climate.

14.2.2.5 Good Design for Safety and Comfort

A circular shape is capable of resisting wind pressure created by sand storms and wave pressure created by earthquakes. The walls are of insulating quality and are thick, giving good thermal comfort inside the house in both temperature extremes. Roofing is also properly connected to the walling system, giving higher structural safety to the shelter as a unit. The combination of safety and comfort has resulted in a time tested shelter technology that is respected locally for its immediate as well as long term benefits.

14.2.3 Indigenous Knowledge and the Support of Science

SEEDS visited the affected areas immediately after the floods and carried out a damage assessment along with a study of the local natural and built environment. The team assessed and documented the traditional construction practices in the area. These were found very environment friendly as the materials created no ecological or carbon footprint and the houses were very conducive and thermally comfortable in the extreme weather conditions prevalent in the area. Moreover, the circular design protected the structures from strong winds and earthquakes and the construction processes were simple and suited to local skills level.

The programme undertaken included the construction of 300 shelters under the *Barmer Ashray Yojana (Barmer Shelter Programme)*. Research was carried out on appropriate technologies for supporting the traditional construction system, and it led to the Stabilised Compressed Interlocking Earth Block (SCEB) technology, wherein local mud was stabilised with 5 % cement, and compressed into blocks that had high structural strength and water resistant capability.

In partnership with Christian Aid, and with funding from the Humanitarian Aid Department of the European Commission, the shelters were built using this appropriate technology, which was a mix of indigenous knowledge, and limited scientific inputs to make it further resilient in the face of new threats. Village Development Committees (VDC) were formed in each village to make decisions and to guide and monitor the construction process. The VDCs comprised of men, women, local leaders, school teachers, NGO representatives and project team personnel working closely with local government officials. The traditional circular designs and the "breathing" thatch roofs were retained. An efficient system was established to mass-produce the SCEBs very quickly to provide housing to the affected families in a span of 6 months. The house-owner families mainly did the construction with limited support from the project team. The knowledge and skills were left with local construction workers so that they can be replicated and scaled up in the region. Upon completion, local families preferred these traditional structures far more than the modern concrete technology based houses provided by other sources, which turned into ovens under the scorching desert sun. A review of the programme in 2012 revealed a high level of acceptance in the community for the improved Dhani structures as opposed to modern material and technology based houses that had been built in recent times.

14.3 Cloudburst and Flash Flood in Leh, 2010

During the night of 6th August 2010 a cloudburst hit the area of Leh, Ladakh, claiming hundreds of lives while several hundreds are still missing. Buildings were destroyed, communication lines broken and highways leading to Srinagar and Manali washed away. The current casualty figure as of 13th August stands at one hundred and eighty seven with approximately four hundred people missing. A total number of thirty four villages have been affected by the disaster with some regions still inaccessible. Foreigners have been airlifted back from such remote areas and food has been dropped. However, as these regions remain inaccessible accurate details of any casualties are still missing. Rescue work still continues and the final death toll is expected to increase over the coming days.

In terms of rainfall, the India Meteorological Department (IMD) recorded precipitation between 1.30 a.m. and 2 a.m. A weather station near Leh recorded 12.8 mm of rainfall in 24 h across an area of only 30 km^2. In August, the normal rainfall in Leh is 15.4 mm and the previous highest record was 51 mm. However this August's total rainfall is expected to be higher still. Ladakh Autonomous Hill Development Council chairman Chering Dorjay earlier estimated that 40 % of the infrastructure, which includes bridges, link roads and irrigation canals, has been destroyed. Another 40 % has been partially damaged and perhaps 20 % has remained intact.

Of the 34 villages Choglamsar, situated 6 km from Leh, was worst hit. A mud slide has passed the village either flooding the buildings or swiping them away. As time passes and rescue teams are still excavating and cleaning, the risk of disease outbreaks rises here. A secondary problem for several villages is the loss or blockage of irrigation canals, critical during this period some weeks before the autumnal crop harvest.

The Leh district, situated at an altitude ranging from 2,900 to 5,900 m is the most elevated and habituated district of the earth. Sparse population settled around the banks of major rivers/streams earns their livelihood through agriculture and cattle rearing. The district has inherent constraints for example prolonged winter, scanty rainfall, rugged terrain's limited availability of productive land.

14.3.1 Damages to Shelter

There is been different kind of damages in the existing houses due to heavy rains in short amount of time. Different kinds of damages have been observed based on the kind of rain water flowing from different areas. One of the major reasons of the damages to houses is the flowing water along with mud and rocks which exerted the lateral load on the buildings. Another reason of the large damages to houses is the use of mud blocks for walling as mud dissolves once it comes in contact with water. Locally people also use mud as mortar for construction which is also one of the reasons for failure of the walls.

Damages to corners of the walls because of absence of any lateral load resisting features in the houses. This is very common type of damage in the houses where mud blocks is used for the construction of houses. There is also large amount of damages to houses particularly in the rural regions of Leh because of use of mud as mortar for building houses.

There is also complete failure of the walls in some of the areas because of the heavy mass of the mud and rocks flowing in the houses and putting pressure to the out of plane walls which require small load as there were no lateral resisting members in the houses.

In some of the houses there are also damages in the openings because of the mud flowing with the rain water entering in the house through openings and most of the houses which came under the path of the rain water.

There is also complete failure of some of the RC frame buildings because of poor quality of construction coupled with large pressure exerted from rain water on the buildings. As RCC construction is fairly a modern construction there are lots of construction practices followed in the region which increases the vulnerability of structure due to poor quality control during the time of construction.

There are also wide damages to school building in the area as it was come in the path of the flowing water. In some of the places entire room of the school building was swept away due to pressure of water along with mud and rock.

14.3.2 Approach

Based on the past experiences in the field of disaster management and particularly in the field of shelter design and implementation we have followed very holistic approach in order to proposed shelter strategy. Different aspect covered into shelter design and implementation can be understood as per given figure below (Fig. 14.2).

As shown in the above figure different aspects have been covered under strategy of the interim shelter. As you can see there is a combination of both technical as well as social aspect of the affected community. Affected areas are multi disaster prone so we have to design our shelter by considering vulnerability of the area.

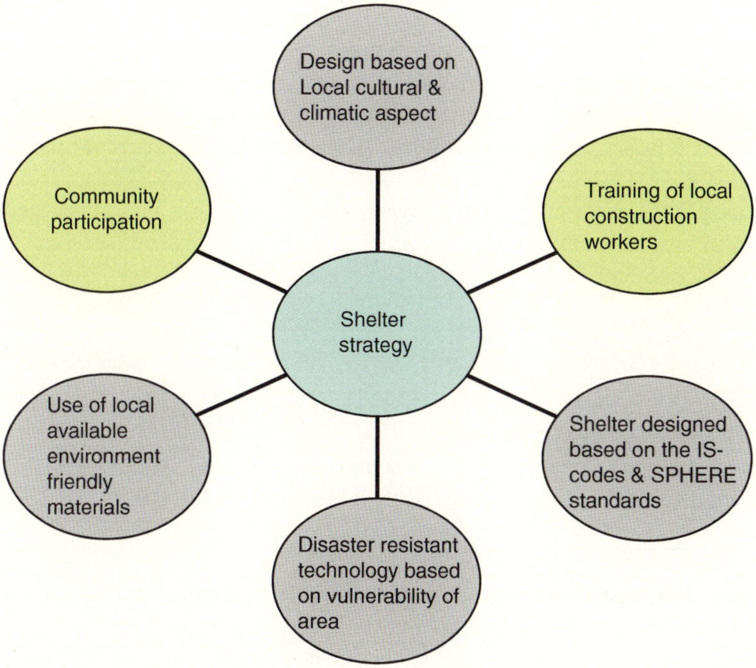

Fig. 14.2 Basic approach of shelter design

As there are lots of activities where affected community can take part actively which can also give them some livelihood options and they can easily overcome the trauma to start a new life again.

14.3.3 Traditional Houses

Traditional vernacular houses in Ladakh have evolved over centuries in response to the harsh climate and thriving Buddhist culture. Over the centuries these houses have allowed Ladakhis to lead a self sufficient and comfortable existence with limited local resources. An understanding of the traditional house, and how it responds to social and environmental factors, is a key starting point for any proposal for a successful shelter intervention.

14.3.3.1 Form/Orientation

Three storeys high
Entrance to house faces east (considered auspicious)
Houses clustered and often adjoining others (reduces heat loss)

14.3.3.2 Layout

Ground level: Stabling for animals (helps to warm upper levels of the house), courtyards surrounding house act as pens for animals
First Floor: Kitchen, bedrooms (husband and wife sleep separately from children), food storage, latrine, washing area
Top Floor: Roof courtyard, family chapel (elevated position important, steps up), guest room/bedroom, summer room

14.3.3.3 Walls

Walls up to 3 ft thick (generally 40–50 cm), 4 ft deep foundations (not verified)
Ground floor level: Stone
Upper levels: Sun dried mud bricks plastered with fine clay "*markala*"
Walls whitewashed with limestone (white houses considered desirable)

14.3.3.4 Roofs and Floors

Constructed from poplar beams, willow branches, yagdzas (similar to heather), mud and earth
Flat roofs (generally little precipitation, snow can be removed)
Prayer flags displayed on roof (signifying Buddhist households)
Grass piled in roofs over walls (protects from any precipitation)

14.3.3.5 Openings

Small windows and doors (reduces heat loss)
No openings on north elevation (reduces heat loss)
Entrance door with high threshold (to cope with snowfall)
Ornate carving of timber lintels and window surrounds

14.3.3.6 Kitchen

Heart of house—where the family spend most of their time, place for family gatherings and celebrations
Large (6.5×4.5 m in the case study house, anecdotally informed that this is not much larger than average)
Contains:

- Low tables and mats for sitting
- Wooden shelving for elaborate display of pots

- Stove—focal point of kitchen and elaborately decorated (wood burning stove considered preferable to gas in winter for its space warming properties, even in more modern houses)
- Timber floor
- Screen at door prevents drafts

14.3.3.7 Storage Requirements

Large amount of storage space required due long winters: storage of chang (liquor), spices, milk and yoghurt, barley and wheat

Kitchen storeroom on north side of kitchen with thick walls (keeps cool in summer)

Grain storage tanks under kitchen floor

Roofs used for drying apricots and other fruit/vegetables on blankets in the sun (summer), stacked with alfalfa grass for animals and dung/shrubs for fuel for stove (winter)

14.3.4 Balancing Between Traditional, Improved Technical, and Improvised Local

Stabilized Compressed Earth Block (SCEB) Technology offers a cost effective, environmentally sound masonry system. The product, a stabilized Compressed Earth Block has a wide application in construction for walling, roofing, arched openings, corbels etc. Stabilized Earth Blocks are manufactured by compacting raw material earth mixed with a stabilizer such as cement or lime under a pressure of 20–40 kg/cm^2 using manual soil press. A number of manual and hydraulic machines are available in India. The basic principal of all the machines is the compaction of raw earth to attain dense, even sized masonry. Some of the hydraulic machines can even manufacture interlocking blocks. These interlocking blocks are highly suitable for speedy and mortar less construction. Stabilized Compressed Earth Block (SCEB) Technology helps in offsetting the use of fuel wood that is getting expensive every day. On the other hand, compressed earth blocks are sundried and use cement as stabilization for gaining the required strength.

14.3.4.1 Product

The Stabilized Compressed Earth Block is a masonry unit of cuboidal shape. This may be solid or hollow or interlocking. The shape and size of a block is defined by the equipment used in its manufacture. SCEBs can be used for load bearing construction up to three storeys. The cost of a block depends upon a variety of factors including quality and price of available soil, amount of stabilization, labor productivity equipment and overhead costs. The degree of stabilization has the maximum influence on the cost of the product.

14.3.4.2 Raw Material and Manpower

The primary raw material for the production of SCEB is raw earth or soil. OPC cement in small quantities and water are other constituents. Coarse sand or stone dust may be added depending on soil quality. Soil is made up of grains of various sizes. The grain size distribution of a soil determines its suitability for the manufacture of SCEB. 6–10 persons are required to operate a manual machine. For hydraulic machines the manpower required is 6–8 persons. In both the systems one skilled worker is required while the rest are semiskilled. The workers can be trained to operate any machine in 10–12 days.

14.3.4.3 Application

Stabilized compressed earthen blocks can be used almost all the applications of burnt clay bricks. These blocks can be used for load bearing construction up to three stories

14.3.4.4 Energy Effectiveness

Costs are too often limited only to a monetary value. Another important aspect is the energy consumption involved in the material. The production of earth-based materials consumes much less energy and pollutes much less than fired bricks or concrete. CSEB and stabilized rammed earth are much more eco-friendly. They have these advantages compared to fired bricks (Table 14.1).

14.3.4.5 The Balance

The first phase of work was done with SCEBs and thermal insulation in walls as well as Trombe walls to capture heat. However, the percolation of these technologies in houses being built locally by owners themselves was low. It was realized that the time available to change attitudes and practices through sustained education, training, market changes and social changes is very less. The second phase of work therefore used the local sun dried blocks with seismic reinforcements and thermal features in addition. The third phase was a further improved version of this work using traditional seismic resistance features instead of reinforcements. The final outcome was that improvement and improvisation is important, but even if this intervention is not tempered with local contextual elements, it is difficult to achieve a change in practice.

Table 14.1 Energy and ecological aspects of building materials

Pollution emission (kg of CO_2/m^2)	Energy consumption (MJ)
2.4 times less than wire cut bricks	4.9 times less than wire cut bricks
7.9 times less than country fired bricks	15.1 times less than country fired bricks

Ecological comparison of building materials				
Product and thickness	No of units (per m^2)	Energy consumption (NJ/m^2)	CO_2 emission (kg/m^2)	Dry compressive crushing strength (kg/cm^2)
CSEB-24 cm	40	110	16	40–60
Wire cut bricks-22 cm	87	539	39	75–100
Country fired bricks-22 cm	112	1,657	126	30–100
Concrete blocks-20 cm	20	235	26	75–100

14.3.5 Sanitation Unit

In Ladakh, composting latrines in traditional houses provide a highly efficient and safe method of disposal of human waste. Each traditional Ladakhi house has a composting latrine which is emptied once a year and used as fertilizer on the fields. This eliminates the need for pipe work, saves water and power, and avoids leakages from tanks and thus the chance of water-borne diseases. Such diseases have been on the increase in the area since the introduction of flushing toilets. It was therefore proposed that dry composting latrines are the most appropriate waste disposal system for the shelter proposal and for this region.

The latrine takes the form of a small room with a hole in the floor through which human waste drops down one floor level to a pit below. Earth and ash from the kitchen are shoveled down the hole to cover the waste. This helps to reduce smells, aid decomposition, and enrich the waste for use as fertilizer.

Story 1: Moving from Disaster Management to integrated DRR & CCA thinking in Leh.

Leh is one of the largest and remotest districts of India, which is governed by and Autonomous Hill Development Council (LADHC) and a District Administration that reports to the State as well as the LADHC. Since it is a border district, the district is heavily militarized, resulting in the central government and the army playing a major role in its development and policy. Culturally, Leh Ladakh is akin to the Tibetan Buddhism. Leh is highly vulnerable to climate change with most of its population relying on glacial water sources for drinking, domestic and agriculture purposes. Even small changes in temperature and precipitation can have a major negative impact in a fragile environment like Leh. The flash floods of 2010 was a harsh reminder of the fragility of the region and impact that climate related extreme events might have in the future. The people of Leh have adapted well to its unique and severe weather conditions, yet new disasters and rapidly changing climate conditions mean that they have to reassess risks threatening them and ways to reduce these risks.

As part of its efforts to understand the ability of local multi-stakeholder action to catalyze shifts in program and policy environment towards mainstreaming DRR CCA in Ladakh, SEEDS has conducted a variety of activities since the start of the project to engage local stakeholders in discussions and actions towards integration of DRR CCA and its mainstreaming in broader development plans and policies.

SEEDS conducted a district level workshop in Leh to discuss the response of the district to the 2010 flash floods and ways to build on that for better risk reduction and adaptation to climate change impacts. The participants of the workshop felt that while the government was forced to make a DM plan after the 2010 floods, it was more of a relief management plan and did not address key issues of risk and climate change impacts. Based on discussions held during the workshop, SEEDS decided to collect household level data of about 500 farmers from an area called Sakti or Serthi near Leh. Sakti has about ten villages, which according to the data collected are facing high levels of water stress, near drought conditions in early summer and is flash flood prone. Other potential disasters risk identified were—earthquakes, pest infestation and war. While the area is heavily militarized and the army played a crucial role in relief operations during the previous floods, the government's DM plan seems to ignore the potential of damages in future floods and other impending factors that are contributing to increasing risks. After conducting village level surveys, the SEEDS team worked with local farmers in focus group discussions that created ten village level resource maps and plans for the entire Sakti area. These plans not only highlight the high-risk areas, but also form a basis for a village level DRR and CCA plan. Combining all the village level plans has resulted in a Sakti level integrated DRR and CCA plan, which the local Councilors hope to share with the district administration. This plan will also be a key example and input into the revision of the existing DM plan. The government has approved a revision in the plan but the local administration was not sure how to go about it. The local Councilors feel that the Sakti level DRR and CCA plan will offer a solution.

As part of the development of the resource maps and discussions with the community, it was apparent that a few areas needed immediate attention. One of them was "water scarcity." Currently Sakti is fed by the Warila glacier and its streams. These streams or Tokpos as they are called in the local language are drying up faster than usual. The community had been trying to repair a reservoir near the Warila glacier, in the hope of collecting and diverting water to ameliorate the situation. SEEDS' team of architects inspected the reservoir and found out that its location was wrong and as result most of water from the glacier was not being channeled properly into the reservoir. SEEDS held another round of discussions with the community and the newly formed "Sakti development committee" (this committee was formed during the first district workshop in Leh to oversee the work of SEEDS and the project), in which it was decided that perhaps "snow bunds" was a good way to trap the snow in winter and channel it to the bigger streams or tokpos in summer. This was an effective low cost solution, however, the community wanted to wait for some time and discuss their options for water management next summer. Apart from dealing with water stress, the community felt that erratic weather conditions were becoming a challenge for their traditional cropping practices. They felt that with better weather information and some agriculture extension support, the community could not only know when to sow their crops, but also how to manage pests and make most of the "rising" temperatures. Farmers in Sakti felt that there was potential to grow new kinds of vegetables and fruits. However, with the changing weather conditions, storing the produce and pest management was out of their area of expertise. With very small landholdings and youngsters giving up agriculture for tourism-based jobs, Leh is increasingly becoming more and more dependent on food from outside. This not only increases a family's overall spending budget, but is also instilling a sense of loss of culture and encouraging rapid and unplanned urbanization. The government of Leh is also very concerned about this trend and has tried many ways to subsidize agriculture while encouraging eco-tourism. However, as the chief executive councilor puts it "you can give directives, but you cannot make a law in such a situation." The rapid urbanization and encroachment of areas to build new guesthouses and shops was one of the main reasons for the devastation during the 2010 floods. The encroachment had blocked local streams causing major flooding in Leh.

As a result of the above discussions and suggestions, SEEDS planned to establish a climate field school or "Namze Lopta" in Sakti. As part of this project, SEEDS installed an Automatic Weather Station in a community center build by SEEDS in Sakti. The school will train farmers to monitor weather conditions and use them to generate "expected weather conditions" and forecasts with the help of India Meteorological Department (IMD), Jammu and Kashmir. The department of agro-met services Jammu and Kashmir will help farmers apply weather forecasts to farming practices. The department of agriculture of Leh has also agreed to support the project with agriculture extension workers to guide the farmers. Due to a harsh winter, most of the operations are closed in Leh, but the weather data is being disseminated via SMS service (phone) and through the web to the agro-met department of the State. It is expected that by late Feb and early March, which is also the sowing season, the data will help farmers plan out cropping strategies. With the help of a newly installed automatic weather station farmers would also be able to forecast flood or drought conditions, monitor snow and rainfall patterns.

The Director of Agro-Meteorological services, Jammu and Kashmir who inaugurated the AWS in Sakti, talked about the importance of agro-met services for farming communities. He also said that the impending threat of climate change requires that farmers take initiatives to understand how weather monitoring and forecasting can help them to adapt to the changes. He said that the climate field school started by SEEDS would help farmers in devising climate change adaptation strategies under unfavourable climate conditions and also to take advantage of favourable climate conditions. He said that the climate field school for farmers powered by a community owned AWS system would not only be beneficial for the people of Sakti but also for the entire population of Leh and the IMD Jammu and Kashmir.

SEEDS' representative also discussed the benefits of AWS and highlighted the process of using the data and preparing them for unforeseen situations. The Councillor of Sakti pledged his support to the project. He said that the data generated by the AWS would help in planning community based disaster risk reduction and adaptation strategies in the long run. SEEDS is now working towards planning out a short curriculum for the Climate School.

Story 2: Survey with farmers to evaluate their understanding of weather conditions.

In order to evaluate the perception of the farmers on weather related aspects, a questionnaire was prepared and a household survey was conducted with households of the Sakti village. After the survey the data was analyzed and the observations found are given as below:

In the survey majority of the farmers found snowfall as the most important weather parameter. More than 80 % of the farmer expressed that snowfall and rainfall affects their agriculture considerably, whereas temperature and sunshine is considered to be of lesser importance. This can be because of the location of all the settlements. The water that comes from the melting of glaciers feeds all the villages in the area.

When asked about the use of traditional knowledge or scientific methods for agricultural operations in general, 64 % of total farmers follow traditional method while 36 % follow a mix of scientific and traditional method. This can be because of the implementation of various projects by District Rural Development Authority (DRDA). Mostly people rely on traditional methods only for agricultural activities.

On the question of pests and diseases, 48 % of the farmers believe that weather parameter influences pests/diseases, 28 % feel that crop growth stage is also important for pest and diseases. 8 % of them consider both weather and crop growth as the influencing factor for pest/diseases and 16 % of them had no answer regarding to this question.

When asked about the information disseminated by meteorological department on the forecast weather, only 12 % of the people said that they follow the advisories given by the meteorological department for agricultural practices. 80 % of the people said that they do not follow the advisories of meteorological department, but they have their own regional calendar called "*lotto*" which gives them information about the sowing and reaping time. Also 8 % of the people did not answer to this question. On the general knowledge on climate of their region, about 88 % of the farmers from these centers gave the right answer on identification of summer and winter months. But they were not aware of the exact changes in terms of values of rainfall or temperature. 12 % of them did not respond to this question.

As this study was conducted with people above 30 year of age, their view was taken on change observed in climatic conditions according to their past experience. Alarmingly, it was found that 100 % of the people supported the change in climatic conditions in the area. About 70 % of the people say that there has been a major change in the snowfall pattern. Previously there used to be a heavy snow of about 3–4 ft thick, but now it has reduced to 6 in. to 1 ft thick and also it melts fast. This has caused a shortage of water in the region. Also about 30 % of the people feel that there is a considerable amount of rise in temperature in the area. They say that previously they had never seen fans in the area but now-a-days, in summer they have to use fans.

14.4 Conclusions and Lessons Learnt

The study leads to a number of lessons learnt related to construction technologies, material sciences, as well as social systems and processes. The main lessons learnt are summarized as follows:

- Post disaster recovery programmes must capitalize on existing traditional wisdom since it has been tested over generations and is best suited to the local environment and culture. This is true for construction materials and technologies as discussed in this chapter, but also for agriculture, environment, water management and other facets of development.
- Technology should be introduced where necessary, but in minimalistic ways, so as to add value to the traditional systems and make them more resilient in the face of new threats such as those posed by climate change.
- New approaches should be eco-friendly and local to the extent possible. This keeps the cost low, and also minimizes the carbon footprint of the intervention.
- Participation of the beneficiaries in decision making is critical to their involvement and ownership of the process.
- Participation of house-owner families in the actual development process is very useful in cutting costs, enhancing the sense of ownership, and keeping the process flexible enough for each family to be able to customize small things to suit their preference and convenience.
- Transfer of technology to the local workers is very useful to ensure the sustainability of the approach and its replication and scaling up in the area.
- Linkage with local stakeholders including governments, academia and the private sector is useful for creating a local buy-in for the approach, which will help in its sustainability in the long term.
- Linkage with related sectors such as housing, water, sanitation, livelihood, education helps create a more comprehensive package around shelter, habitat and lifestyle and provides value added benefits to the local community.
- Information is the key to most processes discussed, and the ability of local populations to create and access information, to disseminate it to the larger community, and to use it to influence the policy environment is the ultimate goal of local action on DRR CCA.

References

Anderson J, Bausch C (2006) Climate change and natural disasters: scientific evidence of a possible relation between recent natural disasters and climate change. Briefing paper, European Parliament, Policy Department Economic and Scientific Policy

Behera B, Vaswan RT (2008) Living with change: adaptation and innovation in Ladakh. Retrieved 12 Feb 2012 from http://iasc2008.glos.ac.uk/, http://iasc2008.glos.ac.uk/conference%20 papers/papers/B/Behera_120201.pdf

Daultrey S, Gergan R (2011). Living with change: adaptation and innovation in Ladakh. Retrieved 12 Feb 2012 from http://www.ourplanet.com/, http://www.ourplanet.com/climate-adaptation/Daultrey_Gergan.pdf

Foster JB (2009) The ecological revolution: making peace with the planet. Monthly Review Press, New York

Gernier L (1998) Working with indegenous knowledge: a guide for researchers. IDRC, Ottawa

Ganguly K, Panda GR (2010) Adaptation to climate change in India - a study of union budgets. Oxfam India, Economic Justice. Oxfam India, New Delhi

Holland G (1989) Observations on the international decade for natural disaster reduction. Nat Hazards 2:77–82

ICSU (2002) Series on science for sustainable development no 4: science, traditional knowledge and sustainable development. International Council for Science, ICSU

IPCC (2012) Managing the risks of extreme events and disasters to advance climate change adaptation. A special report of working groups I and II of the intergovernmental panel on climate change. IPCC. Cambridge University Press, Cambridge

Ministry of Environment and Forest India (2002). Retrieved 01 Feb 2012 from http://moef.nic.in/cc/index.html, http://moef.nic.in, http://pib.nic.in/archieve/lreleng/lyr2002/roct2002/23102002/r231020029.html

Mitchell T, Van Alast M, Villanueva PS (2010) Assessing progress on integrating disaster risk reduction and climate change adaptation in development processes. Institute of Development Studies, Brighton

Munich Reinsurance Company (2001) World map of natural hazards – a global view of the distribution and intensity of significant exposures. Nat Hazards 23:443–465

Nyong A, Adesina FB, Elasha O (2007) The value of indigenous knowledge in climate change mitigation and adaptation strategies in the African Sahel. Mitig Adapt Strateg Glob Change 12:787–797

Olhoff A (2011). Retrieved 20 Feb 2012 from http://ccdare.org/Portals/, http://ccdare.org/, http://ccdare.org/Portals/131/PDF/Olhoff_2011.pdf

Schipper EL (2007) Climate change adaptation and development: exploring the linkages. Tyndall Centre for Climate Change Research. Norwich, UK.

Seath F (2009) The human impact of climate change in India. Oxfam India, CLRA, Delhi

Stern N (2006) Stern review: the economics of climate change (Vol. 30). Review. HM treasury, London

Stitger C, Dawei Z, Onyewotu L, Xurong M (2005) Using traditonal methods and indigenous technologies for coping with climate variability. Clim Change 70:255–271

Strigl AW (2003) Science, research, knowldege and capacity building. environment. Dev Sustainability 5:255–273

UN (1994) International symposium on geophysical hazards in developing countries and their environmental impacts. Nat Hazards 9:277–279, Perugia, Italy, USA

UNISDR (2011) Global assessment report on disaster risk reduction. Retrieved 20 Dec 2012 from www.unisdr.org: http://www.unisdr.org/we/inform/publications/19846

Van Aalst MK (2001) The impacts of climate change on the risk of natural disasters. Climate change disasters 30(1):5–18

Webster M, Ginnetti J, Walker P, Coppard D, Kent R (2008) The humanitarian costs of climate change. Tufts University, Tufts University. Feinstein International Center, Tufts University, Medford

Chapter 15
Technological and Innovative Measures to Improve Flood Disaster Recovery Following Mumbai 2005 Mega-Flood

Kapil Gupta and Vinay Nikam

Abstract Mumbai City with 12 million people living in an area of 437.71 km^2 is the commercial and financial centre of India and a base for business outsourcing for major international organizations. Due to the reclamations and it's unique position and setting, Mumbai is most vulnerable to natural hazards like cyclones, floods, landslides and earthquakes. Thus, any disaster in Mumbai not only affects the local community but also extends to national and international communities. Frequent disasters put additional pressures on Mumbai's overstressed socioeconomic systems and urban environment. Over 60 % of the city was inundated due to the unprecedented rainfall of 944 mm during the 24 h starting 08:30 on 26 July 2005. The immediate impact of the heavy rainfall was that there was a total collapse of the rail, road and air transport and communication systems. Disruption of power supply resulted in the failure of sewage pumps and the mixing of sewage with floodwaters resulted in outbreak of epidemics. An estimated 419 people and 16,000 cattle lost their lives due to the ensuing flash floods and landslides in the Mumbai municipal limits. At that time, there was no real time flood warning system in place and the instrumentation with the authorities could give rainfall values only once every 24 h. Following this event, 35 state-of-the-art automatic weather stations comprising of tipping bucket rain gauges were installed at 33 locations in the municipal limits and one ultrasonic flow gauge on the Mithi River. In addition, two innovative flood control structures have also been designed and implemented on the city's main Mithi River. The weather stations have been programmed to relay rainfall intensity data in real time (every 15 min) to the disaster emergency control room and through internet to the public. A standard operating procedure based on real time rainfall data has been developed to respond to incidences of heavy rainfall. This has been of immense help in evacuating people during high rainfall events. To further mitigate the

K. Gupta (✉) • V. Nikam
Department of Civil Engineering, Indian Institute of Technology Bombay, Mumbai 400076, India
e-mail: profkgupta@gmail.com; vinaynikam@live.com

R. Shaw (ed.), *Disaster Recovery: Used or Misused Development Opportunity*, Disaster Risk Reduction, DOI 10.1007/978-4-431-54255-1_15,

flooding, two detention ponds/basins have been designed such that they allow the flows of sewage during the non rainy season and delay the flood waters during extreme rainfall events.

Keywords Disaster management • India • Innovative technology • Mumbai • Urban flood recovery

15.1 Introduction

Mumbai, formerly called Bombay, is the capital of Maharashtra state. It is not only the commercial and financial centre of India but also a base for business outsourcing for major international organizations. Mumbai is located on the windward side of the Western Ghats of India and receives high rainfall due to orographic effect from the Arabian Sea. Due to the reclamations and it's unique position and setting, Mumbai is most vulnerable to natural hazards like cyclones, floods, landslides and earthquakes. Thus, any disaster in Mumbai not only affects the local community but also extends to the national and international communities. The disasters put additional pressures on Mumbai's overstressed socioeconomic systems and urban environment.

The present island of Mumbai (Fig. 15.1) with an area of 437.71 km^2 has been formed due to the continuous reclamation of the spaces between seven separate islands and merger with the larger island of Salsette since 1672. Mumbai developed sequentially, first as a naval base for the East India Company, then as a textile centre and now a global centre for trade and commerce. This attracted large migrations from other parts of India and resulted in the rapid and uncontrolled all round growth of the city. The population increased from 8.19 million in 1981 to 12.49 million in 2011 (MCGM 2012). Due to non-availability of land in higher grounds, a large number of informal settlements (comprising 65 % of Mumbai's population) came up in the low lying areas and on the flood plains of the Mithi River. The city is strongly oriented in a north–south direction. A majority of the population resides in the suburbs in the north and commutes to the city located in the south. The rail network constitutes the lifeline of the city and over six million people are transported daily by Mumbai's suburban railway system alone—this is almost 50 % of the total number of passengers travelling daily by train in India.

Flooding in Mumbai is the most frequently occurring natural disaster due to heavy rainfall during the monsoon months. Monsoon rainfall occurs primarily during June to October and 70 % of the average annual rainfall occurs in July and August and 50 % of this occurs in just two or three events (Gupta 2007). During these two to three events, it usually rains uniformly over the city and severe flooding occurs in many parts of the city. Thus, any disruption in transport services due to flooding results in people being stranded thereby causing economic and social disruption—loss of livelihood to the individuals and loss of business to commerce and industry.

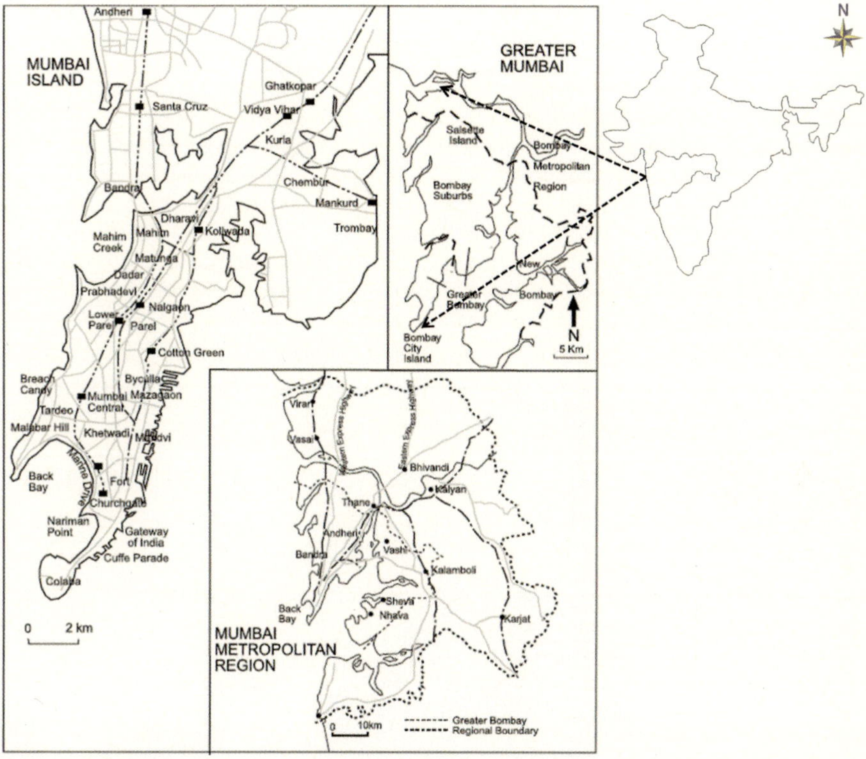

Fig. 15.1 Map showing Mumbai metropolitan region, greater Mumbai and Mumbai island (*Source*: MMRDA)

To address the problem of flooding, a major study was taken up in 1993, called the BRIMSTOWAD project (Brihanmumbai Stormwater Drainage Project, MCGM 1993) and it recommended the augmentation of the drainage network. However, on grounds of economy, it recommended the augmentation of the drainage network for a rainfall intensity of 50 mm/h corresponding to once in 6 months return period. This scheme was implemented only partially due to limited funds.

In addition to the heavy rainfall, the Fact Finding Committee (2006) has identified the main causes of flooding which are summarised in Table 15.1.

Data analysis for the years 1999–2004 showed that the peak rainfall intensity for the time of concentration of 15 min exceeded 72 mm/h over 80 % of the times (Gupta 2007). The average annual rainfall of Mumbai City is 2,050 mm as recorded by the Colaba meteorological station of IMD located at the southernmost tip of the city while that for the suburbs is 2,300 mm as recorded at Santa Cruz, located 27 km away to the north.

A severe flood event, with unprecedented rainfall of 944 mm during the 24 h starting 08:30 on 26th July 2005 caused severe disruption and losses in Mumbai. Over 60 % of Mumbai was inundated to various degrees on 26th July 2005.

Table 15.1 Main causes of flooding in Mumbai (Government of Maharshtra 2006)

	City area	Suburban areas
1	Low ground levels	Low ground levels
2	Low level of outfalls	Siltation of drains
3	Dilapidated drains	Obstructions of utilities
4	Obstructions of utilities	Encroachment along drains
5	Siltation of drains	Slums along outfalls
6	Urbanization and loss of holding ponds	Garbage dumping in drains mainly in slums
7	Increase in runoff coefficient	No access for desilting

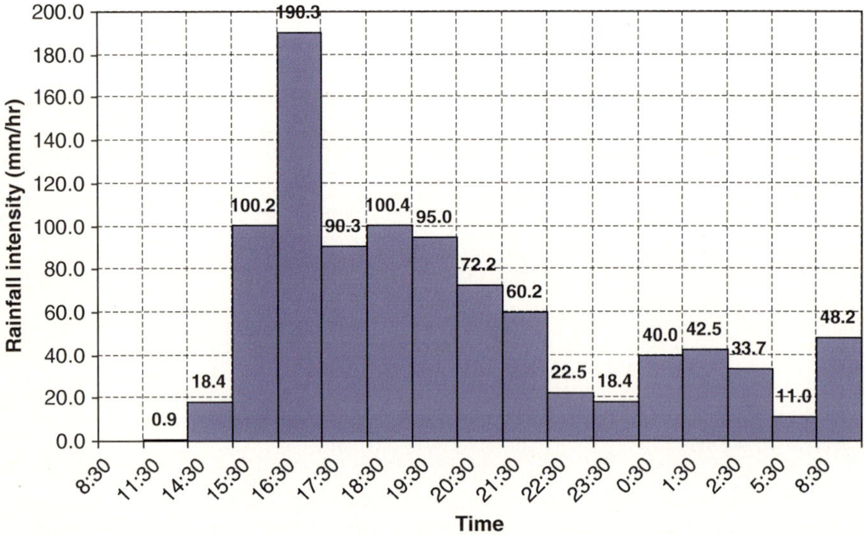

Fig. 15.2 Hyetograph of 26th July 2005 rainfall- 24 h ending 08:30 on 27th July 2005 (*Source*: IMD 2005)

The immediate impact of the heavy rainfall was that there was a total collapse of the rail, road and air transport and communication systems. Disruption of power supply resulted in the failure of sewage pumps and the mixing of sewage with floodwaters resulted in outbreak of epidemics. An estimated 419 people and 16,000 cattle lost their lives in the Mumbai municipal limits. This event has been classified as "very heavy" (>200 mm/day as per the rainfall classification of IMD). The Santa Cruz observatory at Mumbai airport recorded 944 mm during the 24 h ending 08:30 am on 27th July 2006 while the Colaba observatory recorded only 74 mm of rain (Jenamani, et al. 2006). The event was attributed to a highly localized "offshore vortex." The rainfall hyetograph for the 26th July 2005 event is shown in Fig. 15.2. From the figure it can be seen that at Santa Cruz, heavy rainfall started at 14:30 with 481.2 mm falling in just 4 h between14:30 and 18:30 and hourly rainfall exceeding

190 mm/h during 14:30–15:30. The already inadequate drainage system was unable to drain out because of the highest high tide level of the month of 4.48 m at 15:50.

15.2 Enhancement of Flood Response

Following the deluge of 26th July, 2005, it was realized that the implementation of structural measures as envisaged in BRIMSTOWAD (1993) was slow due to non-availability of land and limitations of funding. So immediate alternate measures were considered and these are described below:

15.2.1 Institutional Mechanisms

The institutional mechanisms have been strengthened—a Disaster Risk Management Master Plan is under finalization headed by the Municipal Commissioner. To enable better flood protection and management of the Mithi River, the Mithi River Development Authority has been set up. MCGM has now been entrusted with providing clearance for all construction works in the Mithi River catchment. To reduce runoff to the drains, rainwater harvesting has been made compulsory for areas greater than 300 m^2.

15.2.2 Disaster Control Centre

The disaster control centre of the MCGM has been upgraded at an estimated cost of US$ 2 million and serves as a control room to handle disasters in Mumbai. It is the command and control center for all rescue operations and coordinates with the various agencies. It has state-of-art systems like communications systems with land lines, hot lines, cellular phones, wireless, VHFs, video conferencing, press rooms, emergency water supplies and rations, uninterruptible power supply with standby generators.

15.2.3 Rainfall and Flood Warning System

The Indian Meteorological Department was unable to monitor the rainfall and issue warnings in real time on 26th July 2005. This has been attributed to the lack of state-of-the-art equipment like tipping bucket rain gauges with the IMD. IMD had only two rain gauges in Mumbai of the syphonic type which record data on graph paper attached to clockwork driven drums and were read manually twice a day.

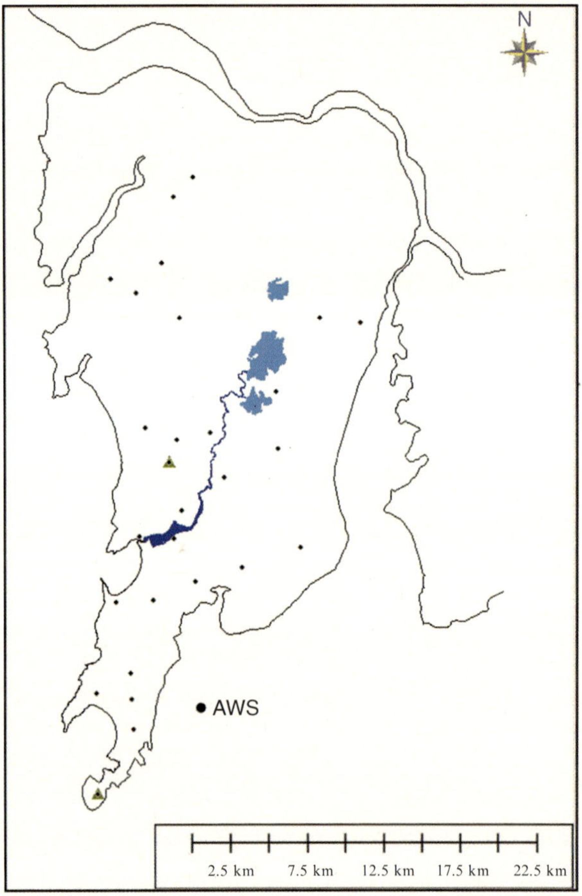

Fig. 15.3 Location of weather stations with tipping bucket rain gauge in the Municipal Corporation of Greater Mumbai (MCGM) area

The MCGM (with IIT Bombay) initiated the installation of 35 automatic weather stations at 33 locations uniformly over the city by June 2006 as shown in Fig. 15.3.

Considering the fact that the fire and rescue services are the first respondents, most of the weather stations have been sited on the top of the fire stations. The other weather stations have been located at the MCGM headquarters, in the catchments of Powai Lake, Vihar Lake and Tulsi Lake. The weather station included tipping bucket raingauges which can give rainfall data every minute directly on a console which can be downloaded to a central computer through intranet. The console has an additional feature, which is very useful for flood warning, namely, an audible alarm at preset rainfall intensity values which has been set to give an alarm for Mumbai when the rainfall intensity exceeds 40 mm/h. The weather stations have been linked to the central control room by internet and the MCGM officials can monitor the updated rainfall every 15 min from each location and issue alerts from the central control room. This has now enabled a much better response mechanism with judicious deployment of resources to the flooded areas. It has also enabled the

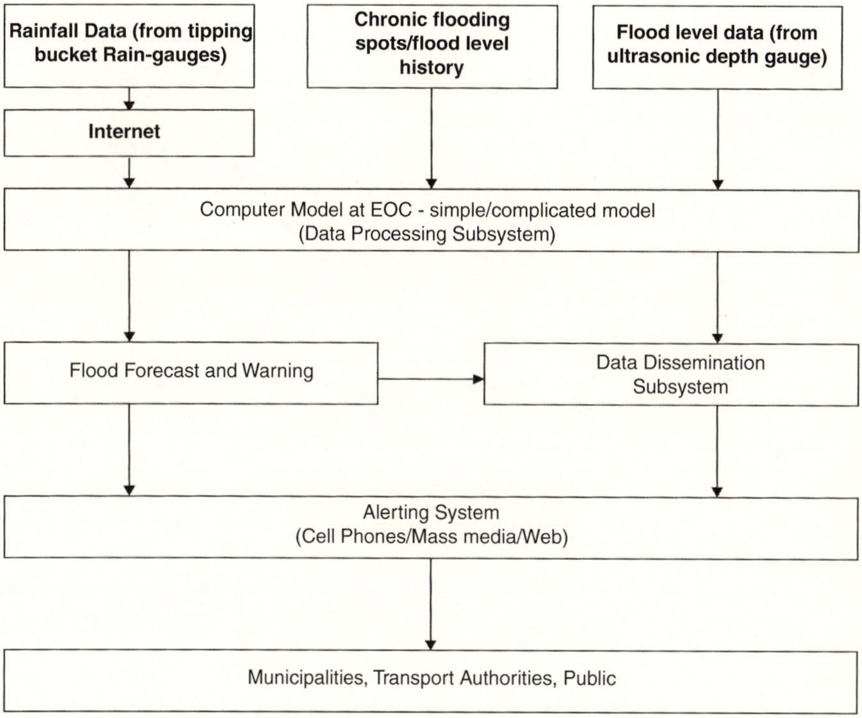

Fig. 15.4 A decision support system for mitigating urban flooding for Mumbai (Gupta 2007)

MCGM to issue warnings to the public through mass media and this is displayed on the internet during the four monsoon months on the website mumbaimonsoon.com. The schematic for the early warning system is shown in Fig. 15.4 while the SOP has been shown in Fig. 15.5.

The data from the AWS has been analyzed and has enabled a detailed of rainfall distribution all over the city for an event. An isohyetal map for the rainfall event during 3–4 July 2006 is shown in Fig. 15.6.

For design and resizing of drainage systems, intensity duration frequency curves (IDF) are required. With the help of the AWS, it is possible to record the rainfall intensity at 5 min intervals and this has enabled the development of IDF curves for Mumbai as shown in Fig. 15.7.

A SVM based model for 15 min ahead rainfall forecast in real-time is presently under development which is proposed to be used to forecast flood levels in the Mithi River using flood simulation models.

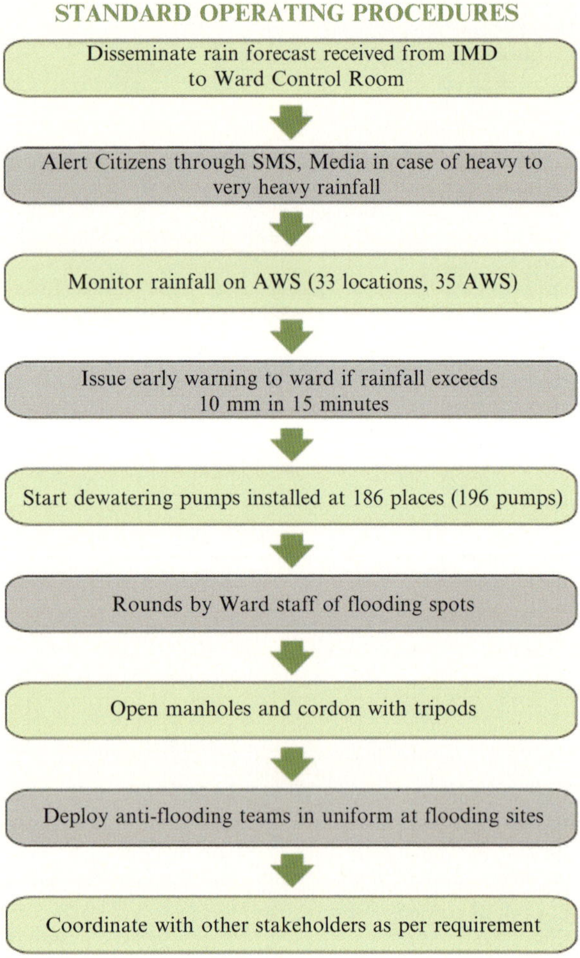

Fig. 15.5 Standard operating procedure based on real-time rainfall information (MCGM Flood Guidelines 2012)

15.3 Innovative Structures on Mithi River to Control the Flooding

Based on the recommendations of the Fact Finding Committee, removal of encroachments on the river banks and widening of the Mithi River was carried out. A flood wall has been completed in most upstream portions and presently is under progress in the downstream reach in the region of tidal-river interaction where mangroves are present. To further mitigate the flooding, two detention ponds/basins have been designed such that they allow the flows of sewage during the non rainy season and delay the flood waters during extreme rainfall events as shown in Fig. 15.8.

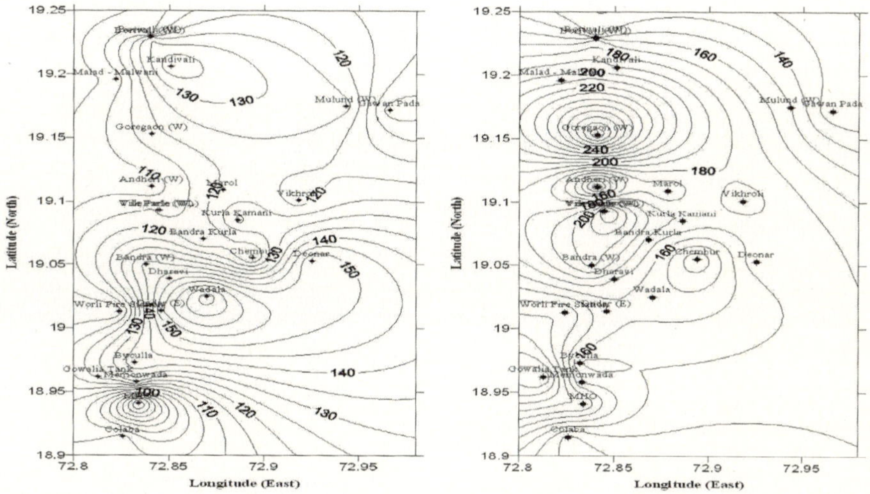

Fig. 15.6 Isohyetal map showing spatial distribution of rainfall in Mumbai (Lokanadham et al. 2009)

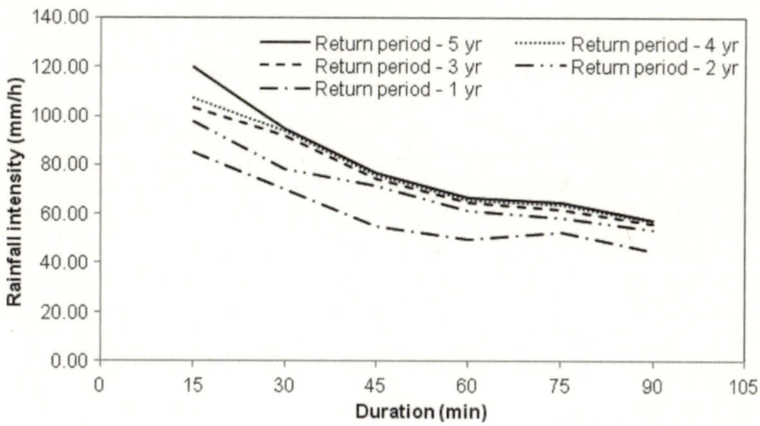

Fig. 15.7 Intensity duration frequency curve for Mumbai from AWS data

15.4 The Way Forward

The various measures described above have shown that it is possible for a city to improve flood disaster mitigation through simple state-of-the-art equipment like automatic weather stations connected through internet. This equipment does not require any special training and once installed, can be operated by office staff. However, the innovative structures described in this chapter have to be designed by trained technical experts, but once properly designed and constructed; they are highly effective in mitigating the flood. Institutional mechanisms can be

Fig. 15.8 Detention pond allows low flows to flow through but stores high rainfall intensity flows and allows overflows

strengthened further by establishing specific river development authorities. Similar measures can be adopted by other cities, especially those facing recurrent flooding.

15.5 Conclusions

The extreme rainfall event of 944 mm on 26th July 2005 has been a lesson for Mumbai and has resulted in Mumbai setting up a much better response mechanism based on real-time monitoring of rainfall at 33 locations in the city to handle recurrences of similar events in the future. This has also enabled a better insight into the rainfall variation over the city, development of design curves for future drainage works and construction of innovative structures on the river to mitigate flooding. However, the present rate of urban development is likely to continue in most of the cities. If all the resources and infrastructure are concentrated in a very small area, the cities must have innovative measures to handle extreme rainfall events and other disasters. Also, developments in any major city need to be accompanied by an adequate water supply, wastewater and stormwater disposal system based on analysis of extreme rainfall events. Under the present global economy, where major call centres and other BPO (business process outsourcing) institutions are located in Mumbai and other cities, disruption in one city has roll-over effects for worldwide business; hence, we cannot ignore flooding in any city as being just a local phenomenon. The Mumbai experience would be helpful for planning response strategies for other large cities to cope with similar events in the future.

Acknowledgements The authors are grateful to the officers of the Municipal Corporation of Greater Mumbai for their help and support.

References

Government of Maharashtra (2006) Report of the fact finding committee on Mumbai floods

Gupta K (2007) Urban flood resilience planning and management and lessons for the future: a case of study in Mumbai, India. Urban Water J 4(3):183–194

Jenamani RK, Bhan SC, Kalsi SR (2006) Observational/forecasting aspects of the meteorological event that caused a record highest rainfall in Mumbai. Curr Sci 90:10

Lokanadham B, Nikam V, Gupta K (2009) Characterization of spatial and temporal distribution of monsoon rainfall over Mumbai, 2009. ISH J Hydraulic Eng 15(2):69–80

Municipal Corporation of Greater Mumbai (1993) BRIMSTOWAD Project, final report, vols 3

Municipal Corporation of Greater Mumbai (2012) Flood preparedness guidelines

Part IV
Cross-cutting Issues

Chapter 16
Mangrove Management and Cyclone Risk Reduction in Kachchh, Gujarat

R. Parthasarathy and Mahima Gupta

Abstract This chapter maps the role of mangrove ecosystem in protecting the coast with the increasing threats from extreme events and also identifies the threats to mangroves from anthropogenic activities along the coast of Kachchh in Gujarat, which is found to be highly vulnerable to cyclones and storm surges. The analysis is based on an earlier empirical study of 176 households covering nine villages along the Kachchh coast and draws cases from other studies too. The empirical study shows that the coastal communities are mostly aware of the protection afforded by mangroves during the cyclones and the multiple benefits of mangrove restoration. However, based on their socio-economic conditions and livelihood structure, the communities have varied perception of the benefits from mangroves leading to differences in the outlook towards conservation of these natural resources.

Keywords Coastal ecosystem • Community perception • Eco-services • Forestry governance • Mangrove management

16.1 Introduction

The coastal ecosystem plays an important role in the Indian economy by the virtue of its resources, habitats, and biodiversity. India's coastline stretches about 7,500 km and supports almost 30 % of its population (NIO 2008). Indian coast can be categorized into east and west coasts. East coast is different from the west coast in many respects. Most part of the east coast is a flat terrain with beaches, coastal sand bars-dunes, and some parts also have backwaters, deltas, mangroves and mud flats. The western coast has narrow rolling plains, a few sandy beaches and several natural

R. Parthasarathy (✉) • M. Gupta
CEPT University, Ahmedabad, India
e-mail: rpsarathycept@yahoo.com; mahimanatural@gmail.com

R. Shaw (ed.), *Disaster Recovery: Used or Misused Development Opportunity*, Disaster Risk Reduction, DOI 10.1007/978-4-431-54255-1_16,

inlets with rocky outcrops. The west coast provides broad continental shelf while the east coast is covered with sediment providing relatively narrow shelf (Patnaik and Sivagnanam 2007). These coasts are also known for their vulnerabilities to various types of natural hazards and disasters. For example, Tsunami 2004 that made landfall on the east coast destroyed large areas causing several deaths and huge losses to property while the same impact was evident in the west coast due to cyclones and earthquakes. However, important feature of coastal zones is a variety of ecosystems especially mangrove habitats that play a vital role in reducing the impacts of extreme events by serving as a natural shield that enables coastal communities to lessen their vulnerabilities against disasters like cyclone, storm surges and tsunamis. Mangrove ecosystems apart from acting as protective buffer are also sources of highly valued commercial products and fishery resources. They can be considered to be an important component of natural resources, particularly in a country like India which is surrounded by sea from three sides (Parthasarathy and Raja 2011).

Though the coastal protection role of mangrove has long being recognized, it has gained more prominence after the 2004 Tsunami in South East Asia and much literature on the protective role of mangroves has emerged. However, there are other cases and examples to refer from other parts of the globe. Ewel et al. (1998), Siripong et al. (2005), Chong (2005), Massel et al. (1999), Mazda et al. (1997), UNEP-WCMC (2006) Alongi (2008) and Barbier and Heal (2006), to mention a few, report the function and value of mangroves in protecting properties, infrastructure, lives and productive economic activities on coastlines by acting as barriers that attenuate wave and storm energy and stabilize the shoreline against erosion.

16.2 Eco-Services of Mangroves

The set of eco-services provided by mangroves includes—natural defense against climate related disasters like cyclones, floods, sea level rise, "green house" effect, wave action, salt water intrusion and coastal erosion. Mangroves form natural barriers which provide shore protection both under normal sea conditions and during tropical cyclones, tidal surges and other extreme events. Evidences from the latest scientific research suggest that mangroves can reduce the height of wind and swell waves over relatively short distance. According to a report by the Nature Conservancy and Wetlands International titled *Reduction of Wind and Swell Waves by Mangroves*, wave height can be reduced by as much as 66 % over 100 m of mangroves (McIvor et al. 2012). With coastal populations particularly vulnerable to the impacts of extreme events such as storms and hurricanes researchers say mangrove forests can be used as a tool in coastal defense strategies making their conservation a key part of protection (McIvor et al. 2012). Besides shoreline protection, the mangroves control the entry of sea water inland and thus protect the ground water systems (Ridd and Sam 1996). Other critical ecological services provided by mangroves include, protection to the coastal biodiversity ranging from small insects like bacteria and fungi, a variety of fish, prawns, shrimps etc, to a variety of birds (Hirway and Goswami 2007). These ecosystems are well known to provide natural adaptation to the effects of climate change and contribute towards the global reduction of greenhouse gas emissions. By limiting carbon dioxide and storing it in its biomass, mangrove forests are able to reduce the amount of excess carbon in the atmosphere, thereby lessening the greenhouse gas contribution to global warming.

Ganesan (2004), Danielsen et al. (2005), Parish and Lee (2005) and IUCN (2005) confirm that during 2004 Tsunami, the coastal areas with dense mangrove cover suffered fewer losses and less damage to property and lives than those areas in which coastal vegetation had been degraded or converted to other land use. M S Swaminathan Research Foundations (MSSRF 2005) has produced report examining the effects of the tsunami on communities living within the Pichavaram mangrove wetland in Tamil Nadu, India. The MSSRF explains that hamlets behind mangroves were physically protected from the tsunami, whereas settlements located on or near the beach and not protected by mangroves were totally devastated. It is reported that mangrove forest reduced the impact of the tsunami by reducing the velocity of the incoming waves (due to friction created by the dense mangrove forest), and by distributing water among the canals and creeks of the mangroves, thus decreasing the level of inundation.

It is important to note that most of the evidence on the protective function of mangroves remains subjective largely perception-based therefore, scientific evidences would be a useful complement. The role played by coastal vegetation in preventing extreme storm surges has been debated by many authors. There seems to be a consensus among many experts that there is no experimental proof to support the fact that vegetation can reduce the effect of long-period waves such as storm surge or tsunami (Kerr et al. 2007; Alongi 2008; Vermaat and Thampanya 2006 and others).

While field based confirmation is scant and debated, the theory behind wave energy dissipation suggests that surface friction (a function of size and density of vegetation structures) disperses wave energy. The hydrological processes involved when waves, tides and currents interact with a mangrove forest are complex, but there are attempts to understand these processes (Furuwaka et al. 1997; Lewis 2005). Harada et al. (2002) have conducted a hydraulic experiment to study the tsunami reduction effect of the coastal permeable structures using five different models—mangroves, coastal forest, wave dissipating block, breakwater rock, and houses. This work reveals that mangroves are as effective as compared to solid seawall structures for reduction of tsunami-hit house damages behind the mangrove forest.

The type of information on the relationship between ecosystems and disasters differs between different regions. Existing literature has also established that some species respond differently to different disasters, and no single species can be used as a response to all natural disasters in coastal areas. Some varieties of vegetations are more wind resistant, whilst others are more wave resistant.

16.3 Study Area: An overview

The coastal and marine environment of the Kachchh district is known to support rich fisheries, coastal reefs and mangrove vegetation on creeks, estuaries and mud flats all along the coast (GEC 2010b). The district falls in the arid tracts and experiences tropical monsoon climate with an average annual rainfall of 350 mm. In spite of being arid, Kachchh district covers maximum area under wetlands and has the largest mangrove stand in the India's west coast (778 km^2 in 2011 as per Forest Survey of India assessment). The Gujarat State Disaster Management Agency (GSDMA 2013a, b) considers Kachchh district as highly susceptible to the various natural disasters such as earthquakes, cyclones, storms and tsunamis. About 4 to 6 tropical cyclones originate in the Arabian Sea every year which are characterized by torrential rain and storm surges (IMD). Historical records show severe cyclonic events in the year 1998 and 1999 over Arabian Sea which made landfall at the coast of Kachchh claiming hundreds of lives and loss of property. Convincing evidences on the role of mangroves has emerged from the field studies along the coastal villages of the district in the aftermath of these cyclonic events. It is known that mangroves play a crucial part in saving lives and property in the areas shielded with dense mangrove cover. Where mangroves were sparse or absent, the waves were able to infiltrate inland destructing property, inundating agricultural farms causing salinity ingress besides impacting livelihoods.

Against this background, this chapter attempts to map the role of mangrove along the Kachchh coast from protection service perspective and looks at the aspect of mangroves management towards conservation of these natural resources. The chapter argues that with widespread conversion of mangrove forest into industrial areas might contribute significantly to the calamitous loss of lives and human habitat in similar future events. The major issue faced by Kachchh today is the conflict between the industrial activities which have been promoted as a part of developmental and reconstruction activities after the earthquake of 2001 and the ecology of the region. It has been observed that there is no correspondence between the industrial policies and environmental policies in the state (Parthasarathy and Raja 2011) which is suspected to pose severe threat to the environment particularly to the mangrove cover of the region.

The paper is divided into three sections. Section 16.1 discusses the extent and characteristics of Kachchh mangroves in Gujarat. With the help of site specific observations, Sect. 16.2 presents the protection services rendered by mangroves. Section 16.3 details the changes in the mangrove areas and the management initiatives. The management initiatives are of two parts: in the first is the Government of Gujarat's initiative to assess the risk and vulnerabilities and in the second are the efforts towards mangrove regeneration by the government, the industries, the NGOs and the people.

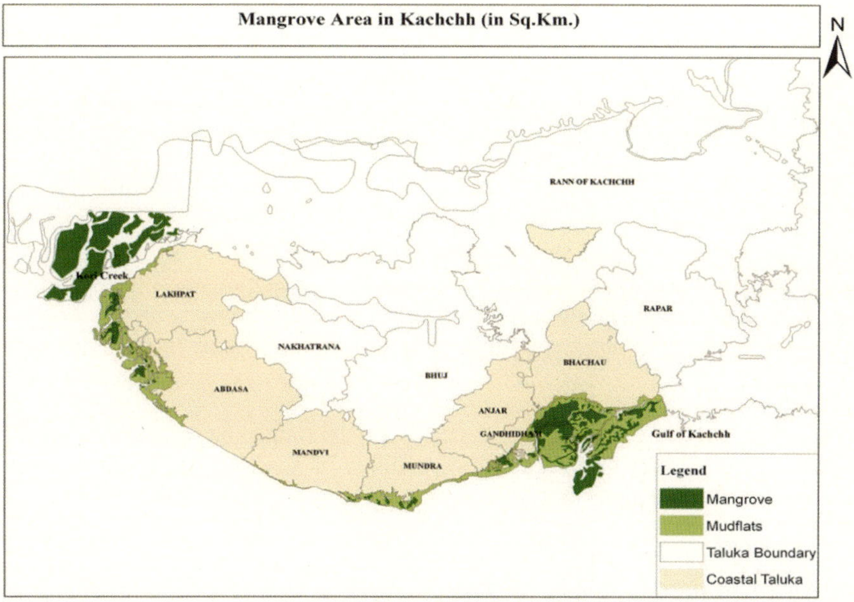

Fig. 16.1 Mangroves in Kachchh. *Source*: Based on GEC (2010a, b)

16.3.1 Mangroves of Kachchh: Extent and Characteristics

Kachchh coast with its 406 km length constitutes about 24.7 % of the 1,666 km coastal stretch of Gujarat. In spite of its high aridity (4 in a scale of 1–4), coastal zone of Kachchh presents diverse ecological entities like mangroves, sandy coasts, mudflats, creeks and other tidal incursions which enhances its coastal landscape and natural resource base. Extensive mangrove formations which constitute the single largest mangrove entity of the west coast and vast continental shelf of 164,000 km^2 facilitate a rich fishery ground. Mangrove cover in the district is nearly 75 % of the total mangrove in Gujarat (FSI 2011). These mangroves could be understood according to three zones based on their geographical location: (1) From Kori creek to Jakhau village along the coast of Abdasa and Lakhpat talukas; (2) Those adjoining the coast of Mandvi and Mundra talukas and (3) Between Anjar and Bhachau talukas (Fig. 16.1). From Kandla to Mundra, the coast is characterized with several creeks inhabiting mangroves. Good mangrove growth is found along the coast of Mundra, Abdasa and Lakhpat talukas which extends up to Kori creek. However, coastal stretches from Mandvi to Jakhau do not support mangrove formations. Table 16.1 indicates that out of the seven coastal talukas of Kachchh, Mandvi talukas has the least mangrove area (0.27 km^2) while dense mangrove patch is present in Mundra, Abdasa, Anjar and highest in Lakhpat taluka (464 km^2). The estimate of FSI (Forest Survey of India) for the year 2011 indicates that of the total 778 km^2 of

Table 16.1 Density of mangroves (in km²) by talukas

	1998			2007		
Talukas	Dense	Sparse	Total	Dense	Sparse	Total
Abdasa	16.50	5.04	21.5	20.57	3.19	23.76
Anjar	7.5	11.85	19.35	5.34	6.43	11.77
Bhachau	2	17.38	19.38	0.49	96.49	96.98
Gandhidham	–	–	0	2.89	59.07	61.96
Mandvi	0.2	0.13	0.33	0.21	0.06	0.27
Mundra	9.51	9.59	19.1	3.55	9.8	13.35
Lakhpat	308.52	339.19	647.71	87.33	346.87	434.2
Total	344.19	383.18	727.37	120.38	521.91	642.29

Source: Singh (2000) and GEC (2011)

mangroves only 118 km² (15 %) are dense and the remaining 660 km² (85 %) are sparse or open. Also, harsh ecological condition of Kachchh coast seems to have resulted in the monotypic prevalence of *Avicennia marina* species of mangroves.

16.3.1.1 Protection Services of Mangrove in Kachchh

Tropical cyclones originating in the Bay of Bengal and the Arabian Sea cross Kachchh in the later part of their tracks. The storms and cyclones governed by rainfall pattern in June–July have severe influence in this region with devastating effects on coastal areas. Generally in June, the storms are confined to the area north of 15°N and east of 65°E. In August, in the initial stages they move along northwesterly course and show a large latitudinal scatter. West of 80°E, the tracks tend to curve towards north. In October, the direction of movement of a storm is westerly in the Arabian Sea. However, east of 70°E some of the storms move towards north-northwest and later recourse northeast to strike Gujarat north Mekran coast. Table 16.2 shows the different cyclones and their intensities that made landfall in Kachchh coastline. The Table points out that number of cyclones hitting Kachchh district has increased in period 1990 to 2010 and a total of eight cyclones were reported during this span of 20 years. Prior to 1990s, the maximum number of cyclones recorded within the same time period was only four.

The cyclones of 1998 and 1999 in Kachchh are known to be the most devastating ones along the western coast in previous 30 years. The damage was greater in the first 10 km from the coast where small settlements were badly affected. Several accounts of the mangroves having a mitigating effect during the cyclone surfaced in the aftermath. Field observations and primary survey conducted along the coastal villages of the district shows cyclone did less damage to property in the regions which were fringed with dense mangroves, than in areas where mangroves were sparse or were absent. Villages from Mundra, Abdasa and Lakhpat talukas with each dense and sparse mangrove cover were interviewed at various points of time for their perception about the protection services provided by mangroves (Table 16.3).

Table 16.2 Cyclones on Kachchh coast (1897–2010)

Year	Month	Intensity	Track route
1897	July	DEPR	Off Jaffrabad–Veraval–Gulf of Kachchh
1909	Sept	DEPR	Surat–Jaffrabad–Kandla–NW
1944	Aug	STR	Ahmedabad–Kandla–Off Jaffrabad–W
1989	June	DEPR	Junagadh–Rajkot–Navalakhi–Vadinar–NW
1996	June	SSTR	Kandla–Rajkot
1996	Oct	SSTR	Kandla–Veraval–Jafarabad
1998	June	SSTR	Porbandar–Jamnagar–Kandla
1999	June	SSTR	Porbandar–Dwarka–Nalia
2001	May	SSTR	N–Saurashtra–Kachchh
2007	June	STR	NW Gujarat–Kachchh–Karachi
2009	June	DEPR	NW Saurastra–Kachchh
2010	June	SSTR	Off Sindh–Karachi–Northern Kachchh

SSTR: 88–117 kmph, DEPR: upto 60 kmph and STR: 60–87 kmph
Source: (Kandla Port trust, GSDMA, GUIDE and GMB)

Table 16.3 Changes in mangrove cover in selected coastal villages of Kachchh (1999 and 2012)

		Mangrove cover	
Taluka	Villages	Before (1999 cyclone)	Present status (2012)
Mundra	Dhrub	Dense	Sparse
	Navinal	Dense	Dense
	Jarpara	Dense	Sparse
	Luni	Sparse	Regeneration going on since 2011
Abdasa	Mohadi	Dense	Sparse—regeneration going on since 2007
	Akri Moti	Dense	Dense
	Budiya	Sparse	Sparse—regeneration going on since 2001
Lakhpat	Gunau	Sparse	Sparse—regeneration going on since 2005
	Lakki	Dense	Dense—regeneration going on since 2001

In October 2012, a primary survey was carried out along the 9 coastal villages of Mundra, Abdasa and Lakhpat talukas of Kachchh district, Gujarat
Source: Village Survey (2012)

Villages in the proximity of dense mangrove forest were physically protected from the incoming sea waves entering into the settlement and villagers perceive that mangrove forest reduced the impact of cyclonic wind by reducing its velocity and by distributing water among the channels and creeks along the mangroves, thus reducing the level of inundation. Villagers reported that mangroves and coastal vegetation helped protect the coast and saved lives. Many fishermen for example took shelter in the mangroves during the cyclone and survived. Whereas villages with comparable distance (to the ones mentioned above) from the sea shore having sparse or no mangrove cover were seriously affected. Loss of life and huge damages to agricultural farms, livestock and property has been reported from the unshielded villages. Also channels and creeks linking sweet water flows to the sea were reported to have

Table 16.4 Loss to buildings by end use due to Cyclonic storm and wind (100 year return period)

		Loss to buildings [Rs. in million (2002–2003 prices)]					Mangrove area in 1998 (km²)		
S.N.	Taluka	Residential	Industrial	Commercial	Total loss	Capital loss	Dense	Sparse	Total
1	Abdasa	37	7	9.2	53	2,517	16.46	5.04	21.5
2	Anjar	85	3.2	3.2	92	4,922	7.5	11.8	19.3
3	Bhachau	100	2.1	2.5	105	3,942	2	17.3	19.3
4	Gandhidham	45	2.1	3.1	51	7,466	–	–	–
5	Lakhpat	33	0.5	0.5	34	1,049	308.52	339.2	647.7
6	Mandvi	630	14.1	21.3	666	5,847	0.2	0.13	0.3
7	Mundra	372	8.4	9.7	390	2,881	9.51	9.6	19.1

Source: GSDMA (2005) and GEC (2011); and, http://www.gsdma.org/hazard/cy/100_year.jpg

funnelled water inland causing flooding and turning the sweet water saline and unfit for any purpose. Changes in topography, soil salinity and the flow of freshwater from upstream has been reported in many of the villages post cyclone especially where mangrove cover was sparse and absent However, massive loss of fishing boats, agricultural produce as well as salinity in agricultural land was reported in all the villages which have badly impacted the livelihoods of the communities along the coastal belt. Villages which were just 10–15 m above the sea level and had no layer of mangrove protection reported losing their drinking water wells permanently after cyclone. Huge damage to mangrove forest was also reported due to cyclone, however the extent of the damage is not known. The direct force of the waves and silt deposition by wave action is said to be the cause of suffocating the mangroves.

A study carried out by GSDMA in the year 2005 on Hazard Risk and Vulnerability assessment of Gujarat, appears to corroborate the above anecdotal evidence. Their results clearly show that where mangroves are, they have offered protection to property, infrastructure and lives. However, mangroves that are threatened ecologically suffered severe damage and their protective properties were diminished. Table 16.4 shows that estimated loss to buildings is maximum in Mandvi taluka with had the lowest cover of mangrove whereas Lakhpat and Abdasa talukas suffered relatively less loss which can be attributed to the extensive mangroves in these areas. Similarly, Capital loss is estimated to be highest in Gandhidham followed by Mandvi where mangrove covers were too small to protect the buildings.

Table 16.5 highlights the loss of life due to cyclone with probability of 50 and 100 year return period across each coastal talukas. In Lakhpat taluka which has the highest dense mangrove cover, the estimated number of lives lost is 40 people—the lowest number of estimated fatalities in any coastal taluka of Kachchh. Whereas in Gandhidham followed by Mandvi highest casualties are estimated where mangrove cover is found to be sparse and negligible.

Impact of cyclone on infrastructure has also been estimated by GSDMA specifically for electricity sector. The estimated capital loss to GEB due to cyclone, presented in Table 16.6 is for a return period of 50 and 100 years. The estimation shows that maximum loss to be in Mandvi taluka and the minimum in Lakhpat.

Table 16.5 Estimated loss of life due to cyclone (probabilistic 50 and 100 year return period estimated based on 2001 taluka population)

		Estimated number of lives lost		
S.N.	Taluka	Cyclone (50 year return)	Cyclone (100 year return)	Census 2001 population
1	Abdasa	5	69	97,508
2	Anjar		50	160,292
3	Bhachau		170	147,891
4	Gandhidham		1,035	201,569
5	Lakhpat		40	50,120
6	Mandvi	13	175	170,573
7	Mundra		134	83,010

Source: GSDMA (2005) and Census (2001)

Table 16.6 Total capital losses to GEB due to cyclone (100 year return) (million 2002–2003 prices)

		Capital loss to GEB (Rs. in million)	
S.N.	Taluka	Cyclone (50 year return)	Cyclone (100 year return)
1	Abdasa	21.1	50.3
2	Anjar	30.7	62.4
3	Bhachau	31.6	46.2
4	Gandhidham	11.8	27.1
5	Lakhpat	10.2	18.1
6	Mandvi	43.2	122.3
7	Mundra	12	32

Source: GSDMA (2005)

The evidences from the field and damage assessments however need to be supported by detailed ecological studies since the estimates of protection services of mangroves is still relatively poorly understood as they are dependent on a number of influencing factors like wave energy and height; topography of the coastline and the characteristics of the mangrove forest. Many of the areas that were hit hard by the cyclone like Kandla in Gandhidham were already under serious threat from increasing anthropogenic activities. Over the past decade, hundreds of hectares of mangroves have been converted to industrial land use which has consistently under-valued mangrove resources of the region.

16.4 Changing Status of Mangroves in Kachchh

According to the Gujarat Ecology Commission, the district had 239 km^2 area under mangroves in 1991 which has increased to 854 km^2 in 1999. During 1999 to 2007, however, this area has continuously declined reaching 672 km^2 in 2007 (Fig. 16.2). Hirway and Goswami (2007) have argued that this decline is due to several natural

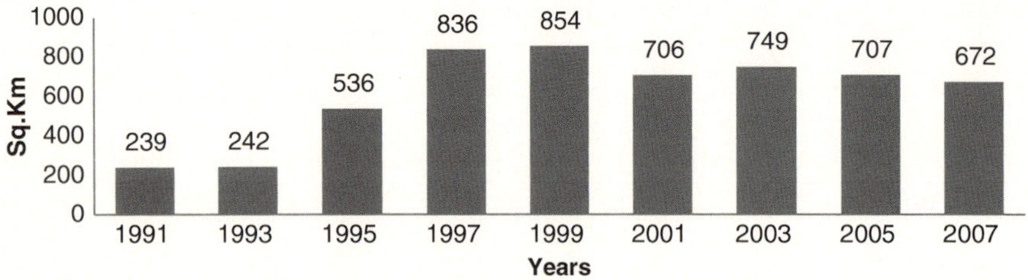

Fig. 16.2 Mangrove area in Kachchh district (in km²). *Source*: GEC (2011)

Table 16.7 Dense and sparse mangroves area in Kachchh (in km²)

1998			2007		
Dense	Sparse	Total	Dense	Sparse	Total
344.19	383.18	727.37	120.38	521.91	642.29

Source: Singh (2000) and GEC (2011)

factors as well as human factors. Though data on the quality and density of mangrove are important there are no time series data on the share of dense mangrove in the total mangrove area. The 1998 data of GEER foundation are the first such estimates. Table 16.7 shows change in the area under dense and sparse mangrove from 1998 to 2007 in the district. It needs to be noted that there has been significant decline in the area under dense mangrove (from 344 to120 km²). There seems to be several reasons for this decline which are discussed below.

16.4.1 Depletion and Degradation of Mangroves

Historically, Kachchh had an extensive mangrove ecosystem which had been depleted over time due to various development activities as well as natural disasters and anthropogenic interactions. Overexploitation of mangroves for fodder and fuel-wood by local communities along with reduced natural regeneration and diversion of mangrove lands for industrial development are largely the causes of degradation. Ports, jetties, thermal power stations during their construction and operation cause both direct and indirect impacts on mangroves. Most of the ports and jetties in Kachchh are located in mangrove lined creek systems or in its close proximity. Creek systems are preferred locations for ports and jetties and other industries due to their operational convenience, less vulnerability to storm and sea surges and easy maintenance of draft for vessel movement. Degradation of this fragile ecosystem in Kachchh is well recorded. This can particularly be seen in Mundra taluka where the new industries have caused depletion of mangroves (though compensatory mangrove plantations have been made) yet, the industrial wastes are found to contaminate the mangrove vegetation and the marshlands affecting the ecology of the area. In other areas, where there are cement plants located also have caused depletion of mangrove cover. Records also indicate degeneration of dense mangroves along the Kandla creek from 506 km^2 during 1960s to a sparse patch of 49 km^2 (ICMAM 2002). All the proposed, operational and upcoming special economic zones in Kachchh are located in the very close proximity to coastal belt. A vast array of materials like engineering goods, drugs, crude oils and polymers are handled at these SEZs whose discards finally reach the marine environment nearby. These SEZs releases along with solid wastes and sewage forms a complex mixture of pathogens, nutrients, suspended solids (SS), oxygen demanding substances, and many other contaminants—each with different environmental impact. Contamination of marine water by these releases makes mangrove ecosystem vulnerable (Kemp and Spotila 1997). All the study villages which had dense mangrove cover before the industries came up in their vicinity have reported reduction in mangrove area to a great extent making them sparse or totally absent in some places. Mangrove cover near villages like Dhrub, Jarpara and Mohadi have been affected the most due to Mundra Port and Cement industries making them vulnerable to cyclones and other related phenomenon in future.

16.4.2 Management of Mangroves

There has been a sharp decline in the area under dense mangroves over the two decade ending 2010. Dense mangroves have either become sparse or are totally lost in a few places. The reduction in the area under mangroves can be attributed to two major factors: Extreme events especially cyclones and anthropogenic pressures mostly industrialization. However, of late, there have been some initiatives for the conservation as well as management of these valuable resources. Two major modes of initiatives that have been taken up for mangrove management in the state are: (1) by The State Government and NGOs and (2) by the Industries.

16.4.2.1 Mangrove Management by State Government

Plantation of mangroves undertaken by the state government has been going on since 1960s in various parts of Gujarat coasts. One of the major problems with regeneration of mangroves has been the allotment of coastal areas to industries, ports/jetties and other infrastructural development. Potential mangrove regeneration sites were allotted to private ports, cement and salt industries. However, Gujarat Ecology Commission through implementation of the Restoration of Mangroves (REMAG) project with the help of India–Canada Environment Facility had taken up a major initiative. REMAG attempted to introduce participatory management of mangroves through community based approach. However, challenges were faced by this community based approach also. Mangrove regeneration was mainly looked as a source of income generation by the participants. The local communities who worked as wage labourers in the regeneration activity lacked a sense of ownership, motivation and incentives to conserve the resources on a sustainable basis. With the completion of the REMAG project, the community based associations (*Cheriay Samitti*) have become defunct. The lack of motivation amongst the participants could partly be attributed to the reduced dependency on mangroves for fuel, fodder and income over time. The use value and sense of ownership for mangrove amongst the present generation also appears to be lower than the earlier ones. The perceptions about its advantages have changed with changing land use pattern and reduced dependencies. These changing outlooks have an impact on the management of mangrove and its sustainability.

16.4.2.2 Mangrove Management by Industries

As part of legal mandate and social responsibility Kandla Port Trust, Mundra port and SEZ Limited (MPSEZ Ltd.,) and other prominent industries like Tata Power, Shell, Adani etc. have undertaken mangrove regeneration activities in the district. In addition, GEC had initiated a PPP project with the financial support of major industrial houses in 2004–2005 in the district. Under this programme, investments by private sector companies/ industries have also been encouraged with a commitment to have community participation in the mangrove development/ restoration efforts. Although there have been various attempts by the industries near the mangrove forests to regenerate mangroves, their quality and the survival rate has been questioned by many.[1] Moreover, claiming to compensate for destruction of more than 100 year old mangroves towards industrial development, promising to restore double the area destroyed needs to be critically looked into for consistency in terms of quality, quantity of the mangroves and the environment (Parthasarathy and Raja 2011).

16.5 Strategies for Managing Mangrove Resources

Even as mangrove regeneration through community participation has been touted as a panacea to the problems related in conservation and management of mangroves; it is evident that firstly, the changing mangrove-livelihood linkages amongst the coastal communities make it appear that communities are no longer interested in the conventional activities. Importantly, field based studies show that the sense of ownership for mangroves amongst the newer generations has reduced and the perceptions about their advantages have changed with changing economic activities and reduced dependencies. If at all, it appears that the mangrove regeneration is only looked upon as an additional source of income among the daily wage labourers. Evidently they do not resemble even close to being co-managers of the resource. Taking into account the changing approaches and attitudes of community towards

[1] Discussions held with various NGOs and Community Leaders.

natural resources, there is a need to revisit the relevance of community based management of mangroves at least in this area. A possible management strategy seems to the one which would integrate all the stakeholders such as State Government agencies like Disaster Management Authority, Ecological Commission and Pollution Control Board, NGOs, Industries and local community. In fact, some of the examples cited above has elements of involvement of the varied agencies. In order to attain an appropriate balance between environment, economy and societal expectations following measures for conservation of mangroves could be explored

1. Afforestation of mangroves by Industries that cause mangrove degradation should be made compulsory by the State Government. State should ensure that the laws and regulations for mangroves afforestation provide for effective penalties against violations, which are sufficient in severity to be effective.
2. Forestry governance systems (with emphasis on monitoring of rehabilitation of degraded mangroves) with the respective roles and responsibilities of Government agencies, communities and civil societies. Monitoring of mangroves should be carried out at two levels (a) routine monitoring by local people; (b) more intensive but periodic monitoring by governmental agencies and NGO's. School children can also be involved in monitoring activities at regular intervals. This will help children grow up with a feeling of stewardship towards natural resources and an understanding of environmental issues and probable solutions.
3. Zoning existing mangrove areas for preservation, conservation or sustainable utilisation based on their location, ecological characteristics and values would also be effective to manage. Mangrove zoning plan should be integrated with development plans in order to designate regions for industrial development. For example, Vietnam has enacted a zoning plan featuring a Full Protection Zone (FPZ) for coastal protection, a Buffer Zone for controlled economic activities (40 % by area), but retaining 60 % forest cover, and an Economic Zone where there are no forest conservation restrictions.

References

Alongi DM (2008) Mangrove forests: resilience, protection from tsunamis, and responses to global climate change. Estuar Coast Shelf Sci 76(1):1–13

Barbier EB, Heal GM (2006) Valuing ecosystem services. Economists Voice 3(3), Article 2, http://www.bepress.com/ev/vol3/iss3/art2

Census (2001) Kachchh District Census Handbook, Office of the registrar general & Census Commissioner, Ministry of Home Affairs, Government of India

Chong J (2005) Protective values of mangroves and coral ecosystems: a review of methods and evidence. IUCN, pp 1–5

Danielsen F, Sorensen MK, Olwig MF, Selvam V, Parish F, Burgess ND, Hiraishi T, Karunagaran VK, Rasmussen MS, Hansen LB, Quarto A, Suryadiputra N (2005) The Asian tsunami: a protective role for coastal vegetation. Science 310:643

Ewel et al (1998) Different kinds of mangrove forest provide different goods and services. Global Ecol Biogeogr Lett 7:83–44

FSI (2011) State of forest report 2011. Ministry of Environment and Forest, Dehradun
Furuwaka K, Wolanski E, Mueller H (1997) Currents and sediment transport in mangrove forests. Estuar Coast Shelf Sci 44:301–310
Ganesan S (2004) Point calimere escapes tsunami fury. The Hindu. http://www.hindu.com/200412/30 stories/2004123006870500.htm
GEC (2010a) Mangrove atlas of Gujarat. Gujarat Ecological Commission, Gandhinagar
GEC (2010b) Socio-economic and ecological benefits of mangrove plantation: a study of community based mangrove restoration activities in Gujarat. Gujarat Institute of Development Research and Gujarat Ecological Commission, Gandhinagar
GEC (2011) Trends of changing climate and effects on eco-environment of Kachchh district, Gujarat. Gujarat Ecological Commission, Gandhinagar
GSDMA (2005) Gujarat hazard risk and vulnerability atlas. Gujarat State Disaster Management Authority, Gandhinagar
GSDMA (2013a) http://www.gsdma.org/hazard_cy.htm. Accessed 25 January 2013
GSDMA (2013b) http://www.gsdma.org/hazard_ss.htm. Accessed 25 January 2013
Harada K, Imamura F, Hiraishi T (2002) Experimental study on the effect in reducing tsunami by the coastal permeable structures. In: Final proceedings of the annual international offshore and polar engineering conference, USA, 2002, pp 652–658
Hirway I, Goswami S (2007) Valuation of coastal resources, the case of mangroves in Gujarat. Academic Foundation, New Delhi
ICMAM (2002) Geographical information system for Gulf of Kachchh. Department of Ocean Development, ICMAM Project Directorate, Chennai
IMD http://www.imd.gov.in/services/cyclone/tropical-cyclone.htm. Accessed 30 October 2012
IUCN (2005) Early observations of tsunami effects on mangrove and coastal forests. www.iucn.org
Kemp SJ, Spotila JR (1997) Effects of urbanization on brown trout salmo trutta, other fishes and macroinvertebrates in Valley Creek, Valley Forge, Pennsylvania. Am Midl Nat 138:55–68
Kerr AM, Baird AH, Campbell ST (2007) Comments on Coastal mangrove forests mitigated tsunami. Estuar Coast Shelf Sci 67:539–541, Kathiresan and N. Rajendran [Estuar Coast Shelf Sci 65: 601–606 (2005)]
Lewis RR (2005) Ecological engineering for successful management and restoration of mangrove forests. Ecol Eng 24:403–418
Massel SR, Furukawa K, Brinkman RM (1999) Surface wave propagation in mangrove forests. Fluid Dyn Res 24:219–249
Mazda Y, Magi M, Kogo M, Hong PN (1997) Mangroves as a coastal protection from waves in the Tong King delta, Vietnam. Mangroves Salt Marshes 1:127–135
McIvor AL, Moller I, Spencer T, Spalding M (2012) Reduction of wind and swell waves by mangroves, natural coastal protection series: report 1. Cambridge coastal research unit working paper 40, Published by The Nature Conservancy and Wetlands International, p 27, ISSN 2050–7941
MS Swaminathan Research Foundation (MSSRF) (2005) Tsunami & Pichavaram mangroves. http://www.mssrf.org/
NIO (2008) Coastal and marine pollution. National Institute of Oceanography, Goa
Parish F, Lee D (2005) Preliminary information on impacts of the 24th December 2004 tsunami on selected coastal ecosystems in Aceh province, Indonesia. Global Environment Centre, Malaysia. www.riverbasin.org
Parthasarathy R, Raja S (2011) Role of mangroves in livelihoods and climate change: the case of semi-arid Kachchh in Gujarat. A report submitted to CESS, Hyderabad
Patnaik DC, Sivagnanam N (2007) Disaster vulnerability of coastal states – a short case study of Orissa. Department of Geography, University of Madras, Chepauk Campus, Chennai
Ridd PV, Sam R (1996) Profiling groundwater salt concentrations in mangrove swamps and tropical salt flats. Estuar Coast Shelf Sci 43(5):627–635
Singh HS (2000) Mangroves in Gujarat current status and strategy for conservation. Gujarat Ecological Education and Research (GEER) Foundation, Gandhinagar

Siripong A, Yumuang S, Swangphol N (2005) The 26 December 2004 tsunami: impacts on mangroves along some Thai coasts. In: The importance of mangrove and other coastal ecosystems in mitigating tsunami disasters. Symposium held in Kuala Lumpur, Malaysia, 23 August 2005

UNEP-WCMC (2006) In the front line: shoreline protection and other ecosystem services from mangroves and coral reefs. UNEP-WCMC, Cambridge

Vermaat JE, Thampanya U (2006) Mangroves mitigate tsunami damage: a further response. Estuar Coast Shelf Sci 69:1–3

Chapter 17
Adaptive and Transformative Capacities of Communities After Disaster: The Case of Oil Spill in Guimaras, Philippines

Andrew Eusebio S. Tan and Juan M. Pulhin

Abstract Disaster recovery from industrial pollution such as major oil spill presents formidable challenge for governance and disaster management considering their adverse environmental and socioeconomic impacts. The community's social dynamics could explain how responses to such perturbation are made and what inherent traits of the households and communities contribute to their adaptiveness to disasters. The collective decisions and actions leading to a desired and improved end likewise plays considerable role in their adaptive capacity. Using both qualitative and quantitative approaches in research, this chapter examines the ecological and sociological impacts as well as the adaptive and transformative capacities of selected communities affected by oil spill in Guimaras, Philippines—one of the worst cases in the country's history. Questionnaire elicits data on community's quantitative level of adaptive capacity while focus group discussion, in-depth interviews, observations and narrative analysis were employed for qualitative data. Justification for the different parameters of adaptive and transformative capacities of communities was also explored. Results showed that communities are fairly adaptable to the oil spill disaster as revealed by the household adaptive capacity index developed for the study namely economic well-being and stability, demographic structure, interconnectivity to higher level processes and natural resource dependence. Each of these indicators' result was analyzed why the community is fairly adaptable during perturbations. In the same manner, the desired trajectory towards transformative capacity was not evident as there are individual and group decisions and actions needing redirection. The spill incident was not seen as an opportunity by the residents to further improve the existing social institutions. Instead, the decisions made were

A.E.S. Tan
West Visayas State University, Iloilo City, Philippines

J.M. Pulhin (✉)
College of Forestry and Natural Resources, University of the Philippines, Los Baños, Philippines
e-mail: jpulhin@yahoo.com

R. Shaw (ed.), *Disaster Recovery: Used or Misused Development Opportunity*, Disaster Risk Reduction, DOI 10.1007/978-4-431-54255-1_17, © Springer Japan 2014

generally based on their intuition and expediency without consideration of future and long term consequence. The study concluded that for communities to better respond to future spills, there should be an institutionalization of processive and systemic change. Prescriptive approach by outsiders e.g. aid and grant by donors and inadequately informed organizations should be discouraged. Spontaneous actions and those attributed to informal organizations must be effectively harnessed with the understanding that these are for long term benefits and sustainability. Its realization will further strengthen the community's adaptive capacity in addressing future hazards. The chapter concluded with a proposed framework/model with GET (governance, education, technology) matrix in developing adaptive and transformative capacities of communities.

Keywords Adaptive capacity • Guimaras • Oil spill • Philippines • Transformative capacity

17.1 Introduction

Human activities are always linked to benefits and hazards. In the recent years the international community has become increasingly aware of the risks brought about by major technological accidents occurring near populated and environmentally sensitive areas. There is also a growing need to ensure that health, environmental and safety issues are addressed as an integral part of social and financial development.

Some people argued that large scale industrial accidents are unavoidable by-product of economic development. Others dismiss them simply as unintended "side-effects" of technical processes and can be avoided in the future with the aid of man's understanding of how that technology comes about in the first place. Tenner (1996) referred to them as "revenge effects"—the results of technology interacting with people and real environments.

Oil spills endanger public health, imperil drinking water, devastate natural resources and disrupt the economy. In an increasingly technological era, more countries if not all has become more dependent upon oil-based products to maintain its high standard of living. Unfortunately, the recent oil spill incident in Guimaras, Philippines exposed its inadequacy on how to address technological hazards like oil spills (Alabastro 2006; Cosculluela 2006). The country is caught off-guard on how to respond to such disruptions as can be seen in the different pronouncements by different government agencies allegedly tasked to monitor the impact of the spill (Philippine Daily Inquirer Report on Visayan Oil Spill at http://www.inquirer.net/specialfeatures/visayasoilspill/).

What transpired in Guimaras was considered as a disaster. It was considered as a situation or event, which overwhelms local capacity, necessitating a request to national or international level for external assistance. It is an unforeseen and sudden event that causes great damage, destruction and human suffering (Karunasena et al. 2009).

The impacts of the disasters have both human and environmental dimensions (Shaw 2006). Casualties including deaths, injured and misplaced people are the major impacts on human dimension of any kind of disaster. But disasters need not end in itself. The next phase for the community to move forward is to undergo the recovery process.

Most literature classified disasters in two broad categories. The first is natural disasters such as floods, hurricanes, tornadoes or earthquakes. While preventing a natural disaster is very difficult, measures such as good planning which includes mitigation measures and appropriate decision making can help reduce or avoid losses. The second category is man-made disasters. These include hazardous material spills (e.g. oil, toxic substances), infrastructure failure, or bio-terrorism. In these instances surveillance and mitigation planning as well as close monitoring are invaluable towards avoiding or lessening losses from these events.

The interplay of the dynamics in disaster science is attributed to the recovery process. The policies and procedures related to preparing for recovery or continuation of technology infrastructure critical to an organization after a natural or human-induced disaster poised an interesting subject for study. How some communities affected by disasters can easily regain their pre-disaster situation whereas others take longer if not impossible, to return to their pre-disaster situation may be attributed to their adaptive capacity. Some communities find it natural to be visited by disasters while some have a hard time adjusting may be explained by their adaptation technique to such disasters.

Smit and Wandel (2006) made an exhaustive review of the concepts on adaptation and adaptive capacity. They concluded that adaptation whether analyzed for purpose of assessment or practice is intimately associated with the concepts of vulnerability and adaptive capacity. Clarifying it further, they asserted that the vulnerability of any system (at any scale) is reflective of (or a function of) the exposure to sensitivity of that system to hazardous conditions and the ability or capacity of the system to cope, adapt or recover from the effects of those conditions (p. 286).

There is a growing interest in the incorporation of the concept and some literatures associate it to vulnerability, risks, adaptation, resilience and sensitivity, although the same could not be taken as identical (Adger et al. 2007; Brooks 2003; Gallopin 2006). When applied to social systems, however, the existence of institutions and networks that learn and store knowledge and experience, create flexibility in problem solving and balance power among interest groups, play an important role in adaptive capacity (Scheffer et al. 2000; Berkes et al. 2002). Subsequently, systems with high adaptive capacity are able to reconfigure themselves without significant declines in crucial functions in relation to social relations and economic prosperity (Folke et al. 2002). This reconfiguration focusing on the institutions and networks and how it could be identified as having high adaptive capacity needs scrutiny and further study. When there is low adaptive capacity, can it be attributed to the absence of such institutions and networks? If it does, what should be done to these institutions and networks to make it respond effectively, thus leading to high adaptive capacity?

A generic view of adaptive capacity would seem to involve two different components as advanced by Gallopin (2006) namely: (1) the capacity of the community to cope with environmental contingencies [to be able to maintain or even improve its conditions in the face of changes in its environment(s)]; and (2) the capacity to improve its condition in relation to its environment (s), even if the latter does not change, or to extend the range of environments to which it is adapted.

Five years after the disaster, the communities affected by the spill are reeling on how to reconfigure itself and adjust to the devastation the people experienced. Source of livelihood, environmental integrity, institutional arrangements and economic dislocations are among the aspects of their life which are continually affected. The community's post disaster situation is a case worthy of following up as this will galvanize any gains or losses in their well-being. Subsequently, there must be knowledge gained and lessons learned from the experience.

This chapter presents the responses of the affected communities to oil spill, identify the community's dynamics, and determine whether the same lead to the desired trajectories to further improve their situation when compared to their pre-oil spill state of affairs (transformative capacity). It also determines whether the collective actions made by the community's residents lead to their being more adaptive whenever a similar perturbation (of the same intensity and devastation) as that of the oil spill will occur in the future.

The study uses both qualitative and quantitative approach in research. A researcher-made questionnaire eliciting the communities' adaptive and transformative capacity was administered to the 88 respondents (10 % of total household). Other concerns which cannot be obtained through a questionnaire were extracted through a series of focus group discussions with key persons in the communities. Three communities in Guimaras, Philippines directly affected by the spill were chosen as study site. Data gathering was done from July 2007 to February2008 but on March28, 2008, a meeting was convened attended by the communities' key leaders, representatives from the academic community, local government and the researchers' research committee wherein key findings were presented for the communities' validation. As an offshoot of the study, the university of the major author extended its assistance by introducing livelihood opportunities as part of its extension thrusts. Thus continuous visit and interaction with the local communities was done until the end of the extension program on December 2010.

17.2 Oil Spill and Impacts to Recovery Process

The recent Louisiana Deepwater Horizon oil spill in 2010 and the Hebei–Spirit oil spill in Korea are some of the manifestations that industrial accidents create large-scale disruptions to the environment. It affects and created an impact to both the ecological and social systems.

The literature on the impacts of oil spill is not devoid of disturbing results on the well-being of those affected. For example, using the most effective approach for identifying real time exposure and resultant oil impacts have been raised due to past challenges to detecting health and environmental impacts (McCormick 2012). Some of the strongest research on impacts has shown the mental effects of oil spills such as depression, anxiety, and Post Traumatic Stress Disorder (Arata et al. 2000; Sabucedo et al. 2009). Disruptions of social structure was also experienced by an Alaskan community during the Exxon Valdez oil spill when their commercial fishing harvests was abruptly stopped (Picou et al. 2004).

To long term ecological impact on the coastal ecosystem and fisheries habitat remain uncertain. For example, following the Braer spill, the lobster and mussel industries were closed for 7 years because of lingering fears and uncertainties regarding shellfish safety; after the Prestige spill, marine scientists warned that locally harvested shellfish might remain poisonous for up to 10 years (Sutton 2007). In the case of Exxon Valdez some species such as Pacific Herring and pigeon guillemots are not recovering and lingering oil causes suffering of other species such as clams and birds (Baldwin et al. 2000).

Environmental disasters often exceed the local capacity to cope and invite external engagement in response to disasters and extreme events as indicated by inflow of external aid and expertise. Due to the novel and complex nature of oil spills it requires scientific expertise that communities oftentimes do not possess. Both the social and ecological impacts leading to recovery need utmost knowledge in coming up with a workable and effective decisions.

The community suffers large losses in terms of cost in response to disaster as people tend to ordinarily look to the negative consequences. Positive effects however constitute an important part in the disaster recovery processes and may include new skills and occupation, new sources of money, alternative developments and new disaster policies (Cheong 2011). The literature on Asian tsunami of 2004 for example identify some positive consequences like greater equity, economic restructuring, government reform, changed attitudes towards life and the world and increased community solidarity (Olshansky et al. 2008; Rajkumar et al. 2008; Badri et al. 2006). Using the context of adaptation, the long term effects of disaster recovery depend on the nurturing of these positive elements to offset economic loss and social disruptions.

17.3 Context of the Study

17.3.1 The Oil Spill Incident

Late in the afternoon of August 11, 2006, *M/T Solar I,* a vessel owned and operated by Sunshine Maritime and chartered by Petron, sank about 13.3 nautical miles south west off the coast of the island province of Guimaras (Fig. 17.1). It was on its way to Zamboanga from the port of Batangas. It was estimated to have sunk to a depth of 640 m in the Panay gulf area. The vessel was carrying a fuel cargo of 2.19 million

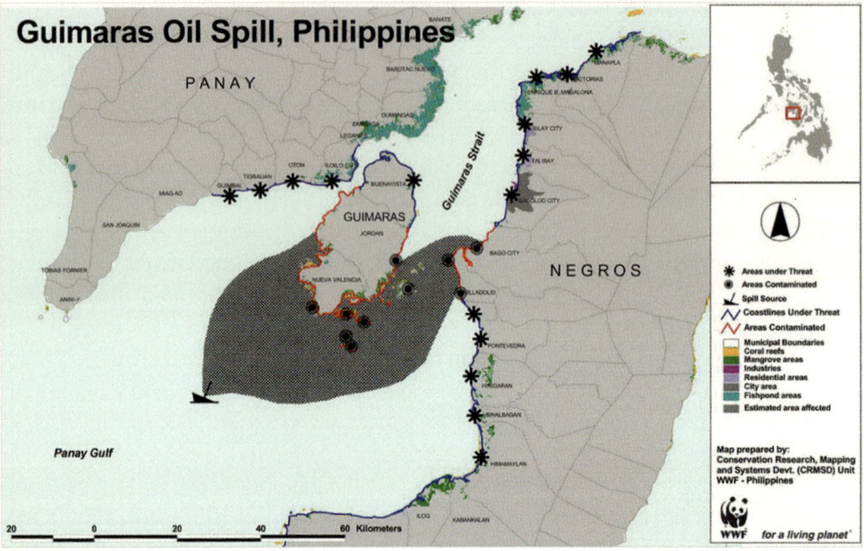

Fig. 17.1 The map of Guimaras and subsequent coastline affected by the oil spill in 2006

liters or 97 % of its carrying capacity of 2.24 million liters. In terms of weight, the fuel cargo was 2,064 metric tons or 96 % of the deadweight tonnage of the vessel. Conflicting reports cannot singly identify how much of the cargo fuel was spilled. Conservative estimates, however, revealed that it could be as low as 500,000 L to as high as 2 million liters of fuel cargo that might have been accidentally spilled out to the strait.

17.3.2 Costs and Damages

The spilled oil affected almost the entire coastline of Nueva Valencia, Sibunag and San Lorenzo in Guimaras Island. As reported by the National Disaster Coordinating Council (Annual Report 2006), 7,870 families were directly affected involving 39,004 persons and displacing 71 families. Said report likewise revealed that in terms of ecosystems, the oil spill affected DENR marine reserves (1,043.45 ha), coastline (234.84 km), coral reef (15.80 km^2), mangrove (478.48 ha), sea weeds (107 ha) and fishponds (974 ha).

The Resources, Environment and Economic Center, Inc. (REECS 2006) pegged the cost of damage to reach Php 9.223 billion). This does not even include the compensatory damages in terms of livelihood costs.

Containing the oil spill and cleaning up the areas were conducted immediately after the incident. During the initial occurrence of the oil spill in the areas, the locals were advised to devise contraptions to contain the oil spill. Spill booms were deployed along coastal waters to protect shallow waters and shorelines of coastal barangays, while the Philippine Coast Guard (PCG) undertook spraying of dispersant to diffuse oil slicks in deeper waters. Petron, the owner of the spilled oil, engaged people under the "cash-for work" scheme to help in the shoreline clean-up.

The National Disaster Preparedness Council created a task force composed of various lead and support national and local government agencies to address the oil spill. The Task Force was mandated to pursue, among others, oil spill response operations, retrieval/recovery operations of bunker fuel and sunken vessel, investigation of oil spill incident, policy review, mitigating further damages, reparation compensation due to economic loss and request for international technical assistance.

Recovery effort consists mainly of (a) coastal clean-up engaged by the community as well as sectors tasked in doing it, (b) provision of alternative livelihood to the affected fisherfolks, (c) education, information and communication on the hazards attributed to oil spill, (d) compensation to the affected sectors and (e) restoration of the degraded environment. The timetable for this recovery effort was not spelled out as some sectors still engaged in coastal restoration e.g. mangrove planting, subject mainly to the availability of whatever technical and financial assistance is at hand.

17.4 Results and Discussion

17.4.1 Adaptive and Transformative Capacities

In general, a species, population or individual may become better adapted by improving the conditions in its environment, even in the absence of changes in the latter. When applied to human systems, learning and technological progress fill that condition thus, the criterion for adeptness goes far "beyond being able to live and reproduce". It includes the viability of social and economic activities, and the quality of human life (Gallopin 2006). Accordingly, adaptability or adaptive capacity of human systems can be defined as the "capacity of any human system from the individual to the humankind to increase (or at least maintain) the quality of life of its individual members in a given environment or range of environments" (Gallopin et al. 1989, p. 377). Responses of biological systems to perturbations are purely reactive; human systems' responses, on the other hand, are both reactive and proactive (Smithers and Smit 1997).

A generic view of adaptive capacity would seem to involve two different components as advanced by Gallopin (2006) namely: (1) the capacity of the community to cope with environmental contingencies [to be able to maintain or even improve its conditions in the face of changes in its environment(s)]; and (2) the capacity to improve its condition in relation to its environment (s), even if the latter does not change, or to extend the range of environments to which it is adapted.

What makes a community better adapted to the changing conditions of the environment? Generally, a community that is exposed to hazards will be more vulnerable when it has less adaptive capacity. This broadly indicates the ways in which vulnerability of communities are shaped. Smit and Wandel (2006) aver that this does not necessarily imply that the elements of exposure, sensitivity and adaptive capacity can or should be measured in order to numerically compare the vulnerability of communities. They advanced that adaptive capacities and their determinants are dynamic (they vary over time), they vary by type, from stimulus to stimulus, and are place- and system-specific (p. 286).

Several studies showed that the forces that influence the ability of the system to adapt are the drivers or determinants of adaptive capacity (Kasperson and Kaspeson 2001; Walker et al. 2002; Adger 2003). In the local level the ability to undertake adaptations can be influenced by such factors as managerial ability, access to financial, technological and information resources, infrastructure, the institutional environment within which adaptation occur, political influence, kinship, networks, among others (Smit and Pilifosova 2003; Adger et al. 2005).

Some determinants of adaptive capacity are mainly local (e.g. the presence of strong kinship network which will absorb stress) while others reflect more general socio-economic and political system (e.g. the availability of state subsidized crop insurance). These, however, are not independent of each other. The illustration cited by Smit and Wandel (2006), for example, shows that the presence of strong kinship network may increase adaptive capacity by allowing greater access to economic resources, increasing managerial ability, supplying supplementary labor and buffering psychological stress. Similarly, economic resources may facilitate the implementation of a new technology and ensure access to training opportunities and may even lead to greater political influence. These individual determinants that are generated by interactions and vary in space and time constitute adaptive capacity.

In social systems, the existence of institutions and networks that learn and store knowledge and experience creates flexibility in problem solving and balance of power among interests groups, and plays an important role in adaptive capacity (Berkes et al. 2002). Subsequently, systems with high adaptive capacity are able to reconfigure themselves without significant declines in crucial functions in relation to primary productivity, hydrological cycles, social relations and economic prosperity. A loss of adaptive capacity is loss of opportunity, constrained options during periods of reorganization and renewal and inability of the system to do different things thus leading the system to emerge for such a long period of undesirable trajectory (Folke et al. 2002).

The determinants of adaptive capacity was made definite as the range of available technological options for adaptation, availability of resource and their distribution across the population, structure of critical institutions and decision-making, human capital including education and personal security, social capital including property rights, a system's access to risk-spreading processes, ability of decision makers to manage information and the public's perceived attribution of the source of stress (Yohe and Tol 2002). It is so encompassing that a single aspect may contribute to the overall adaptive capacity makeup of that particular community.

17.4.2 Adaptive Capacity of Communities

A questionnaire was developed to determine the level of adaptive capacity by modifying the normative indicators of Vincent's (2007) household adaptive capacity index (HACI) (Table 17.1). It consists of items eliciting the scores of a particular household on the identified areas of adaptive capacity namely: economic well-being and stability, demographic structure, interconnectivity to higher level processes, and natural resource dependence. These four components of the index were equally given 25 % each. Scores in these different indices were added to determine their level of adaptive capacity and interpreted in Table 17.1.

17.4.2.1 Economic Well-Being and Stability

It can be inferred that economic well-being and stability of a community are significantly affected when there is absence of infrastructure that could boost their economic well-being. Likewise, communities which are totally dependent on a single resource will obviously suffer the brunt when that resource is obliterated because of a disaster.

17.4.2.2 Demographic Structure

Dependents that are not capable of earning a living will obviously be a disadvantage in the adaptive capacity makeup of that household as more dependents could be viewed as an additional burden. Values of the household and the community in general however, place a significant role in the decision to have so many dependents as possible. This is evident in the failed family planning program. Additional mouths to fedare considered as investment and earners in the future. The minors, the elderly and even the member who is bedridden due to a lingering illness constitute a burden because of their inability to fend for themselves. They are more of a liability than an asset at this point in time. The picture will be different if these mouths to feed contribute to the earning capacity of the household.

Level of education of the respondents can also be a factor in the adaptive capacity makeup of the household. Due to their inability to make decisions that are of far reaching consequence as this requires higher order thinking skills, they will likely be contented with what is immediate and gratifying (Krishna 2003).

17.4.2.3 Interconnectivity to Higher Level Processes

Incidence of poverty was high even before the oil spill, thus, development assistance and access to microfinance was already in place in the area. Further assistance trickled in and organizations were established after the spill. Unfortunately, these did not

Table 17.1 Household adaptive capacity index (HACI) mean scores of the households in the three communities

	Barangay				
Indicator	Tan (n=22)	Suc (n=33)	Seb (n=33)	Mean (n=88)	Interpretation
1. *Economic well-being*					
a. Livestock income	1.18	2.45	2.35	2.10	5,000–9,999
b. Salaries and business	1.05	1.12	1.27	1.16	Below 5,000
c. Agricultural goods production	1.00	1.06	1.39	1.17	Below 5,000
d. Gifts from relatives and kin	1.36	1.12	1.76	1.42	Below 5,000
e. Family expenses	4.59	3.54	3.15	2.76	10,000–14,999
Sub-total score	9.18	9.30	9.93	9.51	
2. *Demographic structure*					
a. Number of dependents	3.59	3.15	3.09	3.24	5–6 dependents
b. Minor dependents	3.27	3.33	3.48	3.37	3–4 dependents
c. Earning capability	2.32	2.24	2.60	2.40	1–2 members
d. Hospitalization cost	4.14	4.24	4.03	4.14	1,001–1,999
e. Educational attainment	2.14	1.64	2.53	2.10	Elementary graduate
Sub-total score	15.45	14.61	15.76	15.25	
3. *Interconnectivity to higher level processes*					
a. Involvement in organizations	3.86	3.15	2.97	3.26	Fairly involved
b. Community connections	3.73	3.52	3.58	3.59	Fairly connected
c. Connections to the government	2.95	2.48	3.09	2.83	Less connected
d. Access to vital information	3.91	2.97	3.12	3.26	Seldom accessible
e. Access to financial resources	3.45	1.91	2.64	2.57	Least accessible
Sub-total score	17.91	14.03	15.40	15.51	
4. *Dependence to water resource*					
a. Dependence to fishing	1.59	1.76	2.42	1.97	Highly dependent
b. Mangrove dependence	2.95	4.48	3.91	3.87	dependent
c. Beach/resort dependence	3.50	4.45	4.24	4.14	Fairly dependent
d. Other employment dependence	2.82	2.67	2.91	2.70	Moderately dependent
e. Contamination dependence	2.04	1.94	2.76	2.27	Moderately dependent
Sub-total score	12.91	15.30	16.24	15.06	
Total score		53.24	57.33	55.33	Fairly adaptable
Score			*Description*		
81.00–100.00			Highly adaptable		
61.00–80.00			Moderately adaptable		
41.00–60.00			Fairly adaptable		
21.00–40.00			Lowly adaptable		
Below 20.00			Not adaptable		

really create a significant change in their economic well-being. Inability to invest in their connections which could have been used as a needed capital (social) to further improve their current plight was observed. It is obvious that level of education, indifference and the low aspiration of the community members do not make this social capital as an asset.

The use of social capital in adaptive capacity has not been maximized in this case. A household whose contacts go beyond the village level and extends to other geographical range and connect with a variety of institutions has more adaptive capacity (Adger 2003).

Memberships to organizations alone do not guarantee a change in their status. A couple of organizations that were established have been serving as milking cows for the hopeful individuals with a promise for benefits in the future. The political interplay of the warring politicians likewise do not serve the common good. It was also observed that local politicians were not serious in alleviating the plight of the affected residents as evidenced by partisan and patronage arrangements (Magramo 2006).

The residents' access to vital information which would otherwise aid them to formulate sound decisions is less accessible. The radios and televisions in their household are mostly used to entertain rather than make them well-informed. Subsequently, their connections to the government are also wanting. It implies that government services through their instrumentalities were not felt by the residents themselves.

17.4.2.4 Dependence on Water Resource

Dependency on the water resource has been determined in this category by extracting the contribution of fishing and other water related resources to household well being. The contamination of the water resource during the oil spill created a dislocation among fisher folks as their livelihood was interrupted for quite some time. When the lifeblood of a household is at risk, it leads to a host of problems which eventually make their existence vulnerable. Conversely, a household that has not totally depended on a single resource will likely overcome whatever losses that resource provide.

The level of adaptive capacity of the three communities is fairly adaptable using the household adaptive capacity index. The indicators dependence to the resource *and* economic wellbeing and stability are rated far above by them compared to community's demographic structure and their interconnectivity to higher level processes.

The community's adaptive capacity can be influenced by their collective action through positive understanding of the reasons why such actions are made and how the same can be sustained. Other than it, collective action has not significantly influenced the community's adaptive capacity.

17.4.3 Disaster Recovery and Transformative Capacity

Recovery efforts after every disaster are not devoid of issues in its realization. Most of these are attributed to the different practices as well as means or manner on how to return to their normal life after the disruption. The decisions made by recovery implementers on different scale at times do not redound to the lower recipients due to variations of doing it. Often though, it is the local-wisdom based disaster recovery model or the people themselves who decides on what is appropriate for them (Kusamasari and Alam 2012). But when outside and indifferent sector will spearhead without local stakeholders' involvement, post-disaster reconstruction is compromised (Cheng et al. 2012). What lessons are learned in the process? Who learned?

Transformative capacity as learning and does concerns self-organization that can transform the system into truly novel strategies and processes (Holling 2001). Making the internet as metaphor, the learning (transformations) represents true invention that can become reality in the kind of situation where the system is deeply responsive (vulnerable) to change or where change is desperately needed. He admits that the consequences of all of these are inherently uncertain and unpredictable. Systems should seize whatever opportunities are in store in the process.

It was Olsson (2003) who exhaustively proposed and extended the relevance of transformative capacity when applied to the social ecological systems when he wrote:

> Management systems contain learning and adaptive capacity but it is based on economic and social feedback and tends to ignore environmental feedback. They become decoupled from the resource base. Due to the co-evolutionary relations of social ecological systems, such mismanagement can cause the linked system to enter unsustainable and undesirable trajection and ecosystems' state and result in social traps. This has been referred to as perverse learning which reduces resilience and increase vulnerability of social ecological systems (pp. 12–13).

Olsson (2003) therefore reiterates that transformative capacity is the ability of an identified social ecological system to move to new or different configuration or create new stability domain; to redefine itself through acquisition of new variables or allowing them to emerge. Such capacity is important since many community development projects today require restoration of degraded ecosystems and reconditioning of social structures and processes.

As an end, transformative capacity will look into the social transformations unfolding in the course of the spill. It is on this assumption that answers for the following questions were sought: Does the community's adaptive capacity have something to do with its transformative capacity? If the needed social transformation will be addressed, how should it be done and to what extent will it lead?

Variables of transformative capacity were included in the questionnaire. Seven aspects of their decision-making were identified namely: clean up operations, compensation claims, environmental law enforcements, rehabilitation of area, organization of people's association, aids and supports and alternative livelihood provision.

Transformative capacity refers to the decisions made in the course of the oil spill. These decisions or actions were analyzed to see if they resulted in a desirable trajectory. In the questionnaire, statements that address the different aspects of transformative capacity were used. The respondents were made to express their views using the three choices namely the same happens (yes), did not happen (no) and uncertain. Table 17.2 summarizes the community's transformative capacity.

17.4.3.1 Clean up Operations

The respondents viewed that there was equity in the choice of participants and that payment was sufficient for those engaged in the clean-up operation. They were rated themselves knowledgeable in matters that concerns clean-up operations. But they were however, not sure if the technology used in the clean-up was efficient and appropriate as this is a new experience for them.

17.4.3.2 Compensation Claims

On the compensation claims made by the residents, some were not knowledgeable or have doubts about the process involved especially on the kind of paperworks to be accomplished. This can be attributed to the ambiguity and complex nature of availing claims. Likewise, some were not informed about it or they were not qualified to apply for compensation. They are, however, sure that the LGU led and spearheaded in the processing of claims.

17.4.3.3 Enforcement of Environmental Laws

The over-all mean in this aspect revealed an uncertain response. These results suggest that there exists indistinguishable construction of whether or not laws were strictly enforced in the area as the residents themselves could not see the difference.

17.4.3.4 Rehabilitation of Damaged Areas

Except for those from Brgy. Suclaran, all respondents believed that there was quick and efficient rehabilitation effort. Respondents from Brgy. Suclaran were uncertain if the rehabilitation efforts did take place in their area since theirs is the farthest from the oil spill site among the three barangays.

Table 17.2 Transformative capacity of the three barangays/communities studied.

Aspect	Tan (n=22)	Suc (n=33)	Seb (n=33)	Over-all
1. *Clean-up operation*				
a. Equity in participation	1.32	1.79	1.48	1.56
b. Sufficient payments	1.36	1.97	1.33	1.58
c. LGU led effort	1.41	1.45	1.12	1.32
d. Efficient technology used	1.77	1.97	1.91	1.90
e. Sufficient knowledge in clean-up	1.41	1.70	1.33	1.49
Sub mean	1.45	1.78	1.43	1.57
2. *Compensation claims*				
a. Compensation equitable	2.04	2.30	2.03	2.14
b. Red-tape in distribution	1.14	1.60	1.60	1.49
c. Transparent criteria used	1.64	2.06	1.64	1.80
d. LGU led in processing	1.23	1.39	1.21	1.28
Sub-mean	1.51	1.84	1.62	1.66
3. *Enforcement of environmental laws*				
a. Strict enforcement of laws	1.41	1.45	1.76	1.56
b. Violators apprehended/penalized	2.41	2.12	2.30	2.26
c. Aware of environmental laws	1.32	1.61	1.94	1.66
Sub-mean	1.71	1.73	2.00	1.81
4. *Rehabilitation of damaged areas*				
a. Rehabilitation quick and efficient	1.60	1.73	1.36	1.56
b. Localized participants	1.09	1.61	1.42	1.41
c. Reason for rehabilitation explained	1.04	1.64	1.40	1.40
Sub-mean	1.24	1.66	1.39	1.43
5. *Organizing of people's associations*				
a. Dues collected with assurance of benefits	1.64	1.70	1.64	1.66
b. Leaders stress responsibility	1.27	1.26	1.29	1.28
c. Objectives well explained	1.32	1.30	1.29	1.30
d. Strength in organization	1.23	1.18	1.03	1.14
d. Sustainability is leaders work only	1.73	2.00	1.48	1.72
Sub-mean	1.44	1.48	1.55	1.49
6. *Aids and supports*				
a. Stringent requirements by benefactors	1.82	2.33	2.09	2.11
b. Recipient expected to comply	1.95	2.15	1.97	2.03
c. System in place to address false claims	2.36	2.24	1.48	1.99
Sub-mean	2.03	2.24	1.85	2.04
7. *Provision of alternative livelihoods*				
a. Well-meaning projects in place	1.82	1.79	1.21	1.58
b. Coordination by LGUs and NGOs	1.32	1.67	1.18	1.40
c. Livelihood projects welcomed	1.32	1.45	1.18	1.32
Sub-mean	1.49	1.64	1.19	1.44
Scale		*Interpretation*		
2.34–3.00		No or absence of such actions/decisions		
1.67–2.33		Uncertain		
1.00–1.66		Yes or presence of such actions/decisions		

17.4.3.5 Organizing of People's Associations

In this aspect, the respondents were certain in the leaders who stressed responsibility, believed that the objectives were well explained, and that there was strength in the organization put up. They were, however uncertain that sustaining the organization was the work of the leaders only.

17.4.3.6 Aids and Supports

In the aids and supports that followed after the oil spill, respondents were uncertain in almost all aspects except those from Brgy. Sebario who believed that a system was in place to address the fictitious claims. This can be attributed to their barangay captain who personally saw to it that everything was facilitated in a transparent manner. Over-all, however, aspects on aids and support were viewed as uncertain by the respondents.

17.4.3.7 Provision of Alternative Livelihoods

Respondents from Brgys. Tando and Suclaran were not certain that well-meaning livelihood projects were in place. Livelihood projects, of which the two barangays were the main beneficiaries due to their demographic make-up were heavily doubted with regards to the implementation of providing livelihood. Over-all, however, provision of alternative livelihood was certain in the three barangays.

The experience had brought the community into trajectories leading to their further need of understanding how the spill can be seen as an opportunity. Otherwise, it will have a significant bearing on how they will be able to respond to future perturbations. Clean-up operation, compensation, enforcement of environmental laws, organizing into associations, alternative livelihoods as well as aids and supports—all exposed the community to experience. The decisions on what course of action to realize is basically anchored on intuition and at times expediency. This is attributed to their inadequacy in knowledge and management on what to do with the spill.

It can be gleaned from the data that the community's adaptive capacity makeup is an instrument for them to create a desired state after the spill. If adaptive capacity is looked upon as a prearranged, certain and definite condition, the oil spill experience subsequently unmasks that pre-arranged condition. It is up to the community how to come across to the experience as something which may appear as either an opportunity grab or a threat to shun away.

The decisions and actions made after the spill reveal the inadequacy of the community to respond. To say that the spill is a new experience for them is an understatement as this is just a common perturbation similar to flood or typhoon. The only difference is that oil spill do not come as often hence, its novelty as a hazard and the manner of responding to it are not easily identified and institutionalized.

All things being equal though, the way how those prearranged condition are in place (Yohe and Tol 2002) and a workable social institutions directed in addressing any perturbation (Smit et al. 2000) will considerably strengthen the community's adaptive capacity. As the community reconfigures to a desired state, the lessons and insights gained and learned in the process may lead them to a more acceptable condition ready and prepared for another perturbation.

17.5 What Insights Are Being Derived?

Visiting the area after years of the incident, the following insights may be derived from the oil spill experience in terms of what the disaster has brought to the affected communities and their households.

17.5.1 Institutionalization of Process and Systemic Change

While disasters wreak havoc to its victims they also offer opportunities for transformation. Lessons from typhoons, flooding and other natural calamities reveal the inadequate preparation and action before, during and after the onslaught. This could be attributed to the affected communities' failure to institute processes and systems that would redound to the adaptive capacity of the community. Among the processes and systems that could be strengthened include a workable disaster preparedness management plan, a sustainable alternative livelihood in case of longer and unsolicited disruptions and a transparent policy making mechanism that exhibits strong political will in realizing the desired outcomes. Collectively, this must be institutionalized and has to be imbibed and accepted by the community to merit utmost support and cooperation.

Because of the spill, significant legislations have been realized both in the local and national sphere. The Oil Pollution Compensation Act of 2007 as well as the establishment of Fund in the clean-up is the significant realizations (Figs. 17.2 and 17.3). Locally, the municipal and provincial government created committees/bodies that will solely address concerns in the light of the spill. An annual educational campaign aimed on the development of awareness about the spill is regularly made. School children were taught as part of their curriculum the hazards of the spill and what to do in case encountering one.

17.5.2 Discouragement of Outsiders' Prescription

The usual outside technical and financial assistance after the disaster is a welcome gesture. Outside assistance here may imply resources, expertise, financial and related supports the affected community avails which are not necessarily sourced out from

Fig. 17.2 Mangroves contaminated by sludge of oil in the area

the affected communities. However, this assistance should be carefully and meticulously planned taking into consideration the existing resources (social, physical and material) before the assistance may be implemented (Tan and Pulhin 2012). An accountability of those responsible in implementing them and the desired changes it wish to carry out may be sought. Likewise, proper consultation with the residents and using the best and paramount knowledge available in coming up with a decision should be strictly and uncompromisingly observed.

As what happened after the spill, outside intervention were relatively ineffective and useless as they only create animosity and indifference to the community's existing situation. To cite a case: a motorized fishing boat was donated for the fisherfolks to use but it was laid idle since they (fisherfolks) cannot agree on the mechanism on how the fishing boat will be effectively utilized by them. Another is when live chicks were donated to augment the residents' lost income after the spill. The supposedly fastidious gesture wiped out all the local chicken they raised because of the introduction of a vaccinated chicken killing the non-vaccinated local population.

There are instances when the donor would prescribe their own "rules" on the assistance that they provide. This is without proper consideration of the local realities and the social interactions going on in the community. It not only made the recipient "enslaved" to the donor's prescription, but it also created a gap that often leads into an unsustainable practice. In one of the donor organization's desire to help, they provided seed capital for the livelihood projects of the residents provided their finished product has to be bought by the organization. At the start, the arrangement seem to be alright but in the long run, the residents, due to their inadequate background in sustaining a business enterprise failed to make it as a profitable venture.

Fig. 17.3 Volunteers doing the clean-up operation

17.5.3 Informal Organizations' Spontaneous Actions Harnessed

Oil spill is a novel industrial disaster for the communities to experience. Droughts and storm surge as well as typhoons are usual encounters but oil spill incident is a new thing for them. Responding to oil spill incident requires another mind frame, technique and even adaptive measures which the community are not prepared to apply. Among the adaptive measures would be the realization of any collective effort in disaster recovery. Since no formal organizations can be easily activated, the informal ones may be tapped to produce results. This include the households, close relatives, community peers and those who are directly at the forefront during disasters of this type. It was found out that formal organizations' interventions like the local and national government as well as established disaster recovery agencies do not come as fast as when these informal organizations acted. What is essential is that they (informal organizations) may be equipped with the needed technology and knowledge to arrive at an appropriate decision making.

In realizing the disaster recovery effort, involvement of the critical mass is crucial. Whatever reasons they (residents) have in mind why they involve themselves, it is imperative that these may be identified and directed to the desired trajectories. Failure in this arena may make the recovery effort a disappointment.

Finally, the one-shot intervention is often criticized for it does not create the needed change in the long run. Processes have to be observed and carefully implemented as intangible results are far more important than one that meets the eyes. Values, reeducation, sustainability thinking—among others, may be considered when interventions after the disaster is made.

17.6 Ways Forward

The oil spill experience provided wealthy and unassailable lessons that can be instituted to similar situations. This affirms that disasters or any other hazards that can befall to a community may be looked both ways—an opportunity to improve the risk reduction approach or it can create further other hazards. To look it as a disaster in its real sense will make one miss the essence of improving further the current approach or practices during disaster recovery.

The lessons may be summarized in the framework below (Fig. 17.4) and how, developmental practitioners and disaster reduction managers can learn when oil spills are inevitable:

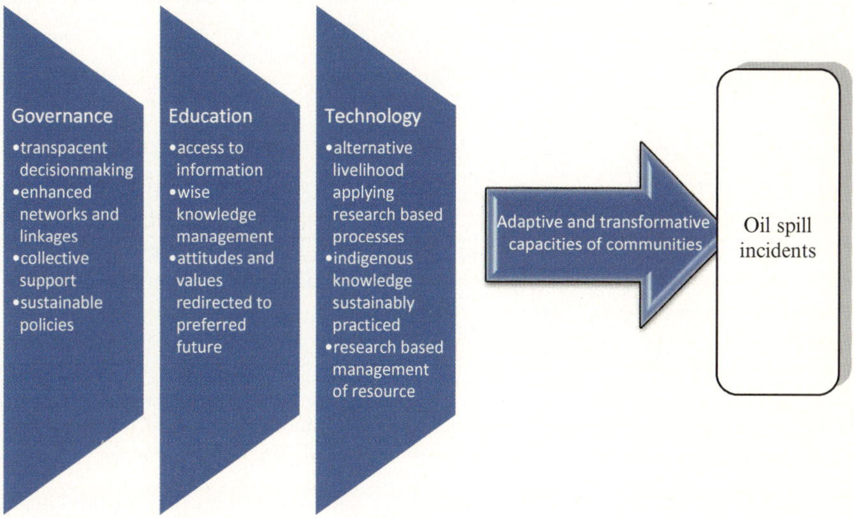

Fig. 17.4 Model/framework with GET matrix

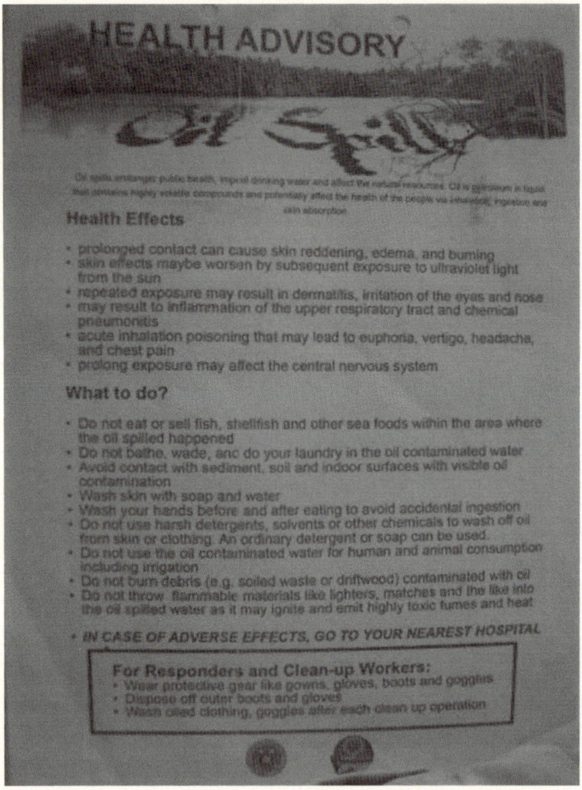

Fig. 17.5 One of the notices to the residents issued by the government's department of health warning them of the possible effects of oil spill

17.6.1 On Governance

Transparency requires the recipients' involvement and active participation in the whole process. The decisions made, involvement of the directly affected, and the benefits the latter may allegedly avail must be done and communicated in a transparent and visible manner. Decision makers must comply these prerequisite if success in disaster recovery is aspired.

17.6.2 On Education

The concept of education for sustainable development (ESD) or for an individual to flourish indefinitely may be ingrained and used as a starting mechanism for the younger generation to follow. This is without prejudice to whatever current strides the community/country formulated on their preferred future (Fig. 17.5).

17.6.3 *On Technology*

The incident is a wake-up call on how to maximize the knowledge available relative to disaster recovery particularly doing oil spills. The clean-up operations which require research-based approaches need further studies as environment differ and technology application cannot be summarily applied to these situations.

References

Adger W (2003) Social capital, collective action and adaptation to climate change. Econ Geogr 79(4)

Adger WN, Arnell NW, Tompkins EL (2005) Successful adaptation climate change across scales. Global Environmental Change Part A 15(2):77–86

Adger WN, Agrawala S, Mirza MMQ, Conde C, O'Brien K, Pulhin J, Bulwarty R, Smit B, Takahashi K (2007) Assessment of adaptation practices, options, constraints and capacity. In: Parry MM, Canziani OF, Palutikof JP, van der Linden PJ, Hanson CE (eds) Climate change 2007: impacts, adaptation and vulnerability, contribution of working group II to the fourth assessment report of the intergovernmental panel on climate change. Cambridge University Press, Cambridge, pp 717–743

Alabastro E (2006) Science-based intervention in solar 1 oil spill. http://www.dost.gov.ph/index.php?option=com_content&task=view&id=160&Itemid=89. Accessed 16 June 2007

Arata C, Picou J, Johnson D, McNally S (2000) Coping with technological disaster: an application of the conservation of resources model in the Exxon Valdez oil spill. J Trauma Stress 12(1):23–39

Badri SA, Asgary A, Elfekhari AR (2006) Post disaster resettlement, development and change: a case study of the 1990 Manjil earthquake in Iran. Disasters 30(4):451–468

Baldwin M, Davis E, Witham B (2000) Recent development: a review of developments in ocean and coastal law 1999–2000. Ocean Coast Law J 5(2):367–368

Berkes F, Colding J, Folke C (eds) (2002) Navigating social-ecological systems: building resilience for complexity and change. Cambridge University Press, Cambridge

Brooks N (2003) Vulnerability, risk and adaptation: a conceptual framework. Working paper 38, Tyndall Centre for Climate Change Research, University of East Anglia, Norwich

Cheng Y, Wilkinson S, Potangarao S, Seville E (2012) Resourcing for post-disaster reconstruction: a comparative study of Indonesia and China. Disaster Prev Manage 21(1):7–21

Cheong SM (2011) A social assessment of the Hebei–Spirit oil spill. Geo J 76:539–549. doi:http//doi.10.107/s10708-010-9368-4

Cosculluela R (2006) Guimaras oil spill a national disaster. http://www.sunstar.com.ph/static/ilo/2006/08/27/news/guimaras.oil.spill.a.national.disaster.html. Accessed 16 June 2007

Folke C, Colding J, Berkes F (2002) Building resilience for adaptive capacity in social-ecological systems. In: Berkes F, Colding J, Folke C (eds) Navigating social-ecological systems: building resilience for complexity and change. Cambridge University Press, Cambridge

Gallopin GC (2006) Linkages between vulnerability, resilience and adaptive capacity. Glob Environ Chang 16:293–303

Gallopin GC, Gutman P, Maletta H (1989) Global impoverishment, sustainable development and the environment: a conceptual approach. Int Soc Sci J 121:375–397

Holling CC (2001) Understanding the complexity of economic, ecological and social systems. Ecosystems 4:390–405

Karunasena G, Amaratunga D, Haigh R, Lill I (2009) Post disaster waste management strategies in developing countries: case of Sri Lanka. Int J Strategic Property Manage 13:171–190

Kasperson JX, Kaspeson RE (2001) Global environmental risk. United Nations University Press and Earthscan, New York
Krishna A (2003) Understanding, measuring and utilizing social capital: clarifying concepts and presenting a field application from India. CAPRi working paper 28, International Food Policy Research Institute, Washington. http://www.capri.cgiar.org/pdf/capriwp28.pdf
Kusamasari B, Alam Q (2012) Local-wisdom based disaster recovery model in Indonesia. Disaster Prev Manage 21(3):351–369
McCormick S (2012) After the cap: risk assessment, citizen science, and disaster recovery. Ecol Soc 17(4):706, http://dx.doi.org/10.5751/ES-05263-170431
Magramo M (2006) Political interplay during the Guimaras oil spill. Paper presented during the international conference on research dissemination, Centro Escolar University, Manila, 12–14 October 2006
Olshansky R, Johnson I, Horne J (2008) Planning for the rebuilding of New Orleans. J Am Plann Assoc 74(3):273–287
Olsson P (2003) Building capacities for resilience in social ecological systems. Doctoral dissertation. Stockholm University, Sweden
Picou J, Marshall B, Gill D (2004) Disaster, litigation and the corrosive community. Social Forces 82(4):1426–1497
Rajkumar AP, Premkumar TS, Tharyan P (2008) Coping with the Asian tsunami: perspective from Tamil Nadu, India on the determinants of resilience in the face of adversity. Soc Sci Med 67(5):844–853
Resources, Environment and Economic Center, Inc. (REECS) (2006) Preliminary cost of oil spill in Guimaras. http://www.rees.org. Accessed 5 December 2006
Sabucedo JM, Constsantino A, Ferraces MJ, Merino H, Duran M (2009) Psychological impact of the prestige catastrophe. Int J clinic Health Psychol 9(1):105–116
Scheffer M, Brock W, Westley F (2000) Mechanisms preventing optimum use of ecosystem services: an interdisciplinary theoretical analysis. Ecosystems 3:451–471
Shaw R (2006) Indian Ocean tsunami and aftermath:need for environment-disaster synergyin the reconstruction process. Disaster Prev Manage 7(1):5–20
Smit B, Pilifosova O (2003) From adaptation to adaptive capacity and vulnerability reduction. In: Smith JB, Klein RJT, Huq S (eds) Climate change, adaptive capacity and development. Imperial College Press, London
Smit B, Wandel J (2006) Adaptation, adaptive capacity and vulnerability. Glob Environ Chang 16:282–292
Smit B, Burton I, Klein R, Wandel J (2000) An anatomy of adaptation to climate change and variability. Climate Change 45:223–251
Smithers J, Smit B (1997) Human adaptation to climatic variability and change. Glob Environ Chang 7:129–146
Sutton P (2007) The environment: a sociological introduction. Polity, UK, pp 78–80. ISBN: 9780745634333
Tan A, Pulhin J (2012) Collective action and adaptive capacity of communities during disasters: the case of oil spill in Guimaras, Philippines. J Environ Sci Manage 15(1)
Tenner E (1996) Why things bite back: technology and the revenge of unintended consequences. Alfred A. Knopft, New York
Vincent K (2007) Uncertainty in adaptive capacity and the importance of scale. Glob Environ Chang 17:12–24
Walker B, Carpenter SR, Anderies J, Abel N, Cumming GS, Janssen M, Lebel L, Norberg J, Peterson GD, Pritchard R (2002) Resilience management in social-ecological systems: a working hypothesis for a participatory approach. Conserv Ecol 6, (online) http://www.ecologyandsociety.org/vol6/iss1/art14/pring.pdf
Yohe G, Tol R (2002) Indicators for social and economic coping capacity – moving toward a working definition of adaptive capacity. Glob Environ Chang 12:25–40

Chapter 18
Environmental Management and Urban Recovery

Akhilesh Surjan and Rajib Shaw

Abstract Disasters are caused when unprepared society met with hazards. Neither there is "one-size-fit-all" solution for disaster risk reduction nor for disaster recovery. Both these aspects are closely intertwined and success of the efforts largely depends of local conditions—social fabric, institutional mechanisms efficiency, environmental considerations, economic base, and above all priority for action. Environment plays an important role in increasing or decreasing the risk of built habitat. For example, ignorance of natural topography and drainage channels while developing infrastructure, buildings and housing in a expanding urban setting can seriously compromise flood safety and also can make disaster recovery extremely expensive and cumbersome. Similarly, due to development pressures, encroachment and destruction of mangroves, forests, floodplains, wetlands, etc., which provides natural buffer against various hazards can potentially expose vast population to the extent where disaster recovery will be extremely difficult and prolonged if not impossible. Environment management especially in the context of urban areas, where high density of population and infrastructure is concentrated, is not an option but a pre-requirement for reducing risk from major disasters and towards swift recovery. This chapter argues that in an urban area, disaster recovery is always a challenging task. However, sound urban planning and environment management practices can help in multifarious ways. For example, it can reduce damage from disasters, enhance response efficiency after the disaster and reduce recovery time and efforts, and even can act as launching pad towards "building back better" towards resilience.

A. Surjan
Global Survivability Studies (GSS), Kyoto University, Kyoto, Japan

R. Shaw (✉)
International Environment and Disaster Management, Graduate School of Global Environmental Studies, Kyoto University, Kyoto, Japan
e-mail: shaw.rajib.5u@kyoto-u.ac.jp

R. Shaw (ed.), *Disaster Recovery: Used or Misused Development Opportunity*,
Disaster Risk Reduction, DOI 10.1007/978-4-431-54255-1_18,

Keywords Environment and disaster linkage • Governance • Housing recovery • Livelihoods • Urban recovery

18.1 Background

> ...throughout the world, we must work harder in the recovery stage to avoid reinstating unnecessary vulnerability to hazards. As I have often said, "building back better" means making sure that, as you rebuild, you leave communities safer than they were before disaster struck.

-Bill Clinton, UN Secretary-General's Special Envoy for Tsunami Recovery, 20 December 2006 (IRP 2007)

The massive wave of urbanization is expanding existing size and swelling densities of urban centers of developing countries of Asia and Africa. While urbanization certainly has potential for all-round development and enhancement of quality of life to its populace, these benefits may not reach to every urban dweller. Institutions responsible for managing cities, social structures which facilitate (or hinders) penetration of urban benefits to those who need it most and physical environment which makes certain parts of the urban area unfit for human settlement—all contribute to shape the present day cities, for better or worse.

Widely used Emergency Management Cycle's includes four phases (1) preparedness/mitigation (2) planning/preparedness (3) response (4) recovery. While recovery always finds mention, its intrinsic aspects remain diluted especially in the urban context. Soon after a disaster occurs, response phase receives utmost attention in which tasks includes search and rescue, humanitarian aid and temporary rehabilitation. Recovery, often considered as collectively as all those efforts, which aimed at bringing the situation towards normalcy i.e. pre-disaster condition.

Returning to normalcy i.e. pre-disaster condition is a contested subject of discussion for various reasons. Firstly, it assumes that pre-disaster conditions were "safe conditions." Whereas the fact is, disaster related losses occur not merely due to intensity of hazard but also due to vulnerable conditions. Vulnerability emanates from combination of factors, most of which have are strongly linked to urban areas. These factors may include but not limited to: lack of awareness of nature of hazard itself, poor quality housing, absence or not adherence to acts, laws and bye-laws to enforce safer construction practices, poor quality infrastructure, land use and housing policies which forces poor people to settle on unsafe locations and informal settlements, and so on.

In this backdrop, this chapter first discusses basic tenets about disaster recovery in general and urban recovery in particular. It progresses further within the context of urban environmental governance and management and its role in better recovery. Through analyzing case-examples of past urban recovery, practical aspects and challenges in long-term recovery will be diagnosed. Finally, the paper will conclude with a note that how urban recovery offers a window of opportunity for urban resilience and sustainable urban future.

18.2 Urban Disaster Recovery

Disaster recovery always finds mention among the important phases of the disaster cycle, but remained least understood (IRP 2007; Mileti 1999; Chang 2010). There is also a vacuum in reliable and standardized database on disaster recovery. Emergency Disasters Database popularly referred as EM–DAT (http://www.em-dat.net/) records and provides quantitative data on immediate losses from disasters (for example, deaths, affected persons or property damage). Routine and thorough examination of disaster-struck locations with consistent data documentation methods can provide deeper insights into recovery especially in complex urban disaster recoveries. However, in practice, most of the recovery related data and other qualitative information is obtained by detailed, one-time surveys only. Recent research on disaster recovery emphasizes that statistical data such as—time series on population or employment, remain a vastly underutilized source of information (Chang 2010). Chang reported, "While advances have been made in measuring and comparing disaster vulnerability (for example, Birkmann 2006; Cutter et al. 2003), a common, robust framework has hitherto been lacking for measuring recovery."

Addressing the needs of swift urban recovery not only requires surveys, case studies and computer modeling but also greater insights of urban ecosystem, governance, planning, etc. Surveys can explain factors responsible for individual businesses recovery. Statistical indicators can provide explanation to differences in recovery in different sectors in the context of regional economic recovery. Detailed case studies of urban disaster recovery often reflect complex processes involved in various stages of recovery. It is difficult to completely standardized disaster recovery process as it varies significantly depending on type of hazard, local context, governance, pre-existing challenges with quality of built and natural environment.

Defining disaster recovery is also a challenge and thus varies widely in literature. One common understanding is that disaster recovery aims at bringing the disaster affected region to a pre-existing "normal" situation. However, in real life situation, it is extremely difficult to bring everything back to pre-existing situation. Depending upon disaster recovery process, affected area usually attain a "new normal" which may be better or worse compared to pre-existing "normal" situation. Reconstruction, restoration, rehabilitation and post-disaster redevelopment are all considered to be part of the recovery process. In the context of developing countries, numerous shades of vulnerability exist in pre-disaster conditions. Disaster recovery, which aims at bringing the community back to pre-disaster levels in the shortest possible duration, may end up recreating similar vulnerabilities, if not given due attention (IRP 2007). Therefore, disaster recovery is now presented as opportunity in which a safer built and natural environment can be created while also addressing some of the pre-disaster condition vulnerabilities.

When the scale of disaster is unprecedented or governance system is weak, disaster recovery process faces conflict of interest among stakeholders involved. In such situations, disaster recovery widens gap between various economic sections of the society. Similarly, top–down approaches to disaster recovery in combination with weak local institutions also lead to slow disaster recovery and uncertainty about future. Disaster recovery is a versatile process, involving regenerating local economy, rebuilding buildings and infrastructure including institutions, households, organizations, businesses, and the society at large.

18.3 Urban Disaster Recovery: Experiences from Asia

Asia is the epicenter of urbanization and thus this paper will specifically dwell on two urban disaster recovery examples from the region, Kobe (Japan) and Aceh (Indonesia). These two examples are carefully chosen to represent urban disaster recovery from a developing as well as developed country perspective. Disaster risk reduction is a precondition to reduce vulnerability of urban community and it also helps is shortening the recovery time. Urban disaster risk reduction is receiving special attention especially since larger share of world's population is now residing in urban areas. Urban risk reduction is a dynamic process, which goes hand in hand with development aspirations of its citizens.

After the disaster, it is often a challenge to recognize when urban disaster recovery phase ends and development phase restarts. Investing during pre-disaster phase in building resilience of disaster affected community and region is very important so that development phase quickly follow disaster recovery phase. Therefore, systematic disaster recovery planning is an essential requirement, especially in an urban area prone to natural hazards, which shall be comprehensively done during the pre-disaster phase. With the increased awareness towards disaster preparedness, many cities in Asia are now investing resources to prepare disaster management plan. However, urban disaster recovery planning is often limited to immediate response, which includes search and rescue, relief, temporary shelter and provision of short-term assistances. In this backdrop, the following section will present a review of recovery experiences from Kobe and Aceh.

18.4 Kobe Earthquake Recovery Experience

Kobe, fifth largest city of Japan, experienced a strong 6.9 magnitude earthquake (magnitude 7.3 on the Japan Meteorological Agency scale) on 17 January 1995. Over 6,400 lives were lost of which about 5,500 lives were lost in building collapse and subsequent fire in Kobe and neighboring region (Government of Japan 2005). Damage from this earthquake is estimated at whooping sum over 10 trillion Japanese Yen (USD 100 billion). Intense shaking of buildings due to earthquake leading to ground failure as well as urban fire resulted in severe damage to housing stock. It is reported that there was complete destruction of 105,000 houses, half-damage to 144,000 houses, and partial damage to 264,000 approximately.

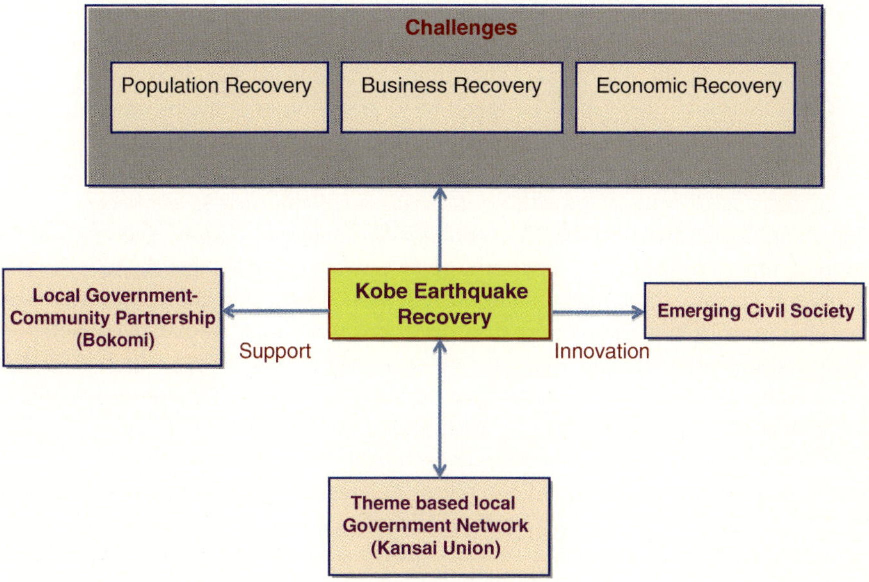

Fig. 18.1 Kobe earthquake recovery: challenges and innovation

Urban infrastructure systems, which serves as lifeline to city's populace, were also seriously damaged—"electric power took nearly a week to restore, telecommunications two weeks, water and natural gas about two and a half months, railway seven months, and highway transportation twenty one months" (Chang and Nojima 2001). Kobe is an important port city. At the time of earthquake, Kobe was sixth large container port in the world. This significantly important Kobe port was not only heavily destroyed but also paralyzed driving force of regional economy (Chang 2000). UNCRD reported Kobe earthquake as "developed world's first experience of a catastrophic urban earthquake" (UNCRD 1995; Chang 2010).

In the following section, recovery of the Kobe region is reviewed which is documented through referring to several published studies (mainly from: Chang 2010; Shaw and Goda 2004; Matsuoka and Shaw 2006; Matsuoka et al. 2012; Nakagawa and Shaw 2004). Discussion is also summarized in Fig. 18.1.

18.4.1 Demographic Recovery

Japan is one of the fastest ageing countries in the world. In the aftermath of Kobe earthquake, city took 10 years to regain its pre-disaster population levels. Significant spatial disparities in terms of physical damage and over the course of recovery are reported in Kobe due to the earthquake. Most wards of Kobe suffered remarkably from loss of population. On the contrary, suburban wards that experienced relatively little damage in the earthquake maintained or exceeded pre-disaster levels and trends. Obviously, there must be temporary relocation of some of the population from heavily damaged wards to housing in these suburban wards, which was partly reversed over time. By 3 years after the earthquake, ward population levels returned to a more stable rate of growth or decline—a new normal level. Economically better-off locations of Kobe, concentrated in suburban, eastern and downtown wards, reported stabilized long-term recovery or exceeded pre-disaster levels. However in many other wards in the western part of Kobe, the 2006 population was varying from 79 % to 93 % of pre-disaster levels. When post-earthquake population changes are considered in light of long-term pre-disaster trends, Chang (2010) found that, to a large extent, the disaster served to accelerate long-term patterns of population decline in the inner city areas and growth in the suburbs.

Differences in recovery experiences have primarily been attributed to physical damage, which is also observed in patterns at a higher spatial resolution. Wards in Kobe city serve as administrative unit and therefore range of published data is available. Typical average population of a city ward ward is around 170,000. Earthquake damage estimated in 165 census districts of Kobe city, which is calculated as the percentage of housing units destroyed, ranged from no significant damage to a loss of 95 % of the housing stock in the areas that were devastated by post-earthquake fires. Study by Chang (2010) noted, "by 6 years after the earthquake, the strength of influence of "housing damage" has diminished dramatically, and housing damage alone provides virtually no explanatory power for differences in population recovery. This suggests that as recovery proceeds, other factors (potentially including differences in income, reconstruction activities, socio- demographic composition and so on) become much more important than simply the extent of physical damage."

18.4.2 Recover of Business Sector

Business sector in Kobe suffered substantially due to earthquake. Chang (2010) noted that "between 1991 and 1996, the number of businesses in Japan as a whole declined by 1 %. If this can be considered the "without-earthquake" trend, the remainder of the 10 % loss Kobe suffered in this same period can be attributed to the disaster." These numbers were further analyzed by business size and revealed "slight decline in small businesses (one to nine employees) and increase in large and very large businesses was occurring in Japan as a whole over this period, these shifts occurred to a much greater degree in the disaster region." However, in Kobe, shrinkage in the number of small businesses was reported by 13 % and on the contrary, 41 % growth was reported in the number of very large businesses. It is often noticed that small businesses not only suffer the most in disasters but also have great difficulty in recovering. Statistical analysis by Chang (2010) shows that this may be part of a larger restructuring effect that is accelerated by disaster. Notably, these effects gradually diminished by 2001, 5 years after the earthquake. Interestingly, "6 years after the earthquake, Kobe had 12 % fewer businesses than in 1991. Half of this loss can be attributed to long-term trends (indicated by the data for Japan), but the remainder can be considered an impact of the earthquake. Moreover, an accelerated loss of small businesses and shift toward very large businesses can be considered a long-term effect of the disaster" (Chang 2010).

18.4.3 Recovery of Economic Activities

For key measures of economic activities, national and sub-national level statistical time series data is extensively available in Japan. From 1991 to 2005, construction sector activities thrived in Kobe due to demand of speedy reconstruction. Temporary boost in per capita income from reconstruction-related activities was observed, including to some degree an inflow of funds from outside the city. However, about 6–7 years after the earthquake, this trend was less prominent. Analysis by Chang (2010) revealed that "per capita income 10 years after the earthquake was only about 89 % of pre-disaster levels. It may be noted that Kobe accounts for only a small fraction (roughly 1 %) of Japan's national population and economic activity. Ten years after the earthquake, Kobe's gross regional product (GRP) decreased by about 9 % while Japan's grew by about 1 %. Per capita income in Kobe can be seen to have decreased in the long run, following the temporary reconstruction boost, from about 5 % higher than Japan's per capita income before the earthquake to about 5 % lower some 10 years after."

This study also observed that "10 years after the earthquake, production activity in many sectors remained considerably lower than pre-disaster levels, including manufacturing (75 % of pre-disaster), construction (65 %) and wholesale and retail trade (80 %). Output in the services sector, however, was 118 % of pre-disaster levels. Stabilization was attained more rapidly in trade and services (1 year) and somewhat more slowly in construction (5 years) and manufacturing (6 years). These disparities reflect in part some differences between export-oriented and locally-oriented sectors. Manufacturers largely produce for export from the region, hence face both more stable post-disaster markets and potentially more intense competition from manufacturers in other regions unaffected by the disaster. Retail trade primarily faces a local market, where customers have been affected by the disaster. Its recovery can be expected to be highly correlated with the economic recovery of the disaster region."

Interestingly, Kobe experienced structural shift away from manufacturing toward a more service-oriented economy. Kobe port was which was severly destroyed by the earthquake took about 2 years to rebuilt. Chang (2010) also revealed that "domestic trade at Kobe port, despite an initial rebound after facilities were rebuilt, stabilised within 4 years at some 40 % of pre-disaster levels. Container trans-shipment cargo, the most vulnerable sector of the port's activities, has levelled off in recent years at less than 10 % of pre-disaster levels." One of the important observation from this study is that "these losses derive from the dynamic context of long-term trade, market and port competition in Japan and Asia, as the disaster served to accelerate pre-disaster structural changes that had already been under way."

18.4.4 Community-Government Partnership in Kobe

Community based organizations (CBOs) played very unique and important role in Kobe's disaster recovery (Shaw and Goda 2004; Nakagawa and Shaw 2004; Matsuoka et al. 2012). Community based organizations are uniquely placed to provide an enabling environment for networking among ordinary citizen. Human networking has tremendous strength to generate social-capital, which is especially noticed in the recovery phase of Kobe earthquake. Kobe's community led initiatives transformed Japanese disaster management regime, which was earlier heavily driven by engineering and technological solutions to prevent (or mitigate) disasters. Matsuoka et al. (2012) examined role of *Bosai Fukushi Komyunithi* (in short—BOKOMI) formed in 191 school districts of the city as community-local government partnership. BOKOMI fostered principles of self-help, cooperation-help and public-help. After rigorous consultation, once a BOKOMI is established, it facilitates: organization of mock-drills, promote disaster education in schools, raising public awareness, seminars on first-aid/emergency kit preparation, town watching from safety and risk perspective. Certain welfare activities are also combined in BOKOMI activities. BOKOMI is an effective way of community participation in an urban area where it is challenging to keep the ties beyond generations and transfer lessons from past disasters to young generation. JICA also attempted similar efforts to build local partnerships in Turkey, Sri Lanka, Costa Rica and Indonesia.

A study done after 9 years of Kobe earthquake also revealed that Kobe earthquake added two very important element in Japan's disaster management system (1) an increase in voluntary and non-government activities, and (2) the enhancement of cooperation between local government and the residents' association (Shaw and Goda 2004). This study asserts that "community participation in the decision-making process was a significant achievement" of Kobe's recovery process. Through the case study of Nishi Suma, one of the worst-affected areas in the Kobe city, this study confirms that "civil societies in urban areas are sustainable if (1) the activities related to daily services are provided by the resident's associations; and (2) these are linked to economic incentives." It also emphasizes vital role of leadership in collective decision-making and also advocates for creation of the support system, which deems essential for long-term sustainability of civil-society activities.

18.4.5 Lessons Drawn from Kobe-Recovery

Kobe disaster recovery provides several major insights. Population of Kobe temporarily declined by about 6 % (according to annual statistics), took 10 years to regain pre-disaster levels following the earthquake. Spatial disparities observed in recovery phase, highlighted not only concentrations of damage but also long-term trends in population shifts away from the older urban core. Recovery in the economy sector revealed "temporary boost from reconstruction activities, followed by settlement at a "new normality" that was roughly 10 % lower than the pre-disaster normal" (Chang 2010). While Kobe's disaster recovery reflects significant variation among economic sectors, long-term trends reveal loss of small businesses (accompanied by an increase in large ones) and an increase in the manufacturing sector's capital-intensity. As a whole, "some indicators regained a "new normality" within a year of the disaster, while others required several years to stabilize" study reveals.

Kobe's recovery also opened a new avenue in Japan's disaster management system. Local community's engagement through BOKOMI is proving a unique experience. Similarly, the concept of sustainable civil society in Kobe's urban setting is an important lesson from a developed country not only to other hazard prone developed countries but also to Asia's developing countries which are also highly prone to natural disasters.

18.5 Reconstruction Experience in Urban Areas of Aceh

On December 26, 2004, a strong earthquake of magnitude 9.0 on the richer scale, hit the Northwest of Sumatra Island, Indonesia, and triggered Indian Ocean Tsunami. One of the most heavily affected areas is Banda Aceh, which is located at the tip of Sumatra Island home to 270,000 people. About 1/4th population of Banda Aceh lost their lives due to Tsunami. All the local government offices experienced severe damage and lost their function. Similarly, all the basic infrastructures and lifelines such as water, electricity, road, and port were destroyed and became dysfunctional. Due to exceptionally high magnitude of the disaster in Aceh, not only large number of death reported but also many residents went missing. Some affected population also became locally displaced persons (LDP). Damage due to Tsunami was mammoth to not only built habitat (such as houses and critical infrastructure) but also impacted social, economical, and environmental spheres in the region. Badan Rehabilitasi dan Rekonstruksi—BRR (in English: Agency for the Rehabilitation and Reconstruction of Aceh and Nias), is an Indonesian government agency which coordinated and jointly implemented the recovery program following the Indian Ocean tsunami that mostly affected Aceh in December 2004, and the Nias earthquake in March 2005. BRR estimated that 130,000 new houses required to be constructed and additional 85,000 houses needed to be repaired. Housing reconstruction

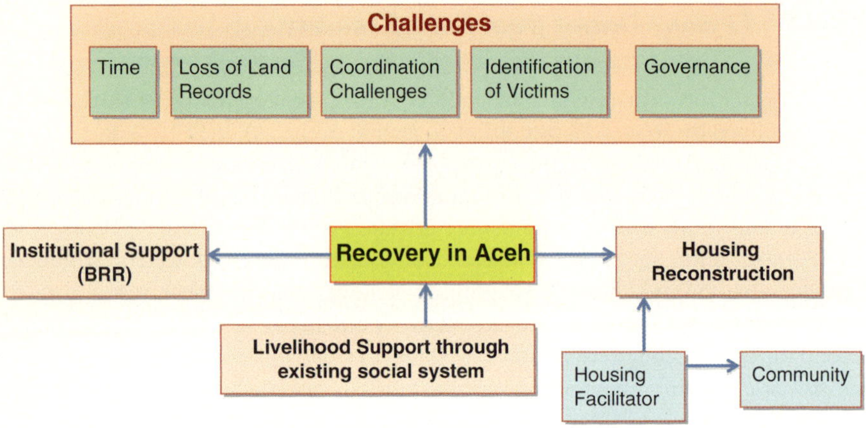

Fig. 18.2 Housing recovery process and challenges in Aceh

and recovery in Aceh was challenging due to various reasons documented in detail in a study done by Ochiai and Shaw (2009), some of which are summarized in the following sections as well as in Fig. 18.2.

18.5.1 Identification of Victims and Land Tenure

The identification of victims was very difficult in Aceh as in many cases; entire family living in a single dwelling was lost. Also, it was found that only one or few family members survived the disaster. Due to this, housing reconstruction process was greatly impacted and cause confusion. Verification of claims for allocation of houses, size of house to be reconstructed and many such decisions couldn't be taken quickly. Moreover, important records of property including maps and land titles were lost which made it further difficult to confirm ownership of specific plot or property. Land tenure related confusion eventually delayed the reconstruction process and also caused other problems such as double counting of beneficiaries.

18.5.2 Coordination Challenges

Following the Tsunami, reconstruction related assistance by national and international organizations poured in Aceh. While all these organizations intended to support housing reconstruction, each of them have their own house design, structure, construction mechanism and method, which resulted in varying delivery time and quality of housing throughout the region. According to a report published by BRR after 1 year of Tsunami, 100 different organizations were involved in housing reconstruction. It was also estimated that 15 large organizations alone reconstructed over 80 % of housing stock. Interestingly, several small organizations were rebuilding houses at a fast pace in the early stages, thus account for more than half of the permanent houses completed in 1 year after the Tsunami. Due to the involvement of large number of agencies in the housing reconstruction process, overall smooth and efficient coordination remained a major challenge not only for the government agencies but also for the international organizations including those work under the umbrella of the United Nations (BRR 2006; Ochiai and Shaw 2009). Due to loss of property records, reconstruction coordination challenges as well as loss of entire or few family members, double-counting also prevailed among survivors who have registered for financial aid. Some of these survivors were registered in the reconstruction needs data of more than one community. In some instances, it was also found that one family received several reconstructed houses, which also caused conflict at times.

18.5.3 Governance

Due to the grave scale of disaster in Aceh, all the government offices were destroyed or their functionality completely came to a halt. Therefore, local reconstruction coordination in the initial stages was provided by the national level government agencies of Indonesia. Five months after the tsunami, Indonesian government established BRR to coordinate the reconstruction process. The delay of 5-months in establishing the responsible reconstruction coordination organization was significant. Moreover, officials working with BRR had limited capacity to carry out clear assessment of housing standards and reconstruction needs. Largely, confusion regarding housing standards remained at the field level which was further deepened when various organizations offering reconstruction assistance followed their own housing construction standards. This can also be attributed to initial delay caused by Indonesian government in establishing BRR. Delay in provision of clear policy guidelines for housing reconstruction to all agencies offering housing reconstruction assistance in Aceh was a major bottleneck, which has hampered overall disaster recovery process.

Abovementioned factors interwoven with each other perplexed all the key stakeholders in the recovery process including government organizations, donor agencies, NGOs, and local community. Delay in establishment of BRR and disjointed institutional coordination mechanism and processes of reconstruction severely affected and delayed housing reconstruction and recovery in Aceh. All these stakeholders experienced mounting pressure from the local people to deliver the assistance quickly and of high quality. However, the devastation by the disaster was such that, no single agency could organize recovery operations smoothly without well-planned recovery plan prepared during pre-disaster phase. It may be noted that although several international donors have assisted Indonesian government and BRR for preparing spatial plans and infrastructure development roadmap of the city, it took some time to complete it, due to large-scale devastation. Aceh exhibited that housing reconstruction can be successful if there exists—well-planned organizational design and processes, trained technical and managerial staff and spatial planning guidelines considering possible damage scenarios in the region.

18.5.4 Housing Reconstruction in Aceh: The World Bank Approach

The World Bank (WB) through Multi Donor Trust Fund was one of the major players in housing reconstruction in Aceh. It supported community participatory approach, which was facilitated through specially appointed and trained staff known as "Housing Facilitator." Housing Facilitators supported communities to conduct the range of activities towards efficient housing reconstruction. These activities included—community organization, formulation of committee and working group, community self-survey, community settlement plan, group implementation plan, sign of community support program (CSP) and housing grant application, opening group bank account, and construction activities. Delivery of WB supported housing was slow compared to other international agencies involved such as UN-Habitat, ADB, etc. However, it has many merits and challenges.

18.5.5 Lessons from Recovery Experience of Aceh

Participatory housing reconstruction process followed by the World Bank in Aceh offers valuable lessons. It also introduced a new culture of adapting participatory approaches in housing reconstruction. Usually community participation is considered to be effective in pre-disaster phase, whereas, post-disaster recovery remained heavily top–down. Aceh experience has proved that community participation in post-disaster reconstruction is equally important. No single participatory approach is recommended as standard approach for all types of disaster recovery, including World Bank's participatory approach in Aceh, as these approaches needs constant improvement through detailed evaluations. The study by Ochiai and Shaw (2009) illustrated that WB's participatory housing reconstruction efforts contributed to five key factors—(1) time/speed (2) quality (technical/safety) (3) socio-cultural concern (4) management (5) cost. These factors together form an important lesson in the field of post-disaster housing reconstruction. Survey conducted as part of this study reveals that building an earthquake-resistant house is main concern of local community. During the recovery phase, level of awareness among the community about building earthquake-resistant house is very high. Building earthquake resistant house can potentially contribute to reduce the future urban risks from similar disasters.

From the case study of World Bank assisted housing reconstruction process in Aceh, Ochiai and Shaw (2009) summarized that: (1) establishment of the organizational structure and process during pre-disaster phase can help ensuring people's participation in housing reconstruction process in the recovery phase (2) role of Housing Facilitator's in supporting community in several ways such as technical, social, and management component is vital (3) in an urban setting, people's participation varies depending on local situation (4) five factors (time/speed; quality (technical/safety); socio-cultural concern; management; cost) are mutually dependent and all stakeholders involved in housing recovery contributes to ensure compliance to these factors.

Aceh's post-disaster recovery experience illustrates that participatory approaches are effective in large-scale housing reconstruction. However, effective and balanced coordination among various stakeholders involved and consideration of five factors is essential and can be adjusted depending on the local context. This can further boost participation of the people with due encouragement. In Indonesia, community-based project implementation approach existed for many years. This also prepared a conducive background for World Bank's housing reconstruction project to adopt the participatory approach in housing reconstruction. Know-how provided through Housing Facilitators and previous experiences helped entire reconstruction project and mechanism to work effectively in the field. Importantly, it is recognized and affirmed further that every-day practice of engaging local community in various developmental projects and pre-planning for recovery are most effective strategies in post-disaster recovery urban areas.

18.6 Sectoral Perspectives on Urban Disaster Recovery

It is not uncommon to observe that response to disaster mainly aims saving lives and offering emergency relief assistance. Urban disaster recovery is often remains a neglected area not mainstreamed with disaster response planning. At the end of humanitarian assistance phase and beginning of reconstruction program, populations affected by disaster remain without long-term support for recovery. Moreover, disaster recovery in urban areas captures narrow perspective and remains limited to the physical aspects leaving apart facilitating social and economic recovery of the region.

In the absence of comprehensive recovery framework, communities themselves proceed to recover. They engage spontaneously in recovery through rebuilding, resettling and thus recreating conditions of disaster risk and vulnerability. Disaster recovery offers tremendous opportunity to reduce risks and guide development. However, in the absence of necessary handholding and guidance to community at the time of recovery, risk can be regenerated. International as well as local assistance is usually very high the immediate aftermath of a disaster. Gradually availability of resources shrinks and recovery phase remains neglected with fewer resources and interest. Interventions for disaster recovery also usually isolated and uncoordinated, resulting in not just duplication of efforts but also wastage of valuable resources. In urban areas, especially in developing countries, integrating risk reduction considerations for sustainable recovery generates win–win situation. Undoubtedly, comprehensive and complex process of recovery needs to be synergized with developmental planning process of cities. To bridge the knowledge gap in disaster recovery at sectoral level, International Recovery Platform recently published "Guidance Notes on Recovery" on a range of existing and emerging themes (IRP, undated). Following sections will draw from these notes and draw some of areas which are of relevance to urban disaster recovery.

18.7 Infrastructure Recovery: Opportunities and Challenges

Infrastructure development and urbanization usually go hand in hand. As a city or urban-region expands, required infrastructure also becomes more complex. Infrastructure, especially lifeline infrastructure is a vital element of an urban system. In case of a large disaster, infrastructure damage can virtually paralyze the urban functions, hamper immediate response and delay recovery. Recovery of infrastructure requires vast financial and human resources. Process of infrastructure recovery is often complex and time consuming. Guidance Notes on Recovery define infrastructure "as the physical and organizational structures, networks, or systems required for the successful operation of a society and its economy." Spatial planners usually divide infrastructure into two categories—physical infrastructure and social infrastructure. Physical infrastructure usually comprise of networks (road, rail and air, water) as well as utilities (water supply, sewage, waste disposal, etc.). Social infrastructure refers to facilities such as hospitals, parks and gardens, community centers, libraries, entertainment and shopping centers, and educational institutions, etc. Infrastructure can be owned, managed and regulated independently or jointly by public and private sectors.

Efficient urban systems not only require well-developed infrastructure but also make its citizens, governments, and businesses heavily dependent on them. When disasters affect infrastructure system in cities, it also creates its own ripple effect. Infrastructure failure may lead to traffic chaos, loss of trade and commerce functions, loss of telephone, radio, television and mobile communication systems, disruption of electricity production and transmission, overwhelm health facilities and hospitals and so on. Cities heavily invest in designing, developing and maintaining infrastructure as it directly relates to offering higher living standards to its populace. IRP (undated) Disasters can cause damage to infrastructure in many ways such as (1) major disruptions to all or part of urban infrastructure systems (2) damage or destruction of the facilities and equipment associated with infrastructure (3) loss in the information upon which infrastructure systems are dependent on (4) injury or death of staff members who runs and manage infrastructure.

Infrastructure is one of the critical component requires restoration on a priority basis in urban recovery regime. In the earliest phases of disaster response, responsible agencies attempt to restore at least most essential infrastructure (partly or make-shift), which is considered crucial to save lives or provide speedy relief. Search and rescue, temporary relief, evacuation, emergency health care, and such critical life saving and sustaining functions can be severely plagued in case of failure of infrastructure. When the infrastructure is created or restored in the midst of disaster response, there is always a risk of compromising with desired level of safety and strength. Thus, comprehensive recovery planning should consider stockpiling of those materials and equipment in the pre-disaster phase, which can restore critical infrastructure with reliable standards of safety. In developing countries, at times, infrastructure built for immediate response also becomes permanent feature of the post-disaster built environment.

Urban disaster recovery requires considering redevelopment in a phased manner. While short-term urban infrastructure recovery may include repair and restoration of partially damaged infrastructure, long-term tasks includes also include replacement of completely damaged infrastructure, and building of new infrastructure depending on needs of the city in post-disaster regime. Infrastructure recovery is not an easy task and requires strong, visionary, quick decision making leadership as well as trained and determined workforce. Following a disaster, there is tremendous pressure to swiftly restore or build critical infrastructure. However, disaster can limit technical capabilities available locally and can hamper availability of material and labor as well.

Infrastructure recovery planning should be dovetailed with both pre-disaster and post-disaster phase. Cities are built over a period of time. Urban areas consist of infrastructure, which is aged, and due for replacement. Post-disaster phase offers opportunity to replace such old infrastructure with modern, hazard resistant infrastructure. Similarly, cities prone to natural hazards also have their developmental aspirations. Such cities, in the pre-disaster phase, must invest in creation of infrastructure, which is built with hazard resistant material, technology and design, so that it remains safe and functional at least from the known risks. Guidance Notes on Recovery characterizes infrastructure recovery components into two primary types—object-oriented and network oriented. It mentioned that "object oriented components of infrastructure tend to be individual, even if multiple units of that infrastructure exist throughout the affected area. For example, hospitals are individual "objects" that together make up a city's health infrastructure. Network oriented infrastructure systems are more interconnected, and often rely upon lines of transmission that traverse great geographic distances. Pipelines, communication wires, transmission lines, and roadways, for examples, are each components of network-oriented infrastructure systems." Guidance Notes also listed key vulnerabilities to infrastructure recovery which are very relevant in urban context such as (1) poor land use planning (2) poor, weak or inappropriate construction materials (3) inappropriate design of buildings and other structures (4) insufficient building codes and inadequate code enforcement (5) poor maintenance, cascading failures (one infrastructure failure leading to other).

Large-scale catastrophic events such as India Ocean tsunami receive great attention from the international donors. Financial assistance offered following such gigantic events also opens a window of opportunity to not only update and improve the infrastructure that existed prior to the event but also create a new one which never existed (such as paved roads in areas which had unpaved or no roads before the disaster). Importantly, infrastructure recovery planning must give due consideration to environmental sensitivity and fragility of the region. It should also complement long-term vision of urban spatial planning so that it can be integrated well in a given urban setting.

18.8 Livelihood Recovery in an Urban Context

Livelihood is one of the most important yet often-ignored dimensions of urban and regional recovery. In urban regions, livelihoods are diverse and interdependent and are shaped within social, economic and political contexts. The natural and built environment also shapes livelihoods and vice versa. Guidance Note on Recovery further states that "quality of soil, air and water; climatic and geographic conditions; availability of fauna and flora; and frequency and intensity of natural hazards—all influence livelihood decisions."

Urban livelihoods comprise of both white and blue-collar jobs. In developing countries, both formal and informal sector opens employment opportunities. Also, many people do not live in the city, but visit the city often or everyday for working in informal or formal sector. In case of a disaster, if city's livelihood options shrink, it can have ripples across the region or even beyond. In addition, developing country cities also attract large number of seasonal migrants who depend on cities for livelihood in those periods when they have no employment in rural areas. IRP (undated) noted, "very few livelihoods exist in isolation. A given livelihood may rely on other livelihoods to access and exchange assets. Traders rely on farmers to produce goods, processors to prepare them, and consumers to buy them. Livelihoods also compete with each other for access to assets and markets. Thus positive and negative impacts on any given livelihood will, in turn, impact others." During disaster recovery phase, this is a particularly important consideration when planning livelihood assistance.

A number of surveys and evaluations done following a disaster (such as—Gujarat earthquake; Indian Ocean tsunami in Indonesia; Haiti earthquake) revealed that local communities persistently stressed the need to restore livelihoods. Sometimes they also rated livelihood recovery much higher compared to receiving relief. Community's capacity to earn their own livelihood also linked to their desire to building confidence and attaining self-reliance and dignity. It also exhibits their willingness not to depend on aid and relief, which is anyway provided for short duration. In urban areas, disasters bring structural change in livelihood options. As noted earlier in the paper, Kobe earthquake was followed by temporary boom in the construction sector. Similar observations are reported from Gujarat earthquake. In Gujarat earthquake, it was also observed that labors, both skilled as well as unskilled, migrated from far flung states (such as Bihar), to fill the additional labor demand in construction sector. Some of these migrated labors decided not to return to their

Table 18.1 Phases of livelihood recovery

Timing	Phase	Constituents
Short term	Livelihood provisioning	Relief-based objective, which relies on swift response and the logistical capacity to deliver critical provisions
Medium term	Livelihood protection	Aligned with the recovery phase and requires careful and complex assessment and benefits from local contextual knowledge
Long term	Livelihood promotion	Transition from recovery efforts to development goals and requires the long term commitment of governments and other development actors

Source: IRP undated

home state and indulge in other incoming earning opportunities. In case of Great East Japan Earthquake and Tsunami, many residents have to move out of their native places, as it is designated as prohibited area due to nuclear radiation. These people are prioritizing on livelihood availability while moving to other parts of Japan. Disasters have potential to create new livelihood options, change livelihood dynamic, or even weaken economic base of affected region.

Livelihood recovery has not yet mainstreamed within disaster response and recovery initiatives, although there are few recent examples where it gained priority. Comprehensively assessing livelihood needs and devising a strategy to achieve economic momentum in urban regions still remain an area of vacuum. Livelihood assistance initiatives offered to replace physical assets lost (such as fishing boats and nets) or providing short-term loans and grants to small businesses. Guidance Notes reflect that "programming and funding for livelihoods support is channeled through multiple sectors, and livelihood practitioners struggle to develop effective coordination mechanisms and tools to assess needs, evaluate impacts, and prevent overlapping and conflicting interventions. Livelihood recovery is also a building process that takes place in a very dynamic environment. Livelihood strategies must be able to adapt or change altogether as the surrounding conditions change. Disaster assistance, across all sectors, also directly and indirectly impacts livelihood recovery, either enabling or impeding it."

Recovery of businesses, be it small or large, cannot be done merely by replacement of physical assets. Businesses are highly interconnected with network of manufacturers, suppliers, distributers, wholesalers, retailers, and so on. Restoration of this network is crucial for businesses to function along with the provision of financial services and the development of markets. To cater to the immediate, short, and long term needs of affected populations, IRP (undated) refers to three corresponding phases viz. livelihood provisioning, livelihood protection and livelihood promotion which are build upon each other (Table 18.1). Scale, intensity and type of disaster dictate the duration of each set of activities in these three phases. Also, it is important to intervene in parallel for community recovery as people will be the consumers who will make businesses grow.

One challenging area is recovery of informal sector, which depends largely on assets (owned or rented, however small asset it may be). Provision of the necessary assets shall be prioritized for informal sector to generate income. After the disaster, people involved in informal sector economy mainly resort to friends, family, and other close-knit social networks for support. Money landers, also working informally, also offer help but charge huge interest rates. As a result of this, affected population may end up into huge debt. In case of insufficient help, people may even sell any remaining productive assets and shift to menial work. Both these conditions may force people to fall in spiraling cycle of poverty. Experience shows that acting swiftly to help in protecting or providing assets is very important to safeguard informal sector livelihoods. External intervention to help in informal sector commonly includes—cash grants, in-kind assistance, and temporary employment.

Large-scale disasters, although devastating, also generate employment opportunity. International and national development assistance require local workforce to implement recovery programs effectively. At times, it is difficult to find suitable workforce locally to cater to various requirements. Another problem is that even when the suitable workforce is available locally, there is gap in communication and hence they don't connect to employers. NGOs, governments, international donors, reconstruction agencies, etc. together forms a new category of employers who offer employment in various recovery related works. When the demand of human resources overweight supply, it may create a situation where limited available workforce demands higher wages and can lead to escalation of project costs. Guidance Notes on Recovery noted an example from Sri Lanka's owner-driven reconstruction project following the 2004 Tsunami. It mention that "the demand for skilled labor outweighed the supply and by August 2005, while the project was still in its earliest phase, the total cost per house rose from rupees 400, 000 to rupees 550, 000. This increase was primarily driven by increased wages of carpenters, masons and other skilled trades-people. Additionally inflation had increased 12.7 %, a significant jump from the pre-tsunami period." Livelihood recovery requires considering all these possibilities and hence comprehensive planning for livelihood recovery in the pre-disaster phase can help conquer some of these challenges.

18.9 Housing Recovery

A house or shelter offers safety and protection to its occupants from external forces (including natural hazards). Housing standards are designed to offer safety from known hazards, however, at a scale accepted in the country's building standards. If an area is prone to certain type of hazard regularly, housing construction mechanism in that area will incorporate these aspects to offer acceptable levels of safety. Disasters serve as reminder to build houses with appropriately stronger materials, innovative hazard resistant design and local environmental and socio-cultural considerations. Guidance Notes on Recovery identified factors such as "poor, weak or inappropriate building material, inappropriate building design, inadequate building code as well as its enforcement, poor land use planning, poor quality access to critical infrastructure

Table 18.2 Distinct phases of shelter recovery

	Type of shelter	Duration of stay	Details
1	Spontaneous shelter	First 72 h	To provide an interim, safe haven while the situation stabilizes.
2	Emergency shelter	First 60 days	To provide emergency shelter and feeding to displaced population requiring shelter.
3	Interim housing	First year and beyond	To provide temporary housing—safe and secure shelter, water, power, and heating—to displaced disaster victims while efforts are underway to make permanent repairs to dwellings, or to find other suitable permanent housing.
4	Permanent housing	No specific time	To provide long-term, permanent housing solutions for disaster victims.

Source: IRP (undated) quoting CUSEC (1998)

(communication, electricity, water, sewerage, transportation, gas, etc.), contributing to increase the vulnerability of shelter and making it difficult to recover."

It is experienced that disasters affect the poor most, who often lives in low quality housing or even in slums and squatters. In addition, mostly in developing countries, construction practices have not matured enough to prioritize "disaster resilient construction." Despite of availability of technical know-how for disaster resistant construction, vulnerable shelters are created ignoring building safety guidelines and codes. Housing recovery begins at pre-disaster phase itself when such building codes can be implemented strictly with massive awareness campaigns among society at large (Table 18.2). Nonetheless, housing recovery by building appropriate shelter not only reduces risk from future hazards, but also has potential to contribute to economic revitalization, densification, rezoning, and modernization, among other factors.

Another important aspect in housing recovery is community participation. There are numerous examples where houses offered after the disasters remain unoccupied or occupied for a very short term. People, mostly those whose livelihoods are dependent on coast, return to previously occupied hazard prone locations or vulnerable construction practices. There is no single recipe to offer a solution to this serious issue as income-generation receives priority over safety among poorer sections of the society. Housing recovery is equally challenging as other aspects of disaster recovery, as noted from the examples of Kobe and Aceh in the earlier section. Reconstruction agencies are in a constant pressure to quickly rebuild or replace the housing. Other challenges to housing recovery may include but not limited to—affordability and availability of building material and labor, loss or lack of land suitable for hazard resistant building construction, pre-existing conflict in community, lack of trust among community towards reconstruction agencies thus making it more top–down, provision of infrastructure and services to make a house livable, etc.

Housing reconstruction has many unique facets that can influence lives and livelihoods of its occupants. For example, selection of site alone can influence social bonding and network amongst community members. Housing layout and design can create spaces, which may facilitate better interaction among residents and thus enhance their mutual understanding. Introduction of children play area, informal spaces at conducive scale, community gardens and similar facilities can help to reduce stress among residents and recover better from the memories of the disaster event. Paying attention to local climate, socio-cultural context, architectural style, occupation and habits, and many such relevant factors is important to housing recovery.

18.10 Local Governance and Urban Recovery

Local government in urban areas essentially includes all those agencies that are responsible for planning, managing or expanding a given urban area or region. Disasters negatively impacts local government functions. Local government organizations are expected to respond quickly, assess damage effectively, and play important role in speedy recovery. Key role of local government organizations is to provide transparent, participatory and efficient environment where citizens enjoy good quality of urban basic services and amenities. In other words, good urban governance is at the core or urban affairs and hence is a pre-condition to urban recovery. Local government organizations together with relevant stakeholders including NGOs and CBOs can provide enabling environment through which sustainability of recovery efforts can be enriched.

Role of local government has become even more important since over half the global population is now concentrated in urban areas. Asia and Africa are urbanizing fast. Cities in developing countries are now becoming denser, housing more people, and tasked to provide better services through sophisticated expanding infrastructure. Urban recovery in these fast-developing cities will be very efficient if attention is paid now to build resilient infrastructure and buildings. This is also the time when comprehensive risk assessment shall be carried out to identify pockets of higher vulnerability or least capacity. Especially drafted policies, plans, acts, building codes, zonal regulations to strategically address identified risks and build resilience will help in effective recovery as well.

Guidance Notes on Recovery clearly points at lack of urban focused research to improve recovery. However, paper by Matsuoka and Shaw (2006) shares important insights on urban governments collation in Japan to offer mutual help as well as regional help. After the 1995 Kobe earthquake, one of the major challenges was to modify existing hierarchical structure of disaster management and putting local governments at the core of the system. Hyogo Prefecture conducts training programs on disaster management for local government officials from other prefectures in Japan. In these courses, Hyogo Prefecture shares it's expertize and lessons from the Great Hanshin-Awaji Earthquake of 1995. This program helps building capacities of local government officials.

Great East Japan Earthquake of March-2011 further boosted cooperation spirit and mechanism among local governments of a region. "Kansai Union" was formed in December-2010 as the first effort towards region-wide cooperation between individual prefectures in Japan. Kansai Union is composed on seven prefectures from Kansai region. The Union identified seven focus areas and disaster prevention is one among them. Hyogo Prefecture, which is also one of the members of Kansai Union, is responsible for disaster prevention and is determined to help the Union members with practical implementation measures of disaster risk reduction. Following the Great East Japan Earthquake, governors of Kansai Union members met to establish a support framework and followed a cluster approach and two prefectures joining as a team to assist one affected prefecture in Tohoku region. In addition, Kansai Governments and Kyushu Governors' Association did another agreement in November 2011, to offer an agreement of mutual assistance. Matsuoka and Shaw (2006) noted that this agreement "constitutes the first inter-regional agreement for cooperation during mega wide-area disasters. The assistance includes (1) dispatch of staff, (2) supply of food, water and daily necessary goods, (3) provision of equipment, (4) acceptance of evacuees, the sick and wounded, (5) provision of transportation such as ships, (6) medical support, and so forth. The counterpart method will be applied to assign the prefectures in charge to support, and at the time of gigantic disasters, the support shall be provided even without requests. This model can be further explored and replicated."

18.11 Urban Environment and Disaster Recovery

IRP (undated) noted that urban environments include both "green" environmental issues (reducing the impact of production, consumption and waste generation on natural resources and ecosystems) and "brown" environmental issues (reducing the environmental threats to health that arise from poor sanitary conditions, crowding, inadequate water provision, hazardous air and water pollution, and local accumulations of solid waste).

Over the years, various studies indicated that well planned and managed mangroves in coastal areas have helped in reducing the impact of tsunami waves, coastal erosion, salinity intrusion, and so on. Mangroves are also considered to protect livelihoods, preserve biodiversity and help reduce recovery time in the aftermath of the disaster. Similarly, study by Surjan and Shaw (2009) established the link between poor waste management leading to localized flooding and water logging. This study examines the case of city of Mumbai, which experienced a mega flood event held in 2005, and many annual local flood events thereafter. It reports that a community-based system of managing waste at the neighborhood level helped keeping the sewers and drains unclogged and facilitated unobstructed flow of rainwater. Therefore avoided flooding or long-duration waterlogging which was the case in other areas where drains were blocked with solid waste (Surjan et al. 2009). This study concludes that in an urban context, the link between environment and disaster issues is very strong and urban environmental issues provides entry points to engage community for long term resilient building agenda.

Post disaster waste management is also one of the often-ignored areas and very little research is available to inform policy makers on this issues. Large-scale disasters such as Tsunami's or earthquakes lead to huge debris. This debris is very difficult to segregate and often mixed with hazardous waste including hospital waste. Post disaster dumping of waste in wetland or poorly planned landfill can pollute soil and groundwater, impact agriculture output, fishery, and other highly ecosystem dependent aspects. Better urban environment management practices creates more conducive environment for better recovery.

18.12 The Road Ahead

Insights form Kobe and Aceh indicate that recovery planning need to be integrated with pre-disaster phase. While preparing disaster management plans, disaster recovery plans should also be prepared as detailed as possible. Well-planned recovery planning helps to take decisions better, smarter and quicker. Recovery planning, should also consider possible scenarios of "new normal" to which the community may need to adapt to follow sustainable development pathway. Figure 18.3 summarizes discussion in this paper by identifying actors and sectors in disaster recovery. This figure also suggests some entry points, which helps in making headway in urban disaster recovery. Learning, especially from the areas, which have previous experience of disasters, will help stakeholders to develop realistic expectations of duration of recovery, difficulties to be faced, and monitoring mechanism required. Pre-disaster risk reduction measures including investments in mitigation, preparedness, good governance, capacity building, environmental quality improvement, community participation etc. pave the way for sustainable recovery and resilient development after the disaster. Recovery is also a gateway for addressing "development deficit." It is very clear that urban disaster recovery is not an easy process but in the developing country context, it brings tremendous opportunity to develop better considering safety, environmental and socio-economic considerations. Planning for disaster recovery well ahead of occurrence of disasters is certainly a necessity of our times. Time has matured to move a step forward from response planning to recovery planning. Urban regions in hazard prone areas can be greatly benefitted with pre-disaster recovery planning. Tohoku triple tragedy is a stark reminder of this necessity and more research will be of immense importance in this area to further broaden understanding on urban disaster recovery.

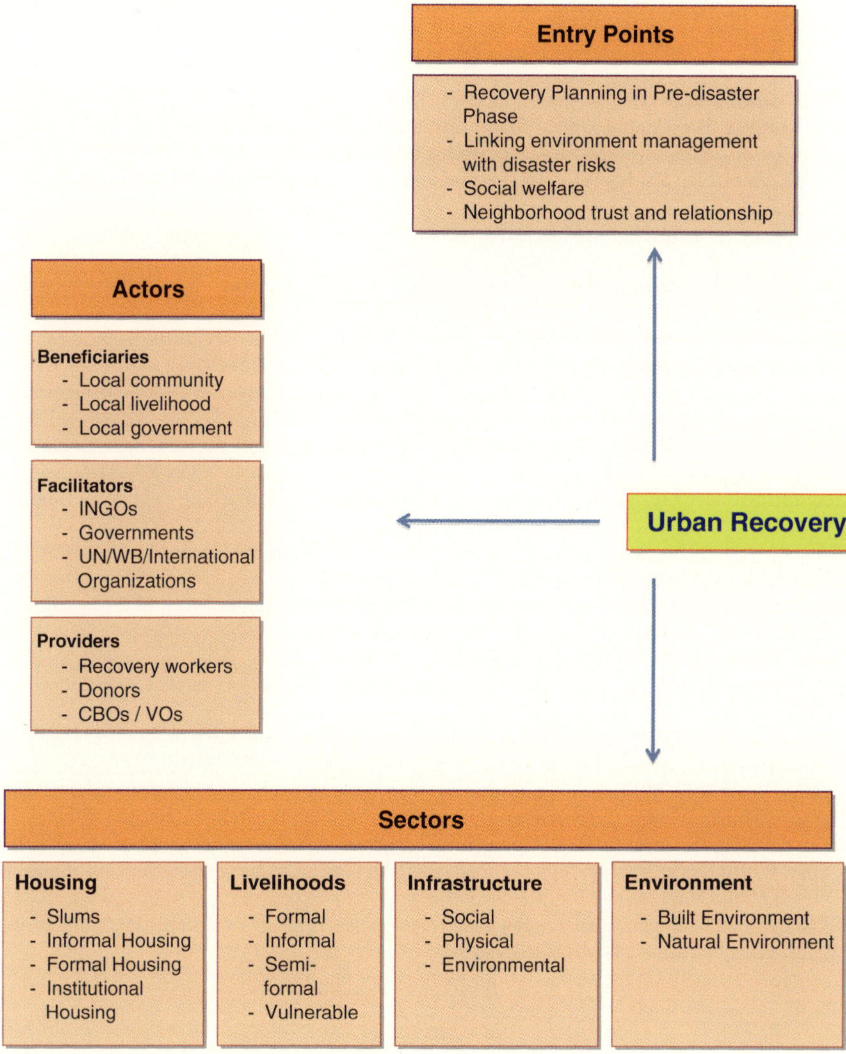

Fig. 18.3 Urban disaster recovery: actors, sectors and entry points

References

Birkmann J (ed) (2006) Measuring vulnerability to natural hazards. United Nations University Press, Tokyo

BRR (2006) Regulation of executing agency rehabilitation and reconstruction agency and the lives of the people of Nanggroe Aceh Darussalam Province and Nias Island, North Sumatra Province, Number: 19/Per/Bp-Brr/Iii/2006 on house rehabilitation assistance for the victims of earthquake and tsunami in Nanggroe Aceh Darussalam Province and Nias Island, North Sumatra Province

Chang SE (2000) Disasters and transport systems: loss, recovery, and competition at the Port of Kobe after the 1995 earthquake. J Transport Geography 8(1):53–65

Chang SE (2010) Urban disaster recovery: a measurement framework and its application to the 1995 Kobe earthquake. Disasters 34(2):303–327

Chang SE, Nojima N (2001) Measuring post-disaster transportation system performance: the 1995 Kobe earthquake in comparative perspective. Transport Res A Policy Practice 35(6):475–494

CUSEC (1998) The Central United States Earthquake Consortium. Spring 5(1):19–22. http://www.cusec.org/publications/newsletter/spring1998.pdf

Cutter SL, Boruff BJ, Shirley WL (2003) Social vulnerability to environmental hazards. Soc Sci Q 84(3):242–261

Government of Japan (2005) White paper on disaster management (Heisei 17 nenban Bousai Hakusho). Cabinet Office, Government of Japan

IRP (2007) Learning from disaster recovery guidance for decision makers. International Recovery Platform (IRP). Preliminary version for consultation, May 2007

IRP (undated) Guidance note on recovery, Jan 2011. http://www.recoveryplatform.org/resources/guidance_notes_on_recovery

Matsuoka Y, Shaw R (2006) Improvements and challenges of Japanese disaster management system. Asian J Environ Disaster Manag 4(2(2012)):119–144

Matsuoka Y, Joerin J, Shaw R, Takeuchi Y (2012) Partnership between city government and community-based disaster prevention organizations in Kobe, Japan. In: Shaw R (ed) Community based disaster risk reduction, vol 10. Emerald Group Publishing Limited, Bingley, pp 151–184

Mileti DS (1999) Disasters by design: a reassessment of natural hazards in the United States. Joseph Henry, Washington

Nakagawa Y, Shaw R (2004) Social capital: a missing link to disaster recovery. Int J Mass Emerg Disasters 22(1):5–34

Ochiai C, Shaw R (2009) Post disaster reconstruction in urban areas in Aceh. In: Shaw R, Srinivas H, Sharma A (eds) Urban risk reduction: an Asian perspective, vol 1. Emerald Group Publishing Limited, Bingley, pp 233–252

Shaw R, Goda K (2004) From disaster to sustainable civil society: the Kobe experience. Disasters 28(1):16–40

Surjan A, Shaw R (2009) Enhancing disaster resiliency through local environment management: case of Mumbai, India. Disaster Prev Manag 18(4):418–423, Emerald Publication, Bingley

Surjan A, Redkar S, Shaw R (2009) Community-based urban risk reduction: case of Mumbai. In: Shaw R, Srinivas H, Sharma A (eds) Urban risk reduction: an Asian perspective, vol 1. Emerald Group Publishing Limited, Bingley, pp 339–354

UNCRD (United Nations Centre for Regional Development) (1995) Comprehensive study of the Great Hanshin earthquake. UNCRD, Nagoya

Chapter 19
Integrated Healthcare as the Future of Disaster Recovery Potential in Tohoku Region

Kenji Isayama and Rajib Shaw

Abstract When disaster first strikes, accidents or emergencies are the most common reasons for people to seek healthcare. In most cases, the rapid response of local providers of care, disaster medical assistance teams, volunteers, and others help to assure that accident and emergency patients receive appropriate intervention. Depending on the nature and extent of the disaster, its aftermath is the time when the primary healthcare system may be least able to provide needed services. A healthcare system, especially one serving the poor, may have been inadequate or overburdened before the disaster, and its ability to respond is often adversely affected by destroyed or damaged buildings and equipment, lack of drugs and other supplies, and increased shortage of healthcare providers. This chapter describes the cases of Kamaishi, Kesennuma and Natori from the affected region of Tohoku by the East Japan Earthquake and Tsunami. Analyzing the past and current healthcare system of these cities, the chapter provides a sustainable integrated model, which needs to be linked to daily health issues, and which is build on the community networks health volunteers, and mutual trusts among the community members.

Keywords Community health center • Community network • Health case impacts • Integrated healthcare system • Tohoku disaster

---

K. Isayama
Department of Medical Science and Technology, Hiroshima International University, Hiroshima, Japan

R. Shaw (✉)
International Environment and Disaster Management, Graduate School of Global Environmental Studies, Kyoto University, Kyoto, Japan
e-mail: shaw.rajib.5u@kyoto-u.ac.jp

R. Shaw (ed.), *Disaster Recovery: Used or Misused Development Opportunity*, Disaster Risk Reduction, DOI 10.1007/978-4-431-54255-1_19, © Springer Japan 2014

19.1 Introduction

On March 11, 2011, a magnitude Mw 9.0 earthquake occurred off the coast of Japan's Tohoku region about 130 km east of Sendai causing catastrophic damage and loss of life (Simons et al. 2011; Fujii et al. 2011). This represents the fourth largest earthquake instrumentally recorded. The earthquake and tsunami caused substantial damage and loss of life along Japan's coastline followed by multiple melt-downs at the Fukushima nuclear power plant (Bowyer et al. 2011). The bulk of the 19,447 fatalities (15,840 dead and 3,607 missing presumed dead) were concentrated in the coastal regions of Miyagi, Iwate and Fukushima prefectures (National Police Agency 2012). The majority at 92.5 % of the fatalities is attributed to the tsunami. The 2011 Tohoku tsunami represents Japan's deadliest tsunami since the 1896 Meiji-Sanriku tsunami earthquake (Tanioka and Satake 1996). For the 2011 Tohoku tsunami we measured a maximum tsunami run up exceeding 38 m along the Sanriku coast in a narrow valley at Aneyoshi, Iwate Prefecture (http://www.coastal.jp/tsunami2011/) (Mori et al. 2011; Fritz et al. 2012).

The East Japan earthquake and tsunami (EJET) produced a devastating tsunami that destroyed many towns and villages near the coast in Iwate, Miyagi, and Fukushima prefectures (Shibahara 2011). Miyagi Prefecture was the area most severely devastated by the tsunami, with extensive loss of life and property; hundreds of thousands of people lost their houses and were forced to move to evacuation areas. In the days and weeks following devastating natural disasters, the threat of infectious disease outbreak is high (Connolly et al. 2004).

Disaster damages were caused both by the disaster itself, as well as secondary reasons including loss of private properties and the health-related public facilities, which affected the local healthcare system. Even for survivors, they feel modulation of the mind and body caused by various disaster stresses that magnifies as time passes. Many residents in the disaster-hit areas lost their family, friends, homes, possessions and jobs in this disaster in which the psychological impact has been immense. Almost "earthquake-related deaths" were elderly. At high rate, the deaths were caused by fatigue in shelter life and with physical and mental fatigue by the tsunami and earthquake. Recovery is not just about the rebuilding of infrastructure in the affected areas, but is also about the rehabilitation of individuals and the process of rebuilding their lives. This study aims to grasp the present condition about the healthcare in Kamaishi, Kesennuma, and Natori City. This study also considers improving future healthcare system for local residents, and proposing future community network lesson from EJET.

19.2 Healthcare and Disaster Recovery

19.2.1 Health Disasters in the World

When disaster first strikes, accidents or emergencies are the most common reasons for people to seek healthcare. In most cases, the rapid response of local providers of care, disaster medical assistance teams, volunteers, and others help to assure that accident and emergency patients receive appropriate intervention. Depending on the nature and extent of the disaster, its aftermath is the time when the primary healthcare system may be least able to provide needed services. A healthcare system, especially one serving the poor, may have been inadequate or overburdened before the disaster, and its ability to respond is often adversely affected by destroyed or damaged buildings and equipment, lack of drugs and other supplies, and increased shortage of healthcare providers (Axelrod et al. 1994).

In Cuba, a growing cadre of family physicians is being trained to work as part of a healthcare team trained to provide health education and preventive services, offer comprehensive medical care, and conduct population based-research. In Cuba, family physicians are required to look at patients in the context of family and community. Medical records are organized by family. Health statistics are recorded and reviewed on a regular basis (Dresang et al. 2005). Cuba's medical system is recognized as a good example of a network of daily medical care and disaster medicine. This system also helps community-coping capabilities when a disaster occurs. Cuba which Hurricane occurs frequently has been focused on disaster medical care as a national policy. Support activities by the medical model of disaster medical care in Cuba was held in Haiti in 2010 earthquake (Appelbaum et al. 2006).

Epidemics do occur in the post-disaster period, but are generally due to respiratory and gastrointestinal disease related to the post-disaster living conditions (Campos 2006). Medical care is important in the immediate post-disaster period however in large disasters such as the 2010 Haiti earthquake medical volunteers will continue to be needed but much more difficult to recruit once the post-disaster media attention fades. Haiti has never had mental health resources. Haitians rely on themselves, their families and their community for support. Culturally appropriate mental health resources should be part of the long-term recovery plan (Jobe 2010).

Recent federal efforts in disaster preparedness, response, and recovery recognize the importance of mental and behavioral health (US Department of Health and Human Services 2009; FEMA 2011). Recognizing psychological support mechanisms as essential elements of “a prepared and responsive health system,” the 2009 US Department of Health and Human Services (HHS) National Health Security Strategy (NHSS) (US Department of Health and Human Services 2009) promotes two goals: (1) building community resilience and (2) strengthening and sustaining health and emergency response systems. The Federal Emergency Management Agency (FEMA) National Disaster Recovery Framework (NDRF) promotes emotional and behavioral health considerations as an essential component of recovery (FEMA 2011). The close interplay between mental health and physical health makes it critical to integrate mental and behavioral health considerations into all aspects of public health and medical disaster management. Successful integration requires mental and behavioral health efforts to be (1) incorporated in assessments and services; (2) addressed in education and training; and (3) founded on and advanced through research. The report used the term "psychological first aid" to describe supportive activities delivered by seven non mental health professionals to family, friends, neighbors, coworkers, and students as well as more sophisticated psychological support delivered by primary care providers to their patients (Disaster Mental Health Subcommittee of the National Biodefense Science Board US Department of Health and Human Services, Office of the Assistant Secretary for Preparedness and Response 2008; Pfefferbaum et al. 2012).

Asian countries are constantly menaced by major natural disasters. Disaster causes a wide range of psychological distress as well as physical and social distress. However, there has been little attention given to the impact of disasters on mental health following disasters throughout Asia. This was considered to be a result of the strong stigma attached to mental health problems. Disaster psychiatry in Asia did not exist as a psychiatric discipline until the 1980s. However, during the 1980s and 1990s, disaster mental health gained an increase in attention as a result of major disasters in Asia. At present, disaster mental health is being recognized as an important psychiatric discipline in many Asian countries (Kokai et al. 2004).

19.2.2 Health Disasters in Japan

In the Hanshin-Awaji earthquake of 1995, the circumstances after the disaster were exceptional. Satisfying even the basic needs of everyday life such as bathing and using the toilet was difficult. Care for the physical and mental health of residents was difficult because of the density of the population living in the area. Privacy was nonexistent for a prolonged period of time. The needs of vulnerable people had to be taken into account and given further consideration. When the lifelines collapsed, support was provided by emergency medical and nursing personnel during the acute phase. However, emergency priority treatment such as triage and treatment of crush victims was insufficient. Damage to the medical buildings caused serious disruption of medical services. Over time, immunity was compromised in the victims of the disaster because of malnutrition and increased stress in shelters. Health providers had to deal with usually rare medical issues, such as early pneumonia following an influenza epidemic. Some challenges remain even now; for example, some victims have residual mental issues from prolonged stay in shelters. Shinfuku, who had the rare experience as a medical doctor of being both a victim and an observer of a major disaster in the Hanshin-Awaji earthquake, indicated that good physical health depended greatly on good mental health, and that good mental health depended on a sound social environment, including housing and community. (Shinfuku 1999; Kuwabara et al. 2008).

Mental stress was the single strongest predictor of life recovery. The elderly earthquake victims received support for their mental health through informal social support network and that paid professionals were hardly used for their mental health needs. This finding may apply not only to the elderly but also to all other generations. Policy measures that strengthen everyday social ties seem to be the key to lower mental stress among the impacted citizens. On this basis, more policy concerns should be paid to assist family systems and to facilitate more active community participation. Programs that encourage family units or parts of family units rather than individual members to participate in community affairs may be the suggested policy direction (Tatsuki and Hayashi 2002).

On 23 October 2004, a severe earthquake with a magnitude of 6.8 and a maximum tremor intensity level of 7 according to the Japanese 7-stage seismic scale occurred in and around the Chuetsu Area, Niigata Prefecture. Approximately 100,000 people took refuge, approximately 120,000 houses were damaged, and the damage amounted to >¥3 trillion (2005.9). Because of the sustained occurrence of aftershocks and delayed reconstruction of community lifelines, 9,160 people who lost their houses were still living in temporary shelter as long as 1 year after the earthquake (Toyabe et al. 2006). Disasters are traumatic events that many people encounter and may cause various psychological or physical health problems (Norris et al. 2002). The impact of the devastating earthquake and subsequent life in an unfamiliar environment would cause psychological distress for almost all people affected by the earthquake (Montazeri et al. 2005; Carr et al. 1997a, b, c; Kenardy et al. 1996). In some people, however, severe mental problems such as depression and/or Post-Traumatic Stress Disorder (PTSD), known risk factors for suicidal thinking, (Blumenthal 1988; Haynes 1990) will occur and they may continue to suffer from these mental disorders for a long time (Ollendick and Hoffmann 1982; Murphy 1986; Rubonis and Bickman 1991).

19.3 EJET Experiences in Case Study Cities

People of cities in northeast Japan, such as Kesennuma and Kamaishi, are enthusiastic about tsunami disaster prevention. They have been very effective in this regard, by organizing workshops, education for tsunami disasters and evacuation practices over the past 20 years. In particular, after the construction of the breakwaters, they also strengthened soft measures such as distribution of tsunami hazard maps to all households. Each city/town government prepared a tsunami hazard map, which overlays the estimated inundation areas of tsunamis and river floods, as well as risk areas for slope failures, based on scientific estimates. Moreover, they also promoted soft measures in local workshops for disaster prevention, such as discussions and practices to find evacuation places and routes. Loud speakers and radio transmissions were installed in each city or town to announce emergency cautions. These prevention activities had a great effect, even in face of the gigantic tsunami (Mimura et al. 2011). On the other hand, EJET hit the northeastern part such as Kamaishi,

Table 19.1 EJET impacts the Kamaishi, Kesennuma, and Natori cities

City	# of deaths A	# of missing persons B	# of deaths and missing persons C (A+B)	Population (2010) D	(%) of deaths and missing people C/D	Current population (2012.8)
Kamaishi	888	153	1,041	39,578	2.63	37,627
Kesennuma	1,204	250	1,454	73,494	1.98	69,476
Natotri	944	43	987	73,140	1.35	72,622

Kesennuma, and Natori Cities of Japan and caused heavy casualties (Disaster Management et al. 2012; Miyagi Prefecture 2012) (Table 19.1), heavy property losses (Okada et al. 2011).

19.3.1 Kamaishi City

Kamaishi produces iron ore and it is a good natural harbor. Consequently, Kamaishi City is the birthplace of the modern iron manufacture in Japan and known as the city of "iron and the fish". The 1896 Meiji Sanriku tsunami devastated Kamaishi. At that time, the population of the Kamaishi area was 6,524 people and 4,985 of them were lost. Its current population reached about 40,000 due to the steel industry. For the mitigation of tsunami disasters, a tsunami breakwater was constructed at the Kamaishi Bay entrance in 1978–2008. There are two breakwaters at the entrance of the bay with 4 m crest elevation, 300 m opening and lengths of 670 and 990 m. These breakwaters were built at a water depth of 63 m, the deepest in the world where a breakwater has been constructed. The tsunami wave reached 6.7 m at 20 km–300 m water depth-off shore Kamaishi. At least four of the town's 69 designated evacuation sites were inundated by the tsunami (Kamaishi Port Office 2011).

The entire coastal areas of Kamaishi are shaped with narrow inlets and steep terrains, which tend to increase the amplitude of tsunami. Unosumai-cho in Unosumai District is located in the north in Kamaishi, right next to another municipality of Otsuchi, where there had the most severe damage in Kamaishi. Tenjin-cho is one of the areas in Kamaishi District, which is the downtown of Kamaishi, with many residential and commercial buildings. The district also had the second largest damage after Unosumai District. Toni-cho in Toni District is a fishing village located in the south in Kamaishi. Many private residences were built at high point as the topography of the area becomes steep right near the port. The level of damage including deaths and missing persons is relatively smaller than other districts (Kawahara 2012).

Special note should be mentioned about the fact that nearly 3,000 children in elementary and middle high schools managed to evacuate safely in Kamaishi. There is a legend of "Tsunami tendenko". As "tendenko" means "scattered" in local language, the legend has been transferred for a long time to teach that, when you feel an earthquake, you should not wait for a unified action but run away individually and immediately. After the Great East Japan Earthquake, most elementary and middle-high schools evacuated students to higher places. Importantly, this action was taken before local authorities issued evacuation alerts. On the other hand, some people regretted that they relied too much on hard structures such as coastal dykes, based on the fact as they thought that these structures had successfully protected against tsunamis until then. There was an instance where raised awareness was of no assistance.

19.3.2 Kesennuma City

Kesennuma City Kesennuma is a City of 75,000 and well-known nationwide in Japan as a harbor city where tons of fresh fish are unloaded daily. Kesennuma has an outstanding catch of Pacific fishes like Saury, Bonito, Tuna, and Shark, shipped all over Japan. Above the fish market, an observation deck had been installed, from which people could watch fish being unloaded below (the roof could also be used as an evacuation site), and tsunami hazard maps are in place throughout the local community and a tsunami height sensor had been installed at the entrance of the Kesennuma Bay to confirm the tsunami size in advance of its arrival. Tide gauge station inside the Kesennuma Bay is totally destroyed. However, during our survey, we found one designated evacuation shelter was inundated. On 11 March 2011, a large part of the city was destroyed by the tsunami. The island of Oshima and its 3,000 residents, included in the city limits, were isolated by the tsunami which damaged the ferry connections. After the tsunami, spilled fuel from the town's fishing fleet caught fire and burned for 4 days (Yalciner et al. 2011).

The emergency manager Ken-Ichi Sato pressed the tsunami alarm button at Kesennuma within 2 min of the earthquake prior to the first official warnings issued by the Japan Meteorological Agency (JMA) and the Pacific Tsunami Warning Center (PTWC). Despite this extraordinary level of tsunami preparedness at Kesennuma, 1,467 residents were killed in the 2011 tsunami (973 confirmed dead and 494 missing presumed dead). The fatality rate was 2.3 % out of a population of 63,841 (http://www.npa.go.jp/archive/keibi/biki/higaijokyo_e.pdf). Some 20,000 houses were destroyed or damaged. In addition, 500 fishing boats were lost at Kesennuma. The tsunami impact peaked with a localized maximum tsunami height of 20 m measured on a tree at Hashikami on the headland to the west of the Kesennuma Bay entrance and run up heights of approximately 20 m on Osaki Cape separating Kesennuma from Hirota Bay. The tsunami heights remained below 10 m within the City of Kesennuma. On the mainland across the narrow channel facing the backside of Oshima Island a localized maximum run up of 20 m was measured in a narrow valley. This is reminiscent of the Babi Island effect observed near Flores, Indonesia (Yeh et al. 1994) The 2011 Tohoku tsunami heights approximately quadrupled the 1960 Chile tsunami heights at Kesennuma (http://tsunami.media.gunma-u.ac.jp/xml_tsunami/xmlindex.php?info=65%20reportMetatab%20reportSectab).

Eyewitnesses at Hashikami at the entrance to Kesennuma Bay reported the first tsunami wave arrival within 30 min of the earthquake and up to three waves (Fritz et al. 2012).

Some of the most severe damage was observed at Kesennuma City near the seashore. All buildings in that area were destroyed by the earthquake, tsunami, and fire; therefore, people were required to go to refuge centers. Control of hypertension, diabetes, and many other health disorders was difficult in these makeshift conditions, and newly developed electronic doctor's bags were used to control these conditions. The electronic doctor's bag, which was invented at Tohoku University, was useful in responding to victims' needs in these emergency conditions. The electronic doctor's bag was first used in the shelter in Kesennuma City, one of the most severely damaged areas, after approval from the ethics committee of Tohoku University Graduate School of Medicine. Use of the electronic doctor's bag facilitated evaluation of the condition of patients at the shelter by doctors located at the University. At the time of the disaster, the personnel shortage precluded on-site evaluation. Therefore, telemedicine and remote medicine were thought to be useful. The electronic doctor's bag enabled electrocardiography, blood pressure measurement, and ultrasonic diagnosis to be performed at the disaster shelters. Using this newly developed system, medical personnel in the shelters were also able to confer with doctors at Tohoku University via Skype. Using this device, control of anticoagulation, blood sugar levels, and blood pressure are able to be achieved, thereby preventing adverse cardiovascular events from occurring (Yambe et al. 2012).

19.3.3 Natori City

Natori City, which is another seriously damaged area, is located at the south of Sendai, in a flat coastal area with no hills or mountains. The March 11th tsunami reached 10 m high in this area. The tsunami reached more than a few kilometers inland from the coast and large areas of rice paddy fields were completely flooded (Sato and Chen 2012). Natori City is located on the Sendai Plains and is predominantly low-lying flat agricultural land, with two main population canters-the main city situated at least 5 km inland and Yuriage District at the coast, adjacent to the mouth of the Natori River. Coastal defense in Yuriage comprised pine coastal forest with concrete harbor walls at the Yuriage Port. The tsunami arrived in Yuriage approximately 65 min after the earthquake and inundated up to 5.2 km inland, close to the embankments of the Tohoku Expressway, where floating and burning wooden houses, cars, boats and other debris were deposited. Natori City officials advised the Earthquake Engineering Field Investigation Team (EEFIT) that Yuriage sustained a greater number of casualties than seaward of Sendai Airport, further south. The fatality rate was high compared to other areas (8.1 %), and as was the case in many other areas in this event, predominant casualties were people over 65 years of age (FDMA 2011). Almost all timber frame residential buildings in Yuriage sustained complete collapse. Steel and RC structures sustained damage ranging from light to collapse.

The RC port building sustained out-of-plane failure of all hollow concrete block masonry infill walls and partial collapse, and the wharf itself suffered significant scour and partial collapse. The majority of reinforcement in the concrete block infill walls was un-deformed rebar. Consistent failure direction of columns immediately inland of the port structure indicates flow direction from the east, which indicates that the wave arrived approximately perpendicular to the coastline, directly into the river mouth and across Yuriage harbor, where there were no significant coastal defenses. Significant scour of sandy soil was observed at a row of four apartment blocks in Yuriage. Scour occurred to up to 2 m deep around the eastern (seaward) ends and northern sides of the two of these buildings, causing severe tilting of one. The position of the most severe scour implies that scouring at this location most-likely occurred during the return flow of the tsunami. Five vertical evacuation structures in Natori City saved 3,285 lives in this event (Iwate Nichi Nichi Shinbun 2011). EEFIT investigated the performance of Yuriage Junior High School; the building suffered non-structural earthquake damage at seismically-designed separation joints, but remained effective as a vertical evacuation structure throughout tsunami inundation. The school was 1.5 km from the shoreline and only 320 m from the river to the north. Constructed on a 1.8 m high embankment, the school was raised above the level of surrounding rice fields, reducing the inundation depth at the eastern (seaward) end of the school buildings to 1.76 m. Flow velocity was low around the school building, with only flood damage to contents, and no debris impact or glazing damage as recorded at other locations. Natori City officials reported that this building provided refuge to approximately 800 people who had to remain there for 2 days following the tsunami (Fraser et al. 2012).

19.4 Future Integrated Healthcare as Lessons from EJET

EJET caused enormous casualties and heavy property losses. Experiences in disasters can be traumatic and affect people not only physically, but also psychologically, socially, and economically. After EJET, problems associated with health and healthcare network were pointed out. For example, in the affected areas, because patient medical records were not shared well among medical and health-related facilities and workers, it took time to assist the victims. In this case, there is room for improvement of healthcare-related network. On the positive side, EJET highlighted the importance of community-based support systems and emergency preparedness. Previous research suggests that primary prevention is the most effective means of reducing disaster damage. However, Japan has the highest proportion of older aging population in the world. The super-aging society is not only an individual issue, but also a salient factor crucial for public policies, such as pensions, health, and long-term care. In addition, although we can expect another mega earthquake anytime soon, it is difficult to maintain awareness of disaster prevention. Indirect disaster prevention countermeasures through the daily activities such as festivals and healthcare will play the important role when disaster happen. Strengthening community

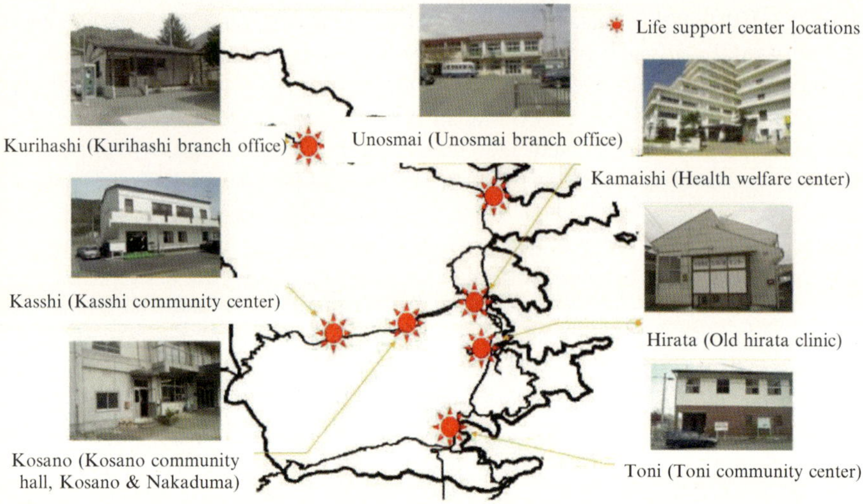

Fig. 19.1 Life support center place of each district. *Source*: Kamaishi City government

network through integrated healthcare may play an important role when an actual disaster happens. Then, Healthcare and the characteristic of each city are shown in the next chapter.

19.4.1 Healthcare Network and Life Support Center in Kamaishi City

Kamaishi City established life support centers in seven places: the Kamaishi, Hirata, Kasshi, Kosano, Kurihashi, Unosumai, and Toni districts) (Fig. 19.1). Community health nurses work in each center to promote health, provide medical care, and ensure the welfare of those who visit the center. Various other activities are offered at this community center. In addition, general administrative services are provided such as distribution of residency cards. In Kamaishi, these life support centers have adopted an attitude of collaboration with local residents to build a network of people. These centers monitor the current situation and problems of the region. Their activities contribute to making the city health conscious and allowing the residents to live in peace.

19.4.1.1 A Local Life Support System

Family support and the assistance system in the region has weakened because of the low birth rate and aging of society. Many problems have occurred as a result of modern lifestyle changes as well. New support structures and methods of service in addition to conventional health education, medical care, and welfare services are required. Therefore, Kamaishi City regards the construction of the community care system as urgent business. Staff such as community health nurses aid residents during consultations in the home or a local facility, depending on the individual situation. In addition, community improvements to assist elderly people are being implemented in Kamaishi. Kamaishi City calls this home visitation program a "local life support system." Its goal is to make Kamaishi a place to live in good health and peace for elderly people who have made it their home for so long. This system has been used as a reference in Chino City, Nagano Prefecture, where a similar program was developed in 2006 and has been implemented since 2007 in Kamaishi City. It was originally scheduled to be implemented for only 3 years.

In terms of directionality of services, cooperation between health providers, medical care workers, and the welfare system is presently insufficient. In generation from childhood to old age, Kamaishi City council desires to provide continuous and consistent service. To accomplish this, services are planned on the basis of real local conditions determined through talks with local residents. Because local connections are weak, and support through local volunteer organizations is lacking, local residents are encouraged to assist themselves. Therefore the city council advocates the education of local community residents and strengthening of cooperation by offering life study sessions.

19.4.1.2 Summary of Systems

1. *Life support center:* Life support centers are located in seven places in Kamaishi City as bases providing health service, medical care, welfare services, and educational sessions for people of all generations.
2. *Support team:* Each center has a support team including a community health nurse. Work related to educating the public, running community center activities, and operating a branch office is also performed at these centers. Staffs trained for supporting the life support teams are available at the city hall.

Connections between schools and communities were strong before EJET in the Toni district. In addition, various groups joined the Toni police box communication meeting and are now active in providing local relief and security in the Toni district. The "Supokon" program was launched in October 2001 in response to the call of the Kamaishi City Board of Education to provide a comprehensive community sports club like those proposed by the Ministry of Education, Culture, Sports, Science and Technology (MEXT). "Supokon" aims to create a healthy and bright cultural environment for citizens to enjoy sports and cultural activities. It also aims to involve more residents in sports and cultural activities, utilizing the rich natural beauty of the Toni area. In a way, the Toni district had this advantage starting before the support centers were established. However, information sharing improved after the life support centers were established, and cooperation after EJET was smooth because of the functioning of this system. Cooperation between healthcare providers has also improved (Fig. 19.2).

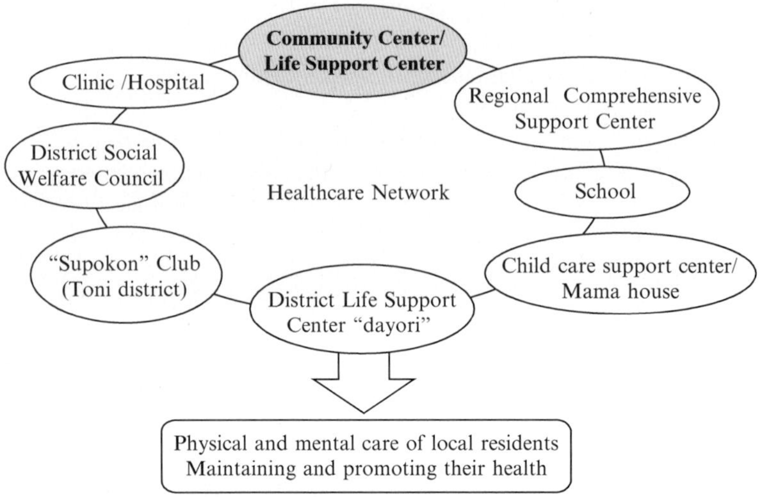

Fig. 19.2 Healthcare network in Kamaishi City

19.4.1.3 Healthcare Related Challenge of Kamaishi City

After EJET, area meetings were organized in each district. An individual living in temporary housing was visited seven times in 1 day because of overlaps in the work of the social welfare council and a Non Profit Organization (NPO). As a result of this incident, weekly area meetings were held from April 2012 in every district. Participant in these meetings include staff of the @Riasu NPO support center, administrative workers, community health nurses, members of the social welfare council, and members of the local residents' association.

Medical expenses are high in Kamaishi City and are even higher in Iwate Prefecture. Kamaishi City once boasted a prefectural hospital, a national hospital, and a municipal hospital. The municipal hospital was then integrated with the prefectural hospital for reasons related to population decline. However, the number of medical facilities remains the same as in the time when the population of Kamaishi City was over 100,000 people. The cost of running all these facilities is very high. In addition, increased aging of the population also increases problems related to medical expenses, and the work burden and low salaries of providers such as community health nurses are increased. Furthermore, local health support differs between disaster-affected areas, where temporary housing was built, and disaster-unaffected areas. Years after the disaster, discrepancies in health, the healthcare system, the role of healthcare providers, and first aid and cardiopulmonary resuscitation training are evident.

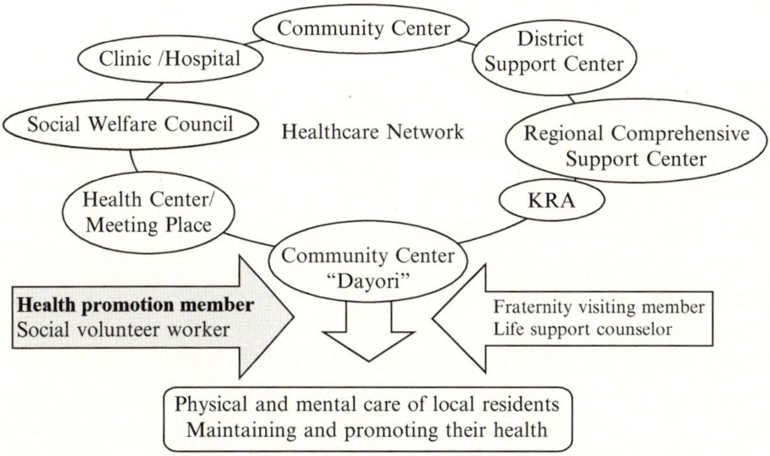

Fig. 19.3 Healthcare network in Kesennuma City

19.4.2 Healthcare and Health Promoting Member in Kesennuma City

In Kesennuma City, 90 emergency temporary housing facilities and 3,459 other housing units were built after EJET. Eighteen months after EJET, victims are still forced to live in inconvenient circumstances. However, various healthcare-related efforts have been made to improve this situation. The city promoted the establishment of a residents' association to promote the formation of new communities in each temporary housing facility. In addition, volunteers from nonprofit organizations have strived to build relationships to promote mutual support. These workers make efforts to interact with residents from outside the city and provide various activities.

Originally the temporary housing and relief efforts through mutual cooperation between the city and NPO were inadequate. Issues related to door-to-door visits to residents by representatives of too many organizations and overlapping support have occurred. Thus, in August 2011, a temporary housing subcommittee comprised city relations department employees involved in temporary housing, members of the social welfare council, and representatives of NPO was established. Since then, consultations have been held roughly once a month and coordination to resolve various issues has increased (Fig. 19.3).

19.4.2.1 KRA

KRA is an abbreviation for Kesennuma Reconstruction Association. The purpose of KRA is to restore and revive the City of Kesennuma through the efforts of Kesennuma citizens themselves. Emergency employment strategies have also been implemented. Cleaning businesses for individual homes and stores that were damaged by EJET have been established. Photograph restoration work is also ongoing for people whose photographs were damaged by the tsunami. Welfare work includes social gatherings and building of new communities in temporary housing facilities. Volunteer coordination is another active employment area.

19.4.2.2 Regional Comprehensive Support Center

Regional comprehensive support centers have been established in two places in Kesennuma City with the aim of helping disaster victims to continue their lives with dignity. Applications for consultation, comprehensive long-term care insurance claims, and welfare services are provided at these centers. In response to the various problems faced by the elderly after a natural disaster, comprehensive support is also provided at these centers. In addition to these more administrative tasks, these centers offer consultations regarding health, welfare, medical care, and life issues.

19.4.2.3 Health Promotion Member

Health promotion managers receive commissions from the city mayor to perform local health promotion activities to maintain and promote health for local residents. One manager is available per province, unless two are deemed necessary by the mayor in particular regions. Members perform the following duties: promotion of necessary activities for the health of residents; cooperation in various screening, health education, and health consultations held by the city; promotion of improved eating habits; and other necessary matters.

19.4.2.4 Emergency Temporary Housing Resident Support Center

Emergency temporary housing resident support centers prevent isolation and social withdrawal through general consultations and interaction with residents in emergency temporary housing. Support centers have been set up in four districts to support residents. Member districts include the Kesennuma district (six life counselors and three nurses), Karakuwa district (two life counselors and one nurse), Motoyoshi district (two life counselors and one nurse), and Ichinoseki district (four life counselors and two nurses).

19.4.2.5 Friendship Visiting Member

Kesennuma City offers the services of friendship visiting members who visit elderly residents to talk, offer advice, and provide support for those living in temporary housing in each region.

19.4.2.6 Healthcare-Related Challenge of Kesennuma City

Liaison meetings of different organizations are currently held in individual districts. However, no area meetings are held in Kamaishi City. Because the residents' associations do not participate in meetings, their needs and ideas may not be adequately reflected. Recently, more and more people have left temporary housing to rebuild their lives. Rebuilding the council will be difficult with a decreased number of people with the ability to manage an autonomous organization. They may have difficulty holding meetings on their own. The collapse of many communities and associated decline in motivation are evident. At present, relationships must be built between neighbors in temporary housing and local governmental organizations. Even in temporary housing, human relations are paramount. In addition, construction of houses from the viewpoint of human welfare, evaluation of disaster-related public housing, and examination of housing transitions will be necessary in future. Cooperation between city hall and various nonprofit and Non Governmental Organizations (NGO) will also be important.

19.4.3 Healthcare Network and Regional Comprehensive Support Center in Natori City

Regional comprehensive support centers are located in three places in Natori City. Nursing-related consultations are provided at these centers. The city plans to expand its family support center with an emphasis on childcare. These regional health centers conduct the following activities (Fig. 19.4).

19.4.3.1 Regional Comprehensive Support Center

Regional comprehensive support centers have been established in three places (east, south, and west) to serve all regions. Senior care managers, social workers, and public health nurses provide comprehensive support as a professional team in these centers.

19.4.3.2 Health Center

AIDS counseling, anti-HIV antibody testing, and registration of bone marrow transplantation are all performed at regional health centers. Infant health checkups, vaccination, and well-baby visits, are performed by public health nurse. Maternity nurses visit all families to whom babies are born. Childcare help services are also offered. Mothers' helpers visit the home to help with housework and childcare, which is difficult without the support of the family immediately after birth. Checkups

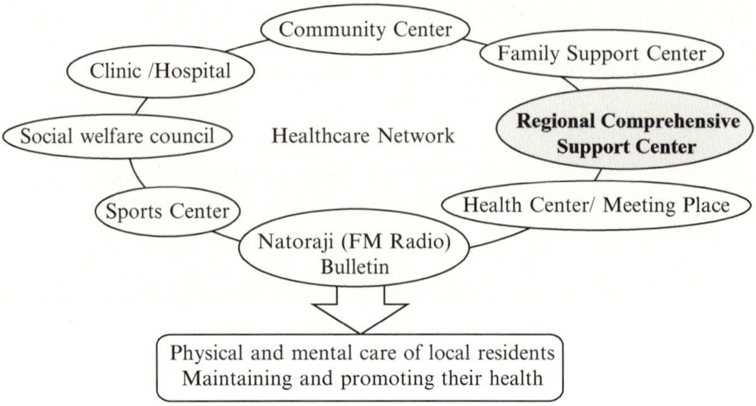

Fig. 19.4 Healthcare network in Natori City

for women after childbirth are also offered. General health checkups for pregnant women and the Cheerful Natori Food Education Plan are also available.

19.4.3.3 Family Support Center

People who require childcare and those who support them must register at the family support center. The center creates an environment where it is easy to find childcare. It also helps families to balance work and childcare, and provides support to childcare facilities in the region.

19.4.3.4 Health, Sports and Recreation in the Community Center

Classes such as healing yoga, easy yoga, Pilates, stretching, tai chi chuan, and walking are held in each community center.

19.5 Common Issues in Three Cities

In Japan as a whole, a decline in population has been recorded since 2005, and the rate of aging is expected to continue to increase in future. The figure above shows changes in population and the number of elderly people in the entire country and in Kamaishi, Kesennuma, and Natori Cities (National Institute of Population and Social Security Research 2008) (Fig. 19.5). The declining population in these cities was evident before the 2011 EJET. A particularly significant decline in the populations of Kesennuma and Kamaishi Cities is predicted in the future.

In addition, the percentage of people over 65 years of age in these cities is predicted to increase to about 40 % in 2020. Although a difference exists between the

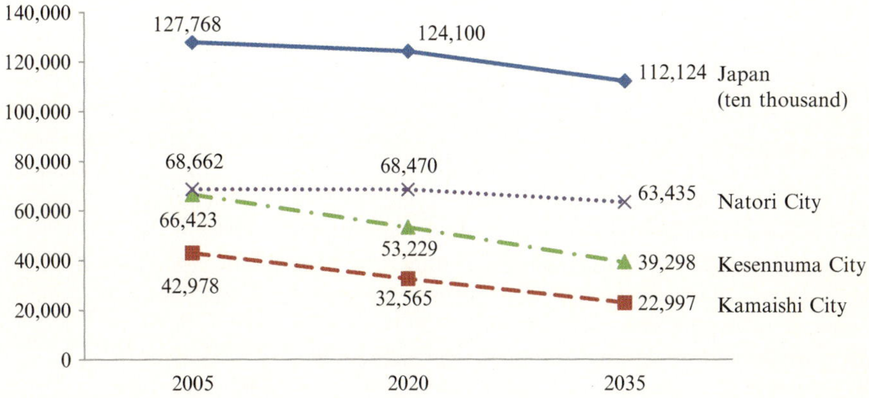

Fig. 19.5 The change of population in Japan and three cities

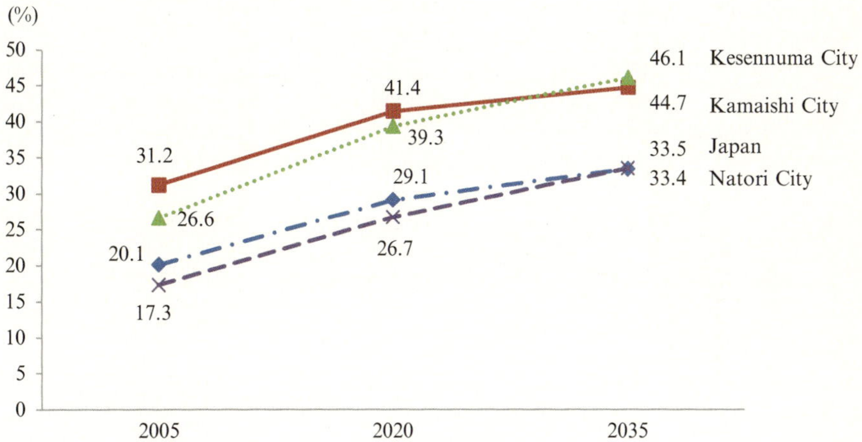

Fig. 19.6 The change of aging rate in Japan and three cities

actual number and the number in the figure because of the effects of EJET and changes in population in each statistical year, 13 years from now, the percentage of people over 65 years of age in Kesennuma City is predicted to reach 46.1 % (National Institute of Population and Social Security Research 2008) (Fig. 19.6).

In addition, the percentage in Natori City will be above the national average. Furthermore, 33 % of the total population of Japan is estimated to be above 65 years of age by 2035. This is similar to the current aging rate in Kamaishi and Kesennuma Cities. Thus, the super aged society in these cities will be a reality in Japanese life 20 years later. The problems of population decline and aging were exacerbated after EJET.

In the chronic phase (from 4 weeks to 5 years) after the Hanshin-Awaji Earthquake, many young people chose to leave shelters and temporary centers, leaving only the elderly. In such circumstances, autonomous management by residents is difficult to sustain. Issues such as loneliness, death, and houseboundedness become a reality for elderly people living in temporary housing. In the three cities most strongly impacted by EJET as well as the Hanshin-Awaji Earthquake, many people with plans and financial means will leave temporary housing in future, leaving people with financial problems and the elderly behind. Their absence limits opportunities for development and hinders the functioning of the council. Decreased motivation because of the difficulties of aging may lead to exacerbation of disease, mental deterioration, and increased houseboundedness. Although the current support system is already overburdened, further cooperation with experts is required.

Health issues for the elderly in temporary housing include lack of exercise because of decreased movement, decreased physical strength because of inactivity or disease, and worsening of chronic disease and dementia because of stress and changes in circumstances. These people need purpose in life and help to improve their health as soon as possible. Despite efforts such as voluntary radio exercise programs, more aid is necessary to form a real community in these shelters. Self-government is vital in these housing facilities. An appointed chairperson considers the needs of inhabitants and acts as a communication coordinator in the temporary housing facilities. Various judgments are required of him or her. Mental stress can result when inhabitants disagree with the chairperson's judgments. Relationships within temporary housing facilities and local autonomous organizations help to connect internal and external networks in these local communities. However, continued support from local organizations is essential.

19.6 Future Efforts and Proposed Actions

Depopulation and aging are serious problems in Kamaishi, Kesennuma, and Natori Cities. These problems make the community more vulnerable against disaster and present financial challenges because of increased medical expenses. Although society depends on the city administration for various services and events, resources are limited. Hence, these municipalities may eventually run out of funds for healthcare services offered by community health nurses, hospitals, and other medical facilities, ant disaster measures from the fire department, and general municipal administration. Communities in local regions must respond to various issues themselves without depending on the government and experts only in future. The healthcare network and system in Kamaishi City offers an excellent model for other municipalities to follow. A life support center functions as a public health center and a community center in Kamaishi City. Because the regional comprehensive center is available as well, depending on the location, the city is able to strengthen its networks. However, public health nurses and healthcare providers may become exhausted because the demand is great for medical care in hospitals and other healthcare facilities.

This study proposes that if members of the health promotion team in Kesennuma City would adopt the Kamaishi City system (model), a better response could be offered to the circumstances of the community. In addition, a shift is necessary in the healthcare system from focusing on public health nurses to focusing on members of the health promotion team in communities. The main role of the public health nurse should be to manage at the local or regional level. If the health promotion team of Kesennuma City and the family support system of Natori City are incorporated into the Kamaishi City system as a base model, support for local residents in affected communities and regions will improve. Furthermore, community networks built through this revised healthcare model will have an important role in the event of a disaster. In addition, the number of victims, post-disaster health impact, and medical costs may be decreased. Daily living for residents can be improved through strengthening of these networks. A disaster-resilient community can be built by further enhancing community networks with participation of residents in their own healthcare.

On the basis of previous disaster experiences, mental and behavioral health must be the focus of healthcare providers. Various institutes and healthcare providers recognize this fact. After a disaster, many medical, psychological, and healthcare teams must provide support for victims in the acute phase. In the chronic phase, many volunteers from NPO/NGO must provide support to local residents in affected areas. However, because disasters impact residents in the long term, and further aging of the Japanese population is expected in the future, a sustainable community network must be established to maintain healthcare for residents that includes their own involvement and that of their communities.

Acknowledgements The author acknowledges JSPS (Japan Society for Promotion of Science), CWS (Church World Services) Asia Pacific, and MERCY Malaysia for financial support.

References

Appelbaum D, Kligler B, Barrett B et al (2006) Natural and traditional medicine in Cuba: lessons for U.S. medical education. Acad Med 81(12):1098–1103

Axelrod C, Killam PP, Gaston MH, Stinson N (1994) Primary health care and Midwest flood disaster. Public Health Rep 109(5):601–605

Blumenthal SJ (1988) Suicide: a guide to risk factors, assessment, and treatment of suicidal patients. Med Clin North Am 72:937–971

Bowyer TW, Biegalski SR, Cooper M, Eslinger PW, Haas D, Hayes JC, Miley HS, Strom DJ, Woods V (2011) Elevated radioxenon detected remotely following the Fukushima nuclear accident. J Environ Radioact 102:681–687

Campos-Outcalt D (2006) Disaster medical response: maximizing your effectiveness. J Fam Pract 55(2):113–115

Carr VJ, Lewin TJ, Webster RA, Kenardy JA (1997a) A synthesis of the findings from the quake impact study: a two-year investigation of the psychosocial sequelae of the 1989 Newcastle earthquake. Soc Psychiatry Psychiatr Epidemiol 32:123–136

Carr VJ, Lewin TJ, Kenardy JA et al (1997b) Psychosocial sequelae of the 1989 Newcastle earthquake: III Role of vulnerability factors in post-disaster morbidity. Psychol Med 27: 179–190

Carr VJ, Lewin TJ, Webster RA, Kenardy JA, Hazell PL, Carter GL (1997c) Psychosocial sequelae of the 1989 Newcastle earthquake: II. Exposure and morbidity profiles during the first 2 years post-disaster. Psychol Med 27:167–178

Connolly MA, Gayer M, Ryan MJ, Salama P, Spiegel P, Heymann DL (2004) Communicable diseases in complex emergencies: impact and challenges. Lancet 364:1974–1983

Disaster Management Office, General Affairs Division, Iwate Prefecture http://www.pref.iwate.jp/~bousai/taioujoukyou/201209051700jintekihigai.pdf. Accessed 5 September 2012

Disaster Mental Health Subcommittee of the National Biodefense Science Board US Department of Health and Human Services, Office of the Assistant Secretary for Preparedness and Response. (2008) Disaster mental health recommendations: report of the disaster mental health subcommittee of the National biodefense science board 18, p 12. http://www.phe.gov/Preparedness/legal/boards/nbsb/Documents/nsbs-dmhreport-final.pdf. Accessed 5 December 2012

Dresang LT, Brebrick L, Murray D et al (2005) Family Medicine in Cuba: Community-Oriented Primary Care and Complementary and Alternative Medicine. J Am Board Fam Pract 18(4):297–303

Federal Emergency Management Agency (2011) National disaster recovery framework.strengthening disaster recovery for the nation. http://www.regulations.gov/#!documentDetail;D=FEMA-2010-0004-0144. Accessed 4 October 2012

Fire and Disaster Management Agency of Japan (2011) Report 137 on the effects of the March 11, 2011 great Tōhoku earthquake. http://www.fdma.go.jp/bn/2011/detail/691.html. Accessed 29 August 2011

Fraser S, Raby A, Pomonis A, Goda K, Chian SC, Macabuag J, Offord M, Saito K, Sammonds P (2012) Tsunami damage to coastal defences and buildings in the March 11th 2011 M_w9.0 Great East Japan earthquake and tsunami. Bull Earthquake Eng 11:205–239

Fritz HM, Phillips DA, Okayasu A, Shimozono T, Liu H, Mohammed F, Skanavis V, Synolakis CE, Takahashi T (2012) The 2011 Japan tsunami current velocity measurements from survivor videos at Kesennuma Bay using LiDAR. Geophys Res Lett 39

Fujii Y, Satake K, Sakai S, Shinohara M, Kanazawa T (2011) Tsunami source of the 2011 off the Pacific coast of Tohoku, Japan earthquake. Earth Planets Space 63:815–820

Haynes MA (1990) Suicide prevention: a US perspective. In: Goldbloom RB, Lawrence RS (eds) Preventing disease: beyond the rhetoric. Springer, New York, pp 129–136

Iwate Nichi Nichi Shinbun (2011) 10,000 people survive after evacuating to a (designated evacuation) building, p 7. Ichinoseki, Japan, 30 May 2011

Jobe K (2010) Disaster relief in post-earthquake Haiti: Unintended consequences of humanitarian volunteerism). Travel Med Infect Dis 9(1):1–5

Kamaishi Port Office (2011) http://www.pa.thr.mlit.go.jp/kamaishi/bousai/index.html. Accessed 10 November 2012

Kawahara T (2012) Systems analysis of the impacts of energy supply shortage in kamaishi for resilience building after Great East Japan earthquake. LUCSUS (Lund University Centre for Sustainability Studies)

Kenardy JA, Webster RA, Lewin TJ, Carr VJ, Hazell PL, Carter GL (1996) Stress debriefing and patterns of recovery following a natural disaster. J Trauma Stress 9:37–49

Kokai M, Fujii S, Shinfuku N, Edwards G (2004) Natural disaster and mental health in Asia. Psychiatry Clin Neurosci 58(2):110–116

Kuwabara H, Shioiri T, Toyabe S, Kawamura T, Koizumi M, Ito-Sawamura M, Akazawa K, Someya T (2008) Factors impacting on psychological distress and recovery after the 2004 Niigata–Chuetsu earthquake, Japan: Community-based study. Psychiatry Clin Neurosci 62(5):503–507

Mimura N, Yasuhara K, Kawagoe S, Yokoki H, Kazama S (2011) Damage from the Great East Japan earthquake and tsunami - a quick report Mitig. Adapt Strateg Glob Change 16:803–818

Miyagi Prefecture Disaster Countermeasures Headquarters http://www.pref.miyagi.jp/kikitaisaku/higasinihondaisinsai/pdf/09071600.pdf. Accessed 31 August 2012

Montazeri A, Baradaran H, Omidvari S et al (2005) Psychological distress among Bam earthquake survivors in Iran: A population-based study. BMC Public Health 5:4

Mori N, Takahashi T, Yasuda T, Yanagisawa H (2011) Survey of 2011 Tohoku earthquake tsunami inundation and run-up. Geophys Res Lett 38

Murphy SA (1986) Status of natural disaster victims' health and recovery 1 and 3 years later. Res Nurs Health 9:331–340

National Institute of Population and Social Security Research (2008) Projected population by city, town and village in Japan (2005–2035). http://www.ipss.go.jp/pp-shicyoson/j/shicyoson08/gaiyo.pdf. Accessed 5 November 2012

National Police Agency of Japan (2012) http://www.npa.go.jp/archive/keibi/biki/higaijokyo_e.pdf. Accessed 4 September 2012

Norris FH, Friedman MJ, Watson PJ, Byrne CM, Diaz E, Kaniasty K (2002) 60,000 disaster victims speak: part I. an empirical review of the empirical literature, 1981–2001. Psychiatry 65:207–239

Okada N, Tao Y, Kajitani Y et al (2011) The 2011 Eastern Japan great earthquake disaster: overview and comments. Int J Disaster Risk Sci 2(1):34–42

Ollendick DG, Hoffmann M (1982) Assessment of psychological reactions in disaster victims. J Community Psychol 10:157–167

Pfefferbaum B, Flynn BW, Schonfeld D, Brown LM, Jacobs GA, Dodgen D, Donato D, Kaul RE, Stone B, Norwood AE, Reissman DB, Herrmann J, Hobfoll SE, Jones RT, Ruzek JI, Ursano RJ, Taylor RJ, Lindley D (2012) The integration of mental and behavioral health into disaster preparedness, response, and recovery. Disaster Med Public Health Prep 6(1):60–66

Rubonis AV, Bickman L (1991) Psychological impairment in the wake of disaster: The disaster–psychopathology relationship. Psychol Bull 109:384–399

Sato M, Chen SW (2012) Polarimetric SAR analysis of tsunami damage following the March 11th East Japan earthquake. In: 9th European conference on synthetic aperture radar 2012, EUSAR, pp 125–128

Shibahara S (2011) The 2011 Tohoku earthquake and devastating tsunami. Tohoku J Exp Med 223:305–307

Shinfuku N (1999) To be a victim and a survivor of the great Hanshin-Awaji earthquake. J Psychosom Res 46(6):541–548

Simons M et al (2011) The 2011 magnitude 9.0 Tohoku-Oki earthquake: mosaicking the megathrust from seconds to centuries. Science 33:1421–1425

Tanioka Y, Satake K (1996) Fault parameters of the 1896 Sanriku tsunami earthquake estimated from tsunami numerical modeling. Geophys Res Lett 23(13):1549–1552

Tatsuki S, Hayashi H (2002) Seven critical element model of life recovery: general linear model analyses of the 2001 Kobe panel survey data. In: Proceedings of the 2nd workshop for comparative study on urban earthquake disaster mitigation, pp 27–46

Toyabe S, Shioiri T, Kuwabara H et al (2006) Impaired psychological recovery in the elderly after the Niigata-Chuetsu earthquake in Japan: A population-based study. BMC Public Health 6:230

US Department of Health and Human Services (2009) National health security strategy of the United States of America, p 11. http://www.phe.gov/preparedness/planning/authority/nhss/Pages/default.aspx

Yalciner AC, Suppasri A, Mas E, Kalligeris N, Necmioglu O, Imamura F, Ozer C, Zaytsev A, Synolakis C, Takahashi S, Tomita T, Yon G (2011) Field survey on the coastal impacts of March 11, 2011 Great East Japan tsunami. American Geophysical Union Fall Meeting, Washington

Yambe T, Shibata M, Sumiyoshi T, Mibiki Y, Osawa N, Katahira Y, Yambe M, Tabayashi K, Yamashina M, Sato E, Sato S, Yagi T, Watanabe M, Akinno Y, Munakata M, Owada N, Akiyama M, Saiki Y, Sugita N, Yoshizawa M (2012) Medical responses following the Sendai quake (East Japan earthquake, march 11, 2011). Artif Organs 36(8):760–763

Yeh H, Liu PL-F, Briggs M, Synolakis CE (1994) Propagation and amplification of tsunamis at coastal boundaries. Nature 372(6504):353–355

Chapter 20
Role of Community Radio in Post Disaster Recovery: Comparative Analysis of Japan and Indonesia

Junichi Hibino and Rajib Shaw

Abstract In the event that communications infrastructure and disaster wireless systems are severely damaged during a natural disaster, the means to pass information on to disaster victims is lost. Furthermore, it is quite common for large scale blackouts to occur during such natural disasters. In such disaster situations, radio acts as a resilient medium of communication that is not influenced by blackouts, and even in cases where broadcasting equipment is damaged it is a medium for which operation can be restored relatively easily. In such situations it is not radio stations that cover a wide broadcast area, but rather low output radio stations (community radio stations being one such example) located in local communities that are capable of playing a major role in collecting, transmitting, and sharing detailed disaster related information specific to these individual communities and the unique needs of disaster victims living in such communities. This chapter provides a comparative analysis of role of community radio in Japan and Indonesia and exemplifies the future role, the community radio should play for effective risk reduction.

Keywords Emergency FM radio • Mega-disaster • Merapi eruption • Role of community radio • Tohoku tsunami

---

J. Hibino (✉)
Radio FM YY, Kobe, Japan
e-mail: hibijun@gmail.com; hibino@tcc117.org

R. Shaw
International Environment and Disaster Management, Graduate School of Global Environmental Studies, Kyoto University, Kyoto, Japan
e-mail: shaw.rajib.5u@kyoto-u.ac.jp

R. Shaw (ed.), *Disaster Recovery: Used or Misused Development Opportunity*,
Disaster Risk Reduction, DOI 10.1007/978-4-431-54255-1_20, © Springer Japan 2014

20.1 Introduction

Natural disasters are a major obstacle to the sustainable development of society. For this reason, strengthening disaster preparedness and facilitating disaster management in local communities located in areas susceptible to the frequent occurrence of natural disasters is integral to sustainable development (JICA 2010). "In order to build disaster-resilient communities, they first need to be empowered so that community members can cope with the adverse effects of natural hazards" (Okazaki et al. 2005).

Data is a source of information; information is a source of knowledge; knowledge is a source of power. Thus, to empower communities, it is essential to ensure access to data and information. Although mass media have played an important role in distributing data and information to society at large, they have not been able to respond to the diverse needs of individual communities at the local level. For this reason, community radio plays an integral role in the empowerment of local communities and in community development in general.

Disaster management and information and communications technology are closely connected. In all phases of a disaster, from "damage mitigation," "preparation," "early-warning," "response" to "restoration and recovery," data, information and communication play important roles. The exchange of information, the sharing of information, and discussion among citizens is essential to develop the disaster management capacity of individual communities. At the same time, it is also essential that neighboring communities exchange information and cooperate with one another. Toward this end, community radio can play an important role in disaster management activities conducted in both times of disaster and in everyday situations, by facilitating the sharing of knowledge and experience and by acting as a tool to foster self reliance and mutual assistance.

In this paper, the role of disaster response radio stations that continue to operate in the disaster affected areas of the Great East Japan Earthquake and Tsunami (as of writing), in the restoration and recovery phases of the disaster is examined. The term "community," as used in this paper, refers to a communal entity composed of individuals that share a geographic or cultural affinity, and "community radio" refers to radio broadcasting that serves such communities. Furthermore, "community broadcasting" refers specifically to FM broadcasting initiated in 1992 as part of Japanese public policy, with the aim of providing broadcasting for a select set of municipalities, and "community media" refers to a broad range of media that includes community radio (in addition to print media, terrestrial broadcasting, digital networks, and physical spaces used for public communication).

20.1.1 The Role of Community Radio in Disaster Situations

In the event that communications infrastructure and disaster wireless systems are severely damaged during a natural disaster, the means to pass information on to disaster victims is lost. Furthermore, it is quite common for large scale blackouts to occur during such natural disasters. In such disaster situations, radio acts as a resilient medium of communication that is not influenced by blackouts, and even in cases where broadcasting equipment is damaged it is a medium for which operation can be restored relatively easily. In such situations it is not radio stations that cover a wide broadcast area, but rather low output radio stations (community radio stations being one such example) located in local communities that are capable of playing a major role in collecting, transmitting, and sharing detailed disaster related information specific to these individual communities and the unique needs of disaster victims living in such communities (Hiratsuka 2011).

20.1.2 What is Community Radio?

In general, a community radio station is radio station that has a low output frequency and relatively small area where it can be listened to. It is said that the fist community radio station was established in Bolivia, South America in 1947 (O'Connor 2004). Sixty-five years later, the number of community radio stations has increased to tens of thousands throughout the world; in North and South America, Europe, Africa, Asia and Pacific Region. To put it briefly, community radio stations were created to reflect and cope with the different needs and characteristics of each country and region. The World Association of Community Radio Broadcasters (AMARC) defines a community radio station under the three following conditions (1) is a non-profit organization, (2) is owned and managed by the community and has community participation, (3) is independent from politics and commercialism. Also, UNESCO defines a community station as follows; a community radio station plays an important role in contributing to solve the great social problems of poverty and alienation of communities, gives power to people who have been left behind in world development, and promotes democracy through citizen participation in the media (Fraser and Restrepo-Estrada 2001).

Such community radio stations broadcast programs on nature conservation, agriculture, regional culture, education, etc. contributing to the development of their communities. Here in Japan, after the Great Hanshin-Awaji Earthquake, and abroad after the Indian Ocean tsunami, the community radio has become a useful tool in disaster management and disaster victim support.

20.1.3 The Role of Community Radio on Disasters

Natural disasters are one cause of preventing the development of a sustainable society. Improving a community's disaster management ability is, for people living in areas where there are many natural disasters, an important issue for them in order to protect their livelihood. On top of this, disaster management is essential for a community's sustained development. In order to create a community that has strong disaster management ability, it is essential that the community develops the ability to improve disaster management on its own.

Data is information, information is knowledge, and knowledge is one kind of power. For communities to be able to improve their disaster management ability on their own, an environment where citizens can obtain data and information must be guaranteed. Although it is true that the mass media contributes to society by spreading data and information, it is also a fact that it is not an information medium that can meet the individual needs of a varied and complex society. For a community to develop its disaster management ability on its own, the community radio is essential.

Disaster management and information transmission technology are closely connected. In all of the phases of a disaster – "damage mitigation," "preparation," "early-warning," "response" and "recovery and revitalization," data, information and communication play important roles. The exchange of information, sharing information and discussion among citizens is essential for the self-improvement and development of community disaster management. The exchange of information and cooperation among neighboring communities is also essential. The equipment and media used for communication through information transmission is a deciding factor with regard to the quality and character of communication. The community radio which promotes the sharing of knowledge and experience and self-help and mutual aid can play an important role in times of disaster and disaster management in normal times.

In times of disaster, the content of the information broadcast by such community radio stations will change depending on the phase of the disaster management process (aid/relief, restoration, and recovery). In the immediate aftermath of a disaster, disaster radio must provide residents with information related to damage caused by the disaster, emergency medical and first aid information, information about directions to evacuations shelters, information on the whereabouts of residents and contact information, as well as information on lifeline utilities directly related to everyday life, such as the provision of electrical power and telephone services.

In addition to providing disaster affected individuals with useful information, community radio can also be used to play a vital role in alleviating the psychological trauma experienced by disaster victims, by broadcasting such things as nostalgic music, folk tales and folk songs that are indigenous to the area and messages from religious leaders. During the relief and restoration periods, information on NGOs and volunteer groups conducting relief and assistance activities is also vital (in addition to the standard information provided by the local government) to enable disaster affected individuals to recover from the disaster. Furthermore, during the recovery period, emergency radio plays a crucial role in the development of a shared recovery plan for the local area, ensuring that the opinions of all individuals are applied to the plan, while at the same time providing information that is necessary to ensure that the recovery activities performed are truly driven and centered around the residents themselves.

Although a large number of organizations will conduct relief efforts in the disaster affected areas, there are cases where such activities do no truly meet the needs of the disaster victims. Community radio plays a vital role in bridging this gap, by encouraging communication between relief organizations and disaster victims. By exercising great care in soliciting the opinions of disaster victims, and by transmitting these voices through radio, it becomes possible for such relief organizations to respond to this feedback by improving their activities in order to better meet the needs of disaster victims.

At the same time, community radio can serve a unique role to monitor if aid/relief, restoration, and recovery policies and activities are being implemented as promised, from the standpoint of the disaster victims. For example, staff members of the local government, disaster prevention agencies, and NGOs that are involved in conducting relief activities to serve disaster victims can be invited to field questions from residents on a call-in radio show program. Such forms of programming can serve to improve the quality of such activities conducted in the disaster affected areas.

In addition to facilitating communication between "relief organizations and disaster victims," community radio can also foster communication by linking "local government and disaster victims," "local government and NGOs," "fellow disaster victims facing differing conditions and issues," and "areas inside the disaster affected region and the outside world." During the long road to recovery, many decisions must be made for the community by the community members themselves. In such cases, it is imperative that a mutual understanding is established between all parties (despite the differences in the roles they assume and the conditions that they face) before any decisions can be made for the community. Community radio can greatly contribute to this process of establishing such a mutual understanding.

Furthermore, the role of community radio in serving those individuals with limited access to information and who are the most in need of such information is a vital one that must not be forgotten. As such disaster situations place such a strain on the lives of all community members, relief activities have a tendency to focus on providing aid that meets the needs of those segments of the population that represent the "greatest common divisor." At the same time, there is a tendency to neglect the needs of foreign and handicapped residents who face special difficulties in overcoming such disaster situations (Yoshitomi 2010).

Community radio acts as more than just a tool to transmit information in a unidirectional manner over a large area, but rather, it is a tool that facilitates interactive communication between the broadcaster and its listeners, as well as between the listeners themselves (Kanayama 2007).

20.2 Outline Overview of Disaster Response Radio Broadcasting in the Aftermath of the Great East Japan Earthquake and Tsunami

As of writing, a total of 18 emergency radio stations have been recently established in the disaster affected areas of the Great East Japan Earthquake and Tsunami. When combined with the existing number of community radio stations, a total of 28 radio stations in four prefectures (Iwate, Miyagi, Fukushima, and Ibaraki prefectures) have contributed in efforts to provide local disaster victims with relief and recovery related information, as well as in efforts to alleviate the damage caused by the disaster and to restore stability to lives of the disaster victims (refer to Fig. 20.1 and Table 20.1).

Of these stations, the 18 newly established stations, as well as 10 of the existing community radio stations (that were facilitated by disaster response measures taken by local government bodies allowing for such things as an increase in broadcasting power), were designated as "temporary emergency FM stations" as defined by government policy. As the broadcasting licenses for such temporary emergency FM stations are issued by the Ministry of Internal Affairs and Communications (hereinafter referred to as the MIC) to the chief executive of the local government, it is first required that the local government submits an application for a license to the MIC before the process can begin. Despite the existence of this system, however, most of the local governments for which temporary emergency FM stations were established in the wake of the Great East Japan Earthquake, were largely unaware of the existence of such a system, and only took steps to establish such stations upon the urging and proposals of both outside and local non-governmental organizations (Kanayama 2012). With that said, it must also be noted that three stations were specifically established by city governments and town offices that had learned of local community radio stations established in other geographic regions in the early stages following the occurrence of the disaster.

Being established under such circumstances, the aforementioned temporary emergency FM stations can now be classified into two separate groups; "publicly established publicly operated radio stations" and "publicly established privately operated radio stations." Publicly established publicly operated radio stations primarily assume the role of "local government radio," mainly focusing on disseminating information provided by the local government. Publicly established privately operated radio stations can be further classified into two types; those stations for which broadcasting operations are a cooperative effort between the local government and an administrative organization (nonprofit or local organization), and those stations for which the local government only performs broadcasting license related paperwork and all operations are conducted by a local organization, in a manner similar to community radio by definition (refer to Table 20.2).

This disparity in the process of establishment and method of operation of the temporary emergency FM stations located in different geographic areas can be attributed to the fact that there is no manual providing an established procedure to start a temporary emergency FM station, as well as the fact that individuals involved in proposing and supporting the establishment of temporary emergency FM stations are often poorly informed or only have access to limited expertise and knowledge (Murakami 2012).

Of the aforementioned temporary emergency FM stations (excluding six stations that have transitioned back to their roles as community radio stations) only two stations have chosen not to renew their licenses and ceased operations. These two stations were established in the early aftermath of the disaster (located in Ohsaki City, Miyagi Prefecture and Sukagawa City, Fukushima Prefecture) and eventually ceased operations after the initial 2 month license issuing period. Both stations were located in

Fig. 20.1 Map showing temporary emergency radio stations which is running as of 1st January, 2013 in affected areas of East Japan and Tsunami

Table 20.1 View of temporary emergency radio stations

Prefecture	Municipality	Establishmental date	Closing date	Type
Iwate	Miyako City	19 March 2011		New establishment
	Ohtsuchi Town	28 March 2012		New establishment
	Kamaishi City	7 April 2011		New establishment
	Ohfunato City	28 March 2011		New establishment
	Rikuzentakata City	10 December 2011		New establishment
	Hanamaki City	11 March 2011	3 April 2011	Existing community radio
	Ohshu City	12 March 2011	29 March 2011	Existing community radio
Miyagi	Kesennuma City	22 March 2011		New establishment
	Minami Sanriku Town	17 March 2011		New establishment
	Tome City	16 March 2011		Existing community radio
	Ohsaki City	15 March 2011	14 May 2011	New establishment
	Onagawa Town	21 April 2011		New establishment
	Ishinomaki City	16 March 2011		Existing community radio
	Shiogama City	18 March 2011		Existing community radio
	Natori City	7 April 2011		New establishment
	Iwanuma City	20 March 2011		Existing community radio
	Watari Town	24 March 2011		New establishment
	Yamamoto Town	21 March 2011		New establishment
Fukusima	Soma City	30 March 2011		New establishment
	Minami Soma City	16 April 2011		New establishment
	Fukushima City	16 March 2011		Existing community radio
	Tomioka Town	11 March 2012		New establishment
	Sukagawa City	7 April 2011	7 June 2011	New establishment
Ibaragi	Iwaki	28 March 2011	27 May 2911	Existing community radio
	Kashima	13 March 2011	1 June 2011	Existing community radio
	Tsukuba	14 March 2011	15 April 2011	Existing community radio
	Takahagi	8 June 2011		New establishment
	Toride City	1 August 2012	31 January 2013	New establishment

inland areas that did not experience any tsunami damage. The remaining 20 temporary emergency FM stations (16 of which were newly established) continue to operate, more than a year and 7 months (as of writing) after the occurrence of the disaster.

Table 20.2 Comparing of temporary emergency radio stations

	Governance	Management	Residents' participation
Miyako	NPO	NPO	Middle
Otsuchi	Local government	NPO	Very high
Kamaishi	Local government	Local government	Middle
Ofunato	Local government	Local government	Very low
Rikuzentakata	NPO	NPO	High
Kesennuma	Local government	NPO	Middle
Minami Sanriku	Local government	Other community Radio	Very low
Onagawa	Resident group	Resident group	Very high
Natori	Local government	NPO	High
Watari	Local government	Resident group	Very high
Yamamoto	Local government	Resident group	Middle
Soma	Local government	Local government	Very low
Minami Soma	Resident organization	Resident organization	Very high
Tomioka	Social welfare council	Social welfare council	Very high
Takahagi	Local government	Local government	Middle

Of the aforementioned 20 stations, 3 stations are of particular interest as they were established more than a year after the actual occurrence of the disaster. In regards to these stations established at such a late phase, the MIC Tohoku Bureau of Telecommunications has publicly stated that, "such temporary emergency FM stations can still be established even in cases where a significant amount of time has passed since the disaster, if there are a significant number of evacuee households living in temporary housing, and where there are no governing bodies available at the temporary housing that can carry out functions such as distributing public announcements for the city/town municipality or disseminating information using a "circulated notice," and where the local government is unable to provide disaster victims with detailed information" (Tohoku Bureau Telecommunications of Ministry of Internal Affairs and Communications).

In the past, as the longest duration of operation of such temporary emergency FM stations has typically spanned 1–3 months (excluding the case of the Mount Usu volcanic eruption where operations continued for a year), the temporary emergency FM stations established in this latest disaster have been in operation for the longest duration in record (Murakami 2012). For this reason it can be observed that the programming contents of these broadcasters has shifted from providing information necessary to reduce risk and damage in the immediate aftermath of the disaster (such as information on the damage caused by the disaster, emergency medical and first aid information, information on the whereabouts of residents and contact information, and relief information), to providing information that places an emphasis on bringing stability to the everyday lives of the disaster victims.

Even among the temporary emergency FM stations there are large differences in the way they are operated, as some stations focus on broadcasting disaster related information mainly provided by the local government, while some stations place an emphasis on securing the participation of local residents in order to facilitate community building and to deepen ties between community members to foster the psychological healing of residents living in the disaster affected areas.

The Rikuzen Takata Emergency FM station was established through a proposal made by a local civic organization to the local city government, and after 8 months of preparation, it began broadcasting in December 2011. In reference to this case, a staff member involved in activities at a temporary emergency FM station in Ofunato City (which is located adjacent to Rikuzen Takata City) commented that "the information provided by such stations is most vital during the phase where residents are still living in evacuation shelters." The staff member went on to further comment that "unless operations at such radio stations can be launched in the early phases where there are a large number of evacuees still living in the shelters, it cannot be said that such radio stations can adequately fulfill their duty." Furthermore, Takayuki Sato, the station director at community radio station Ohshu FM provided comments that were critical of the establishment of such emergency FM stations during the restoration period:

"How can you truly refer to a station launched in December as an emergency FM station? At that stage there will be very little information that is suited for dissemination through radio broadcasting. Such stations are meant to serve a large indefinite number of listeners living in evacuation shelters." Despite such criticisms, it must be noted that the stated purpose for the establishment of the Rikuzen Takata emergency FM station is to connect community members that have been dispersed due to the occurrence of the tsunami (Radio FMYY 2012a). Due to the damage caused by the tsunami, the community originally living in Rikuzen Takata City has sustained devastating damage, and the community members now live dispersed in temporary housing and lack the means with which to share information among former residents. It can be pointed out that the residents themselves took the initiative in this case to utilize radio as a means to bridge this gap.

Another similar example is the case of Minami Soma Emergency FM (a temporary emergency FM station operating closest to the Fukushima Daiichi Nuclear Power Plant), which was established with the aim of creating "a space for communication," considered to be the most pressing need to facilitate the restoration and recovery efforts (Radio FMYY 2012a). For the community of Minami Soma that has been torn apart by the invisible threat of high radioactivity, residents have turned their desperate hopes to radio broadcasting as a means of connecting the thoughts and hearts of the local community.

During such efforts to rebuild and restore a community, the role of community radio broadcasting (as well as temporary emergency radio station broadcasting) is not limited to just transmitting information. Community radio stations play a large role in ensuring that a diverse range of voices (not just the "loudest" voices) are represented, and that individuals playing a variety of roles with a variety of opinions are given a means to connect with each other, and to ensure that such voices are included in the dialogue necessary to perform community building in a disaster recovery context (Buckley et al. 2008).

It must be noted, however, that it is vital to secure funding in order to establish such radio stations and to ensure their sustainable operations. A majority of the temporary emergency FM stations that were established in the aftermath of the Great East Japan Earthquake and Tsunami received assistance from the Nippon Foundation, which provided up to 8.2 million yen in operating funds for each radio station (Matsuura 2011). Most importantly, the radio stations used funding received from the Nippon Foundation to procure broadcast transmission equipment and in-studio equipment. Further funding was provided to cover personnel expenses for the temporary emergency FM stations under the budget drawn up by the Ministry of Health, Labour, and Welfare for the operation of disaster related projects. After receiving funding from the Ministry of Health, Labour, and Welfare for the fiscal year of 2011, a majority of the temporary emergency FM stations will continue to receive funding for the fiscal year of 2012. Although other radio stations, NGOs, and private companies have played a significant role in providing assistance in the form of equipment and knowhow to aid in the establishment of the temporary emergency FM stations, it is most likely that they would not have been able to continue operating for such a long duration, without funding provided by the Ministry of Health, Labour, and Welfare to cover personnel expenses.

During the recovery and restoration phase, temporary emergency FM stations can play a role in contributing to the resolution of issues faced in the community and providing community radio broadcasting that can beneficial to the community. For this reason, there is a great need to support such activities by providing the necessary legal framework and access to public funding.

In the next section, the activities of Minami Soma Emergency FM and Minami Sanriku Emergency FM, two temporary emergency FM stations with very different modes of operation will be examined.

20.3 Case Studies of Disaster Response Radio Stations the Aftermath of the Great in East Japan Earthquake and Tsunami

20.3.1 Minami Soma Saigai FM

In Minami Soma City, Fukushima Prefecture, a vast majority of residents evacuated after the occurrence of the nuclear power plant accident due to fears of exposure to radiation (temporarily reducing the population from 70,000 to 10,000) (JVC 2011). During this time, both residents remaining within the city and those that had evacuated asked a variety of questions, including the following (demonstrating the lack of information provided to the residents):

"What is the current condition of the city?" "How many people still remain inside the city?" "We have heard that relief supplies have not reached the city but what is the current situation?" "What has happened to my neighbors and classmates that I cannot get in contact with?" "We have heard rumors about burglars but are they true?" "What will we do in the event that a fire occurs?" (Radio FMYY 2012b).

Despite all of these concerns there were very few means with which residents could obtain answers to such questions and information about their community in general. There were two reasons for this lack of information, given as follows. One reason for this lack of information was the withdrawal of mass media personnel from Minami Soma City in the wake of the nuclear power plant accident, causing print and television news reports on the Minami Soma area to be virtually non existent. The second reason is that there was no form of preexisting community media in Minami Soma City, such as community radio or cable TV networks. Such conditions continued for roughly a month after the occurrence of the disaster, leading the Minami Soma City Government and local resident organizations to cooperate in establishing the Minami Soma Emergency FM station on April 16th, 2011. A broadcasting license was issued to the mayor of Minami Soma by the national government, and the station began operation as a publicly established privately operated emergency radio station, with the Sakaemachi Shopping District Promotion Association (Haramachi Ward, Minami Soma), an organization with prior experience in operating an event-based FM radio station, in charge of operating the new radio station.

In the initial stages after the radio station was established, programming consisted of reports on disaster related information broadcast at 9 a.m, 1 p.m, and 5 p.m, for an hour at a time, conducted by residents who had previously been involved in the aforementioned event-based FM radio station. Most importantly, these reports were used to announce environmental radiation monitoring results based on daily measurements conducted by the Minami Soma City Government.

A month after establishment, the Japan International Volunteer Center, an NGO involved in the field of international cooperation, began providing assistance to the emergency radio station by dispatching a full time staff member with prior experience in newspaper journalism. Furthermore, in addition to disseminating information provided by the Minami Soma City Government, the scope of information provided was broadened to include civilian information as well. Additionally, the number of residents participating in broadcasting activities began to increase as the station actively called on residents to volunteer, and funding to cover personnel costs became available under the disaster countermeasures budget (6 months after the disaster had occurred).

Exactly 6 months after the occurrence of the earthquake, a special program was aired that allowed residents to reflect over the past 6 months since. Each staff member was given the opportunity to share their experiences in their own words, as well as stories and anecdotes describing the conditions faced by the residents. Furthermore, the leader of a civic action group, the director of the municipal hospital at the Minami Soma City Government, and the chief of the municipal decontamination measures office took part in a live broadcast, allowing the three parties to carry out a live discussion on the air. Starting with the broadcasting of this programming, Minami Soma Emergency FM began shifting from a medium to disseminate information in an unidirectional manner to a medium of interactive communication (Radio FMYY 2012b).

With this shift in programming, the amount of time slots dedicated to broadcasting only music was reduced, and a focus was placed on providing (1) programming that fosters dialogue between residents and (2) programming that is produced to provide residents with healing as well as enjoyment, in addition to the standard information provided in previous broadcasts. Given below are some representative examples of programming provided by Minami Soma Emergency FM:

20.3.1.1 Interview Programming

Live programming was created to present interviews of residents of various backgrounds, to allow such residents to share the stories of their experience in the aftermath of the disaster as well as their current situations. This program is aimed at fostering a feeling of mutual understanding by sharing the voices of people of various backgrounds and conditions with the resident listeners.

Talk Show 1

As the younger generation living in Minami Soma City have difficulty in having their voices heard in the local community, this program provides young people with a forum in which they can freely and frankly share their feelings in their own words, and can ask listeners about "what each of us can do to help in the recovery efforts."

Talk Show 2

This program provides Minami Soma residents of various backgrounds with a forum to share stories of the current conditions they face, as well as the history and their memories of the region before the occurrence of the accident. This program serves to connect the residents of Minami Soma, who have suffered a psychological separation from each other due to the occurrence of the nuclear power plant accident. Furthermore, the station streams the programming over the Internet in order to provide individuals living outside of the city and prefecture with a true picture of Minami Soma, which is often only viewed by the outside world as being a "city that has been contaminated by radiation" due to the effects of the nuclear power plant accident.

20.3.1.2 Entertainment Program

Geared toward senior citizens living in temporary housing, this program features a popular singer from the local region who provides listeners with music and talk segments regarding folk music that is local to the Tohoku region, contributing greatly to the spiritual healing of the disaster victims.

Through the broadcasting of such programming, the amount of emails and fax messages sent to the radio station by listeners began to increase, and by presenting such messages on the air such programs served to further lessen the psychological distance felt between the residents and the radio station.

Additionally, in June of 2012, a 4-h long public live broadcasting event was held in a commercial facility located within the city, and more than 100 residents participated in this event. A majority of the residents who participated in the event listened to Minami Soma Emergency FM on a daily basis as they lived in temporary housing, and all participants interviewed during the show expressed how Minami Soma Emergency FM provided the disaster victims with much needed strength.

During this event, it was decided that Minami Soma Emergency FM would change its station name to "Minami Soma Hibari FM." This name was chosen from a number of entries sent in by local residents. The term "Hibari," which means "skylark" in Japanese, was chosen to represent the hopes that the local community "would be able to recover 1 day and soar freely through the skies again."

In this way, the radio station staff members became deeply involved with the community through such broadcasting activities, and came to identify the specific needs and issues faced by the residents of Minami Soma City in an objective manner. An example of such activities evolving beyond just radio broadcasting is given below.

One of the staff members of the radio station started their own group and is now involved in operating a public space used to promote the interaction of disaster victims (located in the meeting space of a temporary housing development), with the assistance of the Japan International Volunteer Center. Such activities show that emergency FM stations are capable of assuming a role in solving problems faced by the local community in areas that do not involve radio broadcasting. This example is a direct result of efforts made by the emergency FM station in engaging local residents through its reporting activities and in facilitating the exchange of information between residents.

It must be noted, however, that in order to make such community radio related activities sustainable during the restoration and recovery phase, it is imperative to secure the participation of local residents as well as a to provide a sound rationale for financial support.

In regards to the temporary emergency FM stations that were established in response to the Great East Japan Earthquake and Tsunami, the MIC has commented that "(it) will continue to issue and reissue licenses based on the needs of the disaster affected municipalities. We will continue to make such considerations until there is no longer a need for temporary housing." In the case of Minami Soma Emergency FM, although the station plans to continue broadcasting beyond the fiscal year of 2013, in order to be able to continue operations, the station must make a convincing case in order for it to receive the necessary financial support.

Since September of 2011, Minami Soma Hibari FM has been covering the personnel expenses of its staff members through funding provided under the national disaster countermeasures budget. Although the station has also received donations from private companies and civic organizations, 80 % of its operational funding for the fiscal year of 2012 is provided by the national government. Whether the station can continue to broadcast quality programming largely depends on the station providing a sound rationale for financial support. Although the probability remains high that the station will be able to secure funding under the national disaster countermeasures budget for the fiscal year of 2013, in the event that this funding is terminated, the station would have no other choice than to cease broadcasting or sharply reduce the scope of its operations.

As of writing, at Minami Soma Hibari FM, members of the the Sakaemachi Shopping District Promotion Association (the "parent" organization of the station) and the broadcasting staff are making preparations to establish a nonprofit organization that will handle administrative operations of the station, with the aim of making the station more sustainable. By conducting workshops involving local residents to discuss the future of the emergency FM station, and by visiting the other temporary emergency FM stations and community radio stations that serve communities that have experienced disasters in the past, Minami Soma Hibari FM aims to make the transition from a publicly established privately operated temporary emergency FM station to a privately established privately operated community radio station. Toward this goal, Radio FMYY, which was established in the wake of the Great Hanshin Awaji Earthquake in 1995 and continues to broadcast to this day, is providing Minami Soma Hibari FM with assistance based on the station's own unique experiences.

20.3.2 Minami Sanriku Saigai FM

Minami Sanriku Emergency FM was established on May 17th, 2011, becoming the 24th temporary emergency FM station to be established in the aftermath of the Great East Japan Earthquake and Tsunami. Of the 17,000 people living in the Town of Minami Sanriku, Miyagi Prefecture, over 1,000 people were reported dead or missing. The disaster wireless systems that the local government offices had relied on to transmit information during everyday situations was completely destroyed due to the effects of the earthquake and tsunami, leaving the local government with no means to provide residents with disaster related information, relief information, and information for everyday living. As 36 of the 242 staff members working at the local Town office were reported dead or missing, the local government did not have the necessary resources to start its own temporary emergency FM station. A staff member working for the Hyogo prefectural government learned of this dire situation after being stationed at the Minami Sanriku Town office to provide assistance, eventually leading to a joint effort between organizations based in Kobe, such as the Disaster Reduction and Human Renovation Institution, Radio Kansai, Radio FMYY, the University of Marketing and Distribution Sciences, in addition to FM Nagaoka (located in Nagaoka City, Niigata Prefecture) to launch the Minami Sanriku Emergency FM station (Matsuura 2011).

At Minami Sanriku Emergency FM, the local Town office hired staff members using funding provided by the Ministry of Health, Labour, and Welfare's earthquake disaster related project budget, allowing the station to operate as a publicly established publicly operated station. The programming provided by the station placed a priority on disaster related information provided by the local government. As an example of some of the limitations the station experienced, there was a case where it was decided that information on local stores that had resumed operations would not be broadcast, with the rationale that such broadcasts may unfairly benefit specific businesses. In contrast to how this station handled such information, other temporary emergency FM stations did not hesitate in broadcasting such information as they determined that such information was extremely valuable for residents living in the coastal areas where all infrastructure had been washed away.

In another incident, staff members of the FM station tried to report on proceedings that involved discussions of assistance measures and recovery measures at the local town council meeting, only to have the contents of the report screened by local government staff members, with certain items that were considered to be "unsettling" to local residents censored from their reports before they were allowed to be broadcast (Matsuura 2011). By comparison, temporary emergency FM stations in Yamamoto Town (Miyagi Prefecture) and Rikuzen Takata City (Iwate Prefecture), which are both publicly established privately operated stations, broadcast the council meetings of their municipalities on live radio. As it is common knowledge, such council meetings are meant to be conducted in a way where they are open for public observation. Atsushi Takahashi, the director of the Yamamoto Town temporary emergency FM station stresses the importance of providing live broadcasts of council meetings as follows:

"These local municipal councils act as the space in which council members and the municipal government offices negotiate to determine executive decision making related to the recovery efforts, and these negotiations must be disclosed to the local residents. Such broadcasting activities can only be conducted by outlets such as community radio, and it is outside the scope of what mass media is capable of." (Radio FMYY 2012c).

One possible view is that the Minami Sanriku Town office did not fully understand the role that radio stations in disaster affected areas must fulfill and only viewed radio broadcasting as a tool to replace the disaster wireless system that had been destroyed during the disaster. There were even cases where such opinions that the radio programming "did not give an accurate portrayal of the voices of the disaster victims," or other comments of the residents were edited to remove opinions that were critical of the local government office before such comments were aired. Such examples illustrate the limitations of disaster radio stations that are primarily driven and centered around the local government, with programming limited to transmitting disaster related information provided by local authorities.

Starting on April 1st, 2012, Minami Sanriku Emergency FM has entrusted its operations to Tome Community FM a community radio station located in neighboring Tome City, Miyagi Prefecture. As of writing, the station's only original programming is limited to the 1 h live broadcast aired from 9 a.m to 10 a.m, and the rest of its programming consists of rebroadcasted programming provided by Tome Community FM.

20.4 Comparison with Community Radio in Indonesia (Focusing on the Merapi Volcano in Central Java)

Community radio plays an especially crucial role in Indonesia, because disaster risk reduction concepts have not yet reached grassroots communities, despite the fact that such concepts have already been codified in disaster laws at the national level. Under such conditions, community radio holds the potential of spearheading disaster risk reduction related activities both on and off the air. In this sense, community radio can be a strategic agent for the dissemination of knowledge related to disaster risk reduction.

Merapi volcano which is located in the north of the Jogjakarta Special Region in Indonesia attracts attention every 4 or 5 years as it is categorized among some of the most often erupting volcanoes. Merapi volcano has erupted 83 times (on record) before June 2006, and eruptions occur approximately in 2–5 years (short period) or 5–7 year intervals (medium period).

Although the volcano has continued to provide a source of livelihood for local residents, it has also posed a great threat to villagers living in nine sub-districts (consisting of 42 villages and 118 hamlets) located along the slopes of the volcano. The most recent eruption occurred from October to December 2010, causing a significant number of casualties. 353 villagers were killed by pyroclastic flows and more than 350,000 villagers living in the areas within a 20 km zone from the crater were displaced from their disaster-prone homes. Records show that the 2010 eruption was the biggest eruption to occur in the last 100 years (Sumaryono 2011).

The impacts of the eruption can last for more than 2 years after the occurrence of the eruption. The threat of a cold lava flood caused by the eruption will last for 3–4 years because the eruption has produced a great amount of volcanic materials which is estimated to be about 140 millions cubic meters in volume.

20.4.1 Information Network around Mount Merapi (Jalin Merapi)

In 2006, Merapi Volcano erupted after a period of dormancy that lasted about 4–5 years. At that time, there were three community radio stations on the slopes of the volcano: Lintas Merapi FM Community Radio, Radio MMC FM and Radio K FM. Individuals involved in these radio stations felt that the mainstream media's reports on the volcano did not represent the actual conditions faced by the villagers living on the flanks of the mountain (JICA 2012).

Given the situation, COMBINE Resource Institution (COMBINE), a local NGO in Jogjakarta, developed a system by which information was transmitted from walkie-talkies to a repeater radio. This system enabled COMBINE to provide villagers living in and around Mount Merapi with updates on the current situation of the Merapi volcano. COMBINE's role in this system was to write down all information received and to post it on the Jalin Merapi website.

Knowing that the last eruption was greater than that of 2006 and anticipating a more severe impact for future eruptions, COMBINE developed a more complex but integrated information system. COMBINE also involved the communities in broader activities related to managing the information. The three community radio stations became the main media outlet supported by volunteers from many different regions, and the number of volunteers reached roughly 800.

By facilitating the formation of such a community-based information system, COMBINE was able to build networks using various forms of media and to bring a positive effect on relief activities such as facilitating the distributions of aid materials from donors to the victims. Combining Information and Communication Technology (ICT), the level of the villagers' trust and participation in the relief activities increased (Motozuka and Kanki 2012).

Information received from walkie-talkies was posted on the website, meeting the villagers' needs for fast and accurate information. Those residents wishing to ask questions or request for aid could directly send messages via Short Message Service (SMS) or using the Shout Box feature provided on the website. Residents were also free to email the Jalin Merapi website to share information in detail. In order to better coordinate the activities of the roughly 800 volunteers, a Facebook group was created. Jalin Merapi also created a Twitter account to provide information in a quick manner and to reach the broader public, eventually amassing 35,000 followers. All information provided via SMS, Shout Box, Twitter, E-mail and Facebook was later compiled and stored in an organized database which was constantly updated. A database of the evacuees' needs was made available for anyone who wanted to help. Additionally, COMBINE opened a specific channel for monitoring the process of aid material distribution, which could be accessed by the donors as well. It must also be mentioned that the contents of the database were also verified in order to guarantee that the information could be trusted.

Two community radio stations were established with the support of COMBINE to prepare for secondary disasters and to aid in recovery efforts. As of writing, five community radio stations are members of Jalin Merapi (October, 2012). The staff of the radio stations and COMBINE conduct meetings every 3 months to exchange information and to share mutual experience and knowledge, enhancing the disaster management capacity of all network members. In the next section, the disaster management activities of two radio stations that are members of Jalin Merapi will be examined.

20.4.2 Sidorejo Village and Lintas Merapi FM

Sidorejo Village is situated in Kemalang Sub-district, Klaten Regency, Central Java Province. The highest elevated region (900–1,230 m above sea level) of this village is located in Deles Hamlet. From this hamlet, the peak of Merapi Volcano is located only 4 km away. In general, the villagers live by cultivating their lands for planting crops and growing vegetables, while a few work as sand miners in Woro River.

There are currently several social organizations participating in activities to develop the village: Forum Klaster Lereng Merapi (FKLM, Forum of Merapi Slope Cluster), which focuses on developing an ecology-friendly local economy, Lembaga Masyarakat Desa Hutan (LMDH, Association of Forest Village People), which focuses on cultivating the forests together with the villagers, and Pasag Merapi which concentrates its efforts on life saving activities and encouraging villagers to stay alert for threats posed by Mount Merapi.

The villagers rely on radio as their means of communication. This area is served by a community radio called Lintas Merapi FM, which was established in 2001. The station is located 4.5 km on the southern part of the mountain, under the peak of Merapi (Fig. 20.2). The radio station is utilized by people as a source of information on the Merapi Volcano. Through radio, the residents share the latest information on the status of the volcano. For example, if individuals observe a sulfate cloud soaring high, they do not hesitate to inform residents that a volcano eruption is imminent. Residents also listen to the information provided by the BPPTK (Jogjakarta Volcanic Technology Development and Research Centre). The villagers come to a conclusion on the status of the volcano based on what they see and what they hear through such monitoring tools as Lintas Merapi FM and kentongan. When Mount Merapi is safe, this medium is used as a tool for disaster risk reduction education.

In addition to broadcasting, Lintas Merapi FM has formed a children's group called Kelompok Anak Pencinta Lingkungan (KANCING), consisting of a group of children and youths involved in activities on community-based disaster management and environment. Lintas Merapi FM has organized various kinds of activities for children and their parents, which include (1) regular study sessions (e.g. reading activities, presentations, and rescue drills) (2) hiking tours and environmental conservation activities, (3) PPGD (first-aid for emergency situations) training, (4) participation in PPGD exhibitions, (5) completing a painting on the threat of Merapi Volcano, and (5) film screenings related to disasters and the environment. Some activities have been conducted in cooperation with elementary schools in the community and local NGOs. Lintas Merapi FM has also organized a gamelan club (Javanese traditional music club) and has been hosting practice sessions regularly. The members are mostly middle-aged adults and elderly residents. Besides practicing gamelan, they discuss issues related to disasters and the environment.

During the 2010 eruption, the villagers in this area made a conscious effort to evacuate and did not need to ask for as much aid from the government due to the fact that Lintas Merapi FM had made a continuous effort to convey the importance of preserving the environment and to raise awareness on disaster risk reduction.

20.4.3 Jumoyo Village and Lahara FM

Jumoyo Village is situated in Magelang Regency, Central Java Province. It consists of 16 hamlets. The main road of Central Java in Jogjakarta Special Region crosses Jumoyo Village.

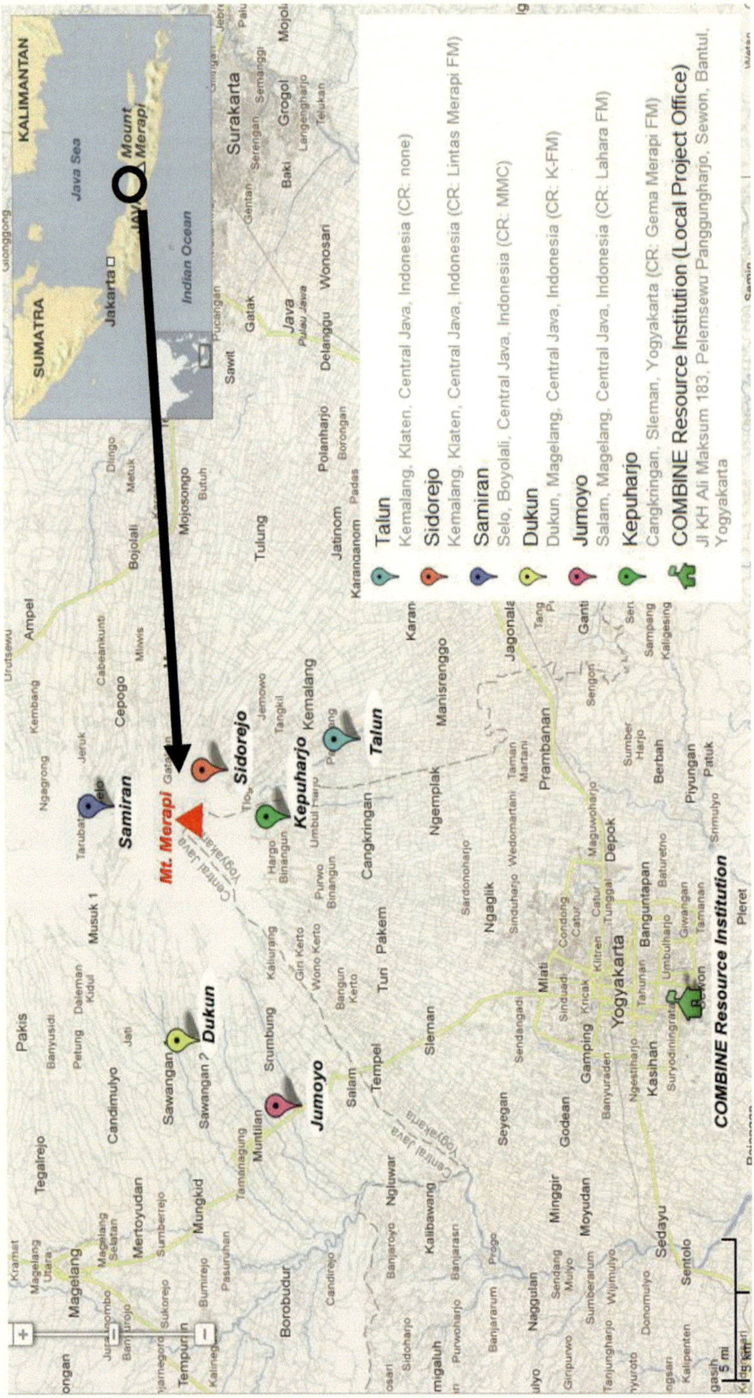

Fig. 20.2 Community radio stations around Merapi volcano in Indonesia

Although the cold lava flood that struck the village (Gempol Hamlet to be exact) did not cause any deaths, about 50 villagers were injured. In this case, the villagers were able to execute a swift and successful evacuation of their homes. As of writing, there are still 165 families from Gempol Hamlet and 6 families from Kadirogo Hamlet staying at two evacuation shelters located in Jumoyo and Nglarangan Village.

Most residents in the village are sand miners and some are farmers. Since the cold lava flood turned their lands into a desert, they can no longer use the lands for farming. As a result, they are forced to earn a living by mining sand, with most of the profits and benefits of the mining business going to the legal owners of the land and not its residents. Some villagers were forced to resort to selling their livestock to make a living. Fortunately, in this case, none of the livestock was lost during the eruptions because the villagers were able to evacuate their livestock quickly and to protect them from drowning in the flood. At the same time, however, they could no longer feed their livestock since the land had become infertile after the eruptions.

To improve the economic status of the villagers, the cooperative office of Magelang Regency has conducted training workshops on making paving blocks and taro chips in June of 2011. Some women from Jumoyo Village also have received training on making doormats, an activity which was conducted by "GP Anshor Anshor," a youth organization located in Magelang Regency.

The residents of Jumoyo village initially lacked knowledge of Mount Merapi. They did not share the knowledge of the volcano for the residents. Since Jumoyo Village is located inside the red zone, it is susceptible to cold lava floods, and it has been reported that floods have hit the villages along Putih River. Initially, most of the villagers did not have adequate information on the volcano. As a result, the villagers conduct sand mining in an area close to the dam that is used to block the cold lava leading to further destruction of the environment.

Since antiquity, the villagers have continued to rely only on mystical signs to determine whether a cold lava flood would occur. One of the residents in Gempol Hamlet told the story about the night before a flood that occurred on Sunday, January 9, 2011. Some of the villagers reported that they had heard the sound of a topeng ireng performance. Such performances usually last until 1 a.m., but the villagers reportedly heard the music until 3 o'clock in the morning, which they considered to be odd. When multiple residents went to check on the performance it was already finished. Since the sound they heard was not from the performance, they concluded that a great flood would occur.

After the occurrence of this flood, monitoring using CCTV was conducted, and monitoring devices and staff were stationed in Jumoyo Village. Images through the monitoring devices were streamed on a website available for anyone to watch live. Jalin Merapi also supported the founding of a community radio station in Jumoyo Village so that it could be used as a means of giving the residents in the community an early warning when a cold lava flood is about to happen. The radio station was named Lahara FM and it broadcasts a variety of information in addition to disaster related information. In the past, it has helped to broadcast information on a program providing free medical therapy for the villagers. For the villager in this locale,

Table 20.3 Comparison with the temporary emergency FM stations of the Great East Japan earthquake and Tsunami

	Temporary emergency radio stations in East Japan	Community radio stations around Mt. Merapi
Ownership	Local government	Local residents' group
Governance	Local government/local residents' group	Local residents' group
Activities	Focusing of community participation to promote dialog among local residents for the recovery	Focusing on information about prevention of next disaster (volcanic eruption)
Networking	Relationship with neighboring community radio stations has not been established to help each other	Strong relationship with neighboring community radio stations to help each other
Finance	Sustainability of the activities is depending on the budget from government	Local residents' contribution as a volunteers sustains their activities

Lahara FM is the only medium available that provides information on the conditions of the Merapi volcano. Residents mainly use information provided by mainstream media for strictly comparative purposes and not as a primary source of information.

20.4.4 A Comparison with the Temporary Emergency FM Stations of the Great East Japan Earthquake

In this section, the disaster risk management activities conducted by a network of community radio stations located in the vicinity of the Mount Merapi volcano located in the central region of Java Island, Indonesia has been examined. A large difference between the temporary emergency FM stations in the disaster affected areas of the Great East Japan Earthquake and Tsunami and the community radio stations in the vicinity of the Merapi volcano can be observed in the relationships between the individual radio stations (Table 20.3). Although some of the temporary emergency FM stations have individual relationships with other stations, there exists no network that crosses geographic and organizational borders that shares a disaster risk management purpose. The information communication network for Mount Merapi, named "JALIN MERAPI" (composed of five community radio stations located in the vicinity of the Merapi volcano) was formed in 2006 after the occurrence of a volcanic eruption, and it serve an important role in facilitating the exchange of disaster related information during the large eruption that occurred in 2010. Furthermore, this network played an instrumental role in the establishment of two new community radio stations that were established in areas that were hardest hit by the disaster, with the aim of dealing with secondary disasters and to facilitate the recovery efforts. Despite such efforts, Jalin Merapi receives no cooperation or support from local governments, and the network relies on the COMBINE Resource Institution to act as its organizational hub(Motozuka and Kanki 2012). Japan, such organizations that offer continuous support of community radio stations in disaster affected areas do not yet exist.

At the same time, the community radio stations in the vicinity of the Merapi volcano primarily focus on providing disaster related information, and do not feature as much programming that involves the active participation of residents, as is seen in the programming broadcast by the temporary emergency FM stations in the disaster affected areas of the Great East Japan Earthquake and Tsunami during the restoration and recovery phases of the disaster.

In October 2012, in a village located in the vicinity of the Merapi volcano, COMBINE Resource Institution and Radio FMYY launched a joint project to share the experiences gained through such efforts in both Indonesia and Japan. By utilizing community media, this project aims to strengthen disaster preparedness of local communities, and this project also receives support from JICA. This project will be conducted to share the past experiences and knowledge gained by organizations in Japan, and the results of this project will be shared with individuals and organizations involved in community radio in Japan.

20.5 Future Suggestion

Broadcasting licenses for temporary emergency FM stations are issued by the local chief executive of the applicable municipality. As mentioned in Chap. 2, there are differences depending on region on how local governments have gone about handling the operations of such temporary emergency FM stations in the wake of the Great East Japan Earthquake and Tsunami, as in some cases operations have been handled by the local governments themselves while in other cases operations are handled by local organizations, and in yet other cases local governments and local organizations have operated such stations as a joint effort. At such radio stations where operations are driven by local organizations, it is common for such stations to broadcast a variety of opinions, including such things as the demands and requests of the residents toward the local government. Although both types of stations share similarities with community radio in that they provide radio broadcasting that serves small local municipalities, it is self evident that they differ greatly as a medium in terms of the degree of community ownership and participation. For example, in order to renew a license for a temporary emergency FM station, unless the local government expresses its desire to renew the license, the MIC will not issue a renewal of the license. In the case of Minami Soma Emergency FM, examined in Chap. 3, although the station has a broadcasting license that is valid until March 31st, 2013, as of the end of October, 2012, the Minami Soma City Government has not yet indicated that it would renew this license. The reason for this unwillingness to renew the license lies in the difficulty in providing a sound rationale for securing funding for the following fiscal year. Compared to the station's activities during the initial phases following the disaster, it is seen as quite difficult to provide proper metrics to quantify the station's activities during the recovery and restoration phase. As a result, a survey of Minami Soma residents will be conducted in November of 2012 regarding the temporary emergency FM station, based on which a decision will be made whether or not to renew the license.

In this case, if the same station had been operating as a community broadcasting station, it would be possible to continue broadcasting during the restoration and recovery phase, regardless of the local government's decision. For this reason, many of the local organizations that operate the temporary emergency FM stations are beginning to take steps toward obtaining a community broadcasting license on their own. In November 2012, the MIC, Tohoku Bureau of Telecommunications distributed a booklet "*Community FM-ka he no tebiki*" (A manual for changing to a community FM) to each temporary disaster FM radio station. On the other hand, the MIC, Tohoku Bureau of Telecommunications sent a notification in writing stating that "negative decisions will have to be made" to local authorities who are considering the shift of temporary emergency FM radio stations to community radio stations. Also, for an application for a license, a letter of agreement from a sponsor to guarantee that there will be financing for 5 years from the time the radio station is opened must be presented. In all cases, however, securing operational funding is most problematic. As the honest opinion of such local organizations, many of them desire to continue operating as temporary emergency FM stations, until they are able to build enough organizational capacity to operate as a community broadcasting station, due to the fact that if they continue to operate under the former designation they will still be eligible to receive funding from the earthquake disaster related budget provided by the Ministry of Health, Labour and Welfare, which can then be allocated toward operating funds.

The MIC, Tohoku Bureau of Telecommunications has been skeptical of the prospects of the emergence of sustainable community broadcasting stations in the coastal areas of the Tohoku region, as these regions lack economic resources, and have been subjected to devastating damage due to this disaster. Despite these views, however, one cannot ignore the case of FM Uken, a community radio station operated by a nonprofit organization, that serves Uken Village, a municipality with a population of less than 2,000 people on Amami Ōshima (located between the Honshu and Okinawa islands). Excluding a single paid staff member, all work to produce and broadcast programming is performed by local volunteers. As this station is located in a region that is susceptible to typhoons and disasters caused by severe rainstorms, an emphasis is placed on providing programming that promotes disaster preparedness in cooperation with the local village office. This station is currently working with two other community radio stations located in Amami Ōshima (FM Amami and FM Setouchi) with the aim of forming a network similar to Jalin Merapi of Indonesia. In order to make this possible, it is imperative that such networks are created in a community radio context, where stations are free to operate independently from local government and private interests.

As it is widely recognized, "participation" is one of the defining characteristics of community radio, and participation as described here refers to volunteer participation of the community members themselves (Fraser and Restrepo-Estrada 2001). The Indonesian community radio stations examined in Chap. 4 exemplify this through the volunteer participation of their local residents. As another example, FMYY, which originally served as an emergency radio station during the aftermath of the Great Hanshin Awaji Earthquake, was able to benefit from the participation of more than 100 volunteers in just its first year of operation, allowing it to serve the community by providing multilingual disaster related information. In the case of FMYY, an entire year passed before the station added it's first paid staff member, around the same time the community radio station received its license a year after the occurrence of the Great Hanshin Awaji Earthquake (Hibino 2010).

As of writing, the disaster affected areas of the Great East Japan Earthquake and Tsunami have begun to transition to the restoration phase, and a large amount of time is still required for the lengthy recovery process. During this long phase, radio broadcasting that allows community members to freely share their voices with their own community can play a vital role.

Rinji saigai hoso kyoku (temporary emergency radio station) system is for FM stations whose goals are originally to broadcast local authorities' disaster information for a short period. However, it is not only information that is required from present disaster FM radio stations in each area of the Tohoku region, but FM radios that will act as a junction for information that will link the hearts of the residents and ensure the continuation of the community. There is a need to create a new framework, which may be called "*Fukko FM*" (revitalization FM), that will meet the needs of the disaster victims as they work for recovery and revitalization after an unprecedented earthquake disaster.

The following comments given at the "Public Meeting to Discuss the Future of Minami Soma Hibari FM" (conducted on September 7th, 2012) by Satoshi Konno, radio station director at Minami Soma Hibari FM, exemplify the kind of role that community radio is expected to play in serving the local community during the restoration and recovery phase of a disaster.

Mass media has many limitations in terms of the space limitations of print media and time slots available for broadcast media. It is impossible to provide a perfectly accurate picture of Minami Soma City through news reporting. Minami Soma FM exists in order to broadcast the actual conditions of the city and the emotions and thoughts of the residents in a fine and detailed manner. Operating in a more flexible manner than the local government, this station exists as a public space and forum that all residents can listen to. Minami Soma FM exists to serve those people who have evacuated and remain separated, as well as those who continue to live in this area, despite their anxieties, and to provide each individual with a means to share their feelings with one another, as they try to cope with this "disaster" for which there is no right or wrong answer. By providing residents with a means to know each other's feelings, better understand each other, and to share in a mutual respect for each other, we hope that we at this station can do our best as "Minami Soma residents" to act as a unifying force for this community.

I hope that our station can continue to grow as a "radio station that is personally close to the community," where more and more residents can feel that they can freely "get involved" and to "go on the air to have their voices heard." By speaking about their thoughts and feelings, and by listening to others, such things lead to a change in consciousness. By connecting with others through speech and by building relationships, we strengthen our community. As we move forward I would like our station to broadcast programming that makes the residents feel that "we can start here to make Minami Soma City a better place."

References

AMARC AP (World Association of Community Radio Broadcasters – Asia-Pacific) (2008) Survey report: practices of community radios in natural disaster management situations. AMARC AP

Buckley S, Ducer K, Mendel T, Siochru SO (2008) Broadcasting, voice, and accountability: a public interest approach to policy, law, and regulation. IBA, Washington

Fraser C, Restrepo-Estrada S (2001) Commnity radio handbook. UNESCO

Hibino J (2010) Community radio to promote social inclusion of minority – management system to sustain activities of radio. FMYY, Kouyo Syobou

Hiratsuka C (2011) Disaster information and media. Liberta Shuppan, Tokyo

Jalin Merapi HP. http://merapi.combine.or.jp. Consulted on 30th September, 2012

JICA Hyogo (Japan International Cooperation Agency Hyogo International Center) (2010) Handbook for the utilization of DMAM (disaster management audio materials) for community based disaster management. JICA Hyogo. Available at http://www.drlc.jp/english/?p=409

JICA Hyogo (Japan International Cooperation Agency Hyogo International Center) (2012) Survey report of investigation of effect of workshop to raise ability of community-based disaster management utilizing disaster management audio material (DMAM). JICA Hyogo

JVC (Japan International Volunteer Center) (2011) "Minami soma diary". http://www.ngo-jvc.net/jp/projects/touhoku-msdiary/index_8.html

Kanayama T (2007) Community media—community FM ga chiiki wo tsunagu. Keio University Press, Tokyo

Kanayama T (2012) Community radio and the Tōhoku earthquake. Int J Jpn Sociol 21(1):30–36

Matsuura T (2011) Temporary emergency radio station in Minami Sanriku town supports victimes. Jounalism: 46–55, Asahi Shinbun Shuppan

Motozuka T, Kanki K (2012) A study of the management support network for the temporary strongholds at wide area evacuation: a case study on activities of Jalin Merapi by Merapi Volcano eruption. J City Plan Inst Jpn 47(3)

Murakami K (2012) An overview of the disaster information transmission system of the local governments in the post-quake Japan. Research for broadcast, NHK Broadcasting Culture Institute, pp 32–53

O'Connor A (2004) Community radio in Bolivua: the miner's radio station. The Edwin Mellen Press, Lewiston

Okazaki K, Narita E, Pandey B (2005) Community-based disaster management for disaster resilient communities. United Nations, Tudor Rose

Radio FMYY (2012a) Radio program "Josef no Megane". 14 November 2012

Radio FMYY (2012b) Report of the workshop of Minami Soma Saigai emergency. FM station

Radio FMYY (2012c) Radio program "Machiha Ikiiki Kiramaki Time. 13 July 2012

Sumaryono A (2011) Managing the Mount Merapi sediments, CRBOM small publication series no. 37

Yoshitomi S (2010) Multicultural society and the power of migrants communities

Chapter 21
Disaster Recovery as a Development Vehicle

Rajib Shaw

Abstract Disaster recovery can be seen as a development opportunity. However, it needs to be linked to local governance system. After the disaster, a significant amount of resources flow in to the affected region. Its proper utilization to suit the local needs need a balance of governance, education and technology management. Apart from a strong governance system, it is also important that the local education and awareness raising mechanism is properly implemented. The application of new technology should be linked to the local needs of the people and communities. Community knowledge is not fully utilized to the potential and should be upgraded with the help of technology and innovation. Low cost technology should be made available to the public for wider impact. Multi-stakeholder approach would enhance the effectiveness of the technological innovation through focused strategies. Government and NGOs need to join hands in formulating disaster recovery plan and translating them to local level actions. Need for judicial arrangements of emergency resources and external aids with respect to economic and social well being of the affected community in disaster recovery.

Keywords Balanced approach of governance • Culture and socio-economic conditions • Development vehicle • Disaster recovery • Education and technology • Local customization

R. Shaw (✉)
International Environment and Disaster Management, Graduate School of Global Environmental Studies, Kyoto University, Kyoto, Japan
e-mail: shaw.rajib.5u@kyoto-u.ac.jp

R. Shaw (ed.), *Disaster Recovery: Used or Misused Development Opportunity*, Disaster Risk Reduction, DOI 10.1007/978-4-431-54255-1_21,

21.1 Introduction

Disaster opportunities can be considered as a development vehicle. If we look at the different disaster situations, especially in the remote areas of the countries, like Aceh in Indonesia, Andaman and Nicobar in India (both affected by the 2004 Indian Ocean Tsunami), mountain areas of Pakistan (2005 Kashmir earthquake), Delta regions of Myanmar (2008 Nargis cyclone), Sichuan province areas (Wenchuan earthquake of 2008). All these areas were not in the development forefront of the respective countries. However, after the disaster, significant resources in terms of financial, technical, human resources came in to the region, and made significant contribution to the development of the region in long term. However, there are exceptions. In several cases, unless the disasters are in large magnitude, it is difficult to encourage the different types of resources to flow in to the remote place. In those cases, local governance issues become very important. We also have several example of failure where the resources flew in, but did not suit to the local socio-economic and cultural context. Therefore, the education and appropriate technology issue also becomes very important.

This chapter analyzes the key findings of the earlier chapters, which are grouped into four different categories: governance, education, technology and cross cutting issues. A workshop was organized in Kyoto in December 2012 with all the chapter authors to discuss the commonalities and differences of respective chapters. Following parts provide the summary of the discussion in the workshop.

21.2 Governance and Recovery

21.2.1 Overview

Governance and Institutional Issues of Recovery is one of the key issues that provide the tools for decision-makers and community leaders to facilitate good policies and strategies, appropriate institutional and legislative systems at national level to grass root level for the designing of effective recovery plans and programs. Five cases that focus on governance issues in disaster recovery process are summarized below.

A study on Post-Disaster Reconstruction and Institutional Mechanisms for Risk Reduction was discussed, where a comparative study of three disasters in India that highlighted an institutional set up named extra ordinary mechanism (EOM) for reconstruction followed in Gujarat, Maharashtra and Tamil Nadu aftermath of earthquake and tsunami disaster was conducted. EOM was successful in Gujarat, because it brought together political executives and bureaucrats in the same platform through Gujarat State Disaster Management Authority (GSDMA). This study further identified that the success and failure depends on the mandate, scope, powers, structure and nature of the EOM.

Collaborative governance and disaster recovery was described with an example on the National Disaster Recovery Framework (NDRF) in the U.S. which addressed that collaborative governance is a form of governing where both public and private entities are involved in collective and consensus-oriented decision making that is being utilized and applied to managing catastrophic disaster effects. NDRF is a flexible structure that is currently being utilized as a resource, which can help engage various stakeholders with each other in the recovery process. Discussions revealed that implementation of NDRF require joint training and common understanding of stakeholders.

A study on Typhoon Morakot and institutional changes in Taiwan was introduced in which typhoon Morakot became the trigger for the National Fire Agency (NFA) to be replaced by the National Disaster Prevention and Protection Agency (NDPPA) through the evolution of Disaster Prevention and Protection Act (DPPA). The newly established DPPA includes several articles that mainly depict the way of implementing the disaster management activities and also gives emphasize on the role of local governments who have to establish Disaster Prevention and Protection Bureaus for operating disaster prevention and recovery system effectively.

Learning from the East Japan Earthquake and Tsunami (EJET) that aimed to propose practical institutional and governance mechanisms of quick recovery for developing countries and development agencies was also discussed. An example from the EJET affected area was given where roads and devastated areas were able to reopen in a week because of robust measures prepared from normal times. It was suggested that (1) advance financial arrangements: clear cost sharing between local and central governments, (2) prior agreements with private sector, (3) mobilization of expert teams through national networks and (4) structural measures to prevent fatal damages etc. are necessary to prompt disaster recovery at normal times.

A study on Institutional response in education sector in Kesennuma city was discussed where it depicted the nature and dimension of massive disaster of EJET, and how educational institutes have implemented strategies for recovering schools and education from that disaster. This study also mentioned that supplying of school lunch was one of the major disaster recovery approaches in Kesennuma city. Furthermore, it also denoted that Kesennuma city board of education started transport for student from living place to school and gave financial support to the students who suffered most by the disaster.

21.2.2 Common Issues

Based on the above description, the following issues are found to be the common institutional issues for the recovery processes. Three specific common issues can be described: institutional transformation, decision and policy-making, and stakeholder engagement.

21.2.2.1 Institutional Transformation

Introduction of new mechanism is found to the most important issues. Regarding governance perspective, the common issues discussed by the group members are as follows: Gujarat State Disaster management Authority introduced new extra ordinary mechanism just after Gujarat earthquake 2001. This extra ordinary mechanism is usually set up in the aftermath of disaster to coordination and speedup reconstruction process

- Leadership at the highest level—political will, commitment and political culture play an important role for good governance
- Avoiding ad-hoc/Short-term arrangements—very limited; hence permanent arrangement—preserve institutional memory
- Cabinet Resolution to Revision of legislation (streamlining of functions to reduce overlap): After typhoon Morakot in Taiwan has changed the idea of National Fire Agency (NFA) replaced by National Disaster Prevention and Protection Agency (NDPPA) through the Disaster Prevention and Protection Act
- Develop comprehensive mandate—reconstruction planning, capacity building, monitoring, reporting

21.2.2.2 Decision and Policy Making

Important issues are as follow:

- Legislation/policies always top-down but space made for collaboration/stakeholder input
- Relief phase: tends to be top-down—reconstruction phase more collaborative
- Nature and structure of mechanism: enables active participation of key-stakeholders
- Reconstruction policies need wider consultation
- Enhanced Fiscal Management
- Facilitate participation of external players (donors/funders)
- Measures to strengthen future capacity for sustainable risk reduction

21.2.2.3 Stakeholder Engagement

Some of the important issues are as follow:

- Enhance stakeholder engagement
- Provide options to the community for meaning full participation
- Trade-offs: between consultation, speed and quality of resources
- Developing schools as safe evacuation centers with adequate facilities

Besides these common issues, the discussion further mentioned the gaps that hinder achievements for better governance in disaster recovery that include:

- Pre-disaster governance conditions
- Social, economical, political and cultural conditions of the country
- Governance structure/political regime
- Pre-disaster policy context; cultural context
- Level of capacity
- Perception of safety/level of awareness

And, based on the above, following are some of the suggestions for institutional issues in recovery process:

- Integration of financial and budgetary systems into governance for disaster recovery
- Need for capacity building targeted to appropriate levels of government and stakeholder groups
- Clarification of the concept of sustainability and resilience in the context of governance systems for disaster recovery
- Inclusion of vulnerable and special needs groups, gender, children etc. in governance mechanisms for disaster recovery

21.2.3 Cases of Used and Misused Opportunities

In a discussion of used and misused opportunities of disaster recovery that determines the success or failure of good governance can be summarized below.

In terms of *used opportunities*:

- Setting up of permanent mechanism: i.e. extra ordinary mechanism (EOM) was successful in Gujarat because single EOM rather than multiple EOM for reconstruction after Gujarat earthquake in 2001 was set up. It brought together political executives and bureaucrats in the same platform through Gujarat State Disaster Management Authority (GSDMA) that continued disaster management activities with budgetary support.
- In Taiwan, National Disaster Prevention and Protection Authority (NDPPA) evolved after Typhoon Morakot in 2009. It is now used as a professional institution in dealing with future disasters and gives emphasize on the role of local governments who have to establish Disaster Prevention and Protection Bureaus for operating disaster prevention system effectively.

On the other hand, in terms of *misused opportunity*:

- After tsunami at Tamil Nadu in 2004, the State Government of Tamil Nadu took disaster recovery program through relocation of fishermen community in inland without taking account of their occupation. The relocated fisherman faced problem to catch fish because the sea was so far from their new place. They argued that, in order to be able to fish, they would need to stay on the beach at all times and keep a watch on the movements in the sea, ready to launch their boats at a moment's notice and rush after a passing school of fish. But the governmental decision impeded the relocated fisherman to continue their jobs. So, this example appeared as a misused opportunity.

- Historically, Japan has suffered severe damage from tsunamis, storm surges, ocean waves, and other natural phenomena. To endure with tsunami, they established dike that are designed to withstand the largest of the predicted tsunami heights and storm surge levels. But in March, 2011 the East Japan Earthquake and tsunami (EJET), however, the height of the tsunami far exceeded predictions. This tsunami swept over this dike before destroying it, leaving a path of death and destruction across the community. In many cases the tsunami was twice the height of the dikes. Therefore latest EJET proved that height of tsunami dyke is not sufficient enough to project from natural disaster to make safety. This can be misused opportunity to promote a false safety.

21.3 Education and Recovery

21.3.1 Overview

Disasters affect education systems and infrastructures. For example, 6,284 public schools were damaged and 733 students and teachers died or are still missing due to the 2011 Great East Japan Earthquake and Tsunami. In the 2010 Pakistan Floods, the education of 1.6 million students was affected either by the destruction of schools or by the extended use of schools as evacuation shelters. The 2001 earthquake in Gujarat, India devastated 1,884 school buildings and 5,950 classrooms and a total of 971 school children and 31 teachers were killed, while 1,051 students and 95 teachers were seriously injured. Given the extent of damages and casualties caused by disasters, there is a strong need for mainstreaming disaster risk reduction (DRR) in the education sector through strengthening school facilities against various natural hazards and through the incorporation of DRR concepts and lessons into the school curriculum.

In the aftermath of disasters, disaster recovery can be an opportunity for building disaster-resistant, safe, and accessible schools and for introducing or enhancing disaster prevention education. This means that aside from fixing what was destroyed by the disaster, recovery should also address the things that already needed improvement even before the disaster. One of the keys to a successful disaster recovery is strengthening resilience by way of recognizing and strengthening community's traditional coping mechanisms rooted in their culture and environments. Stakeholders and actors from outside the education sector have contributed toward this end. Civil Society Organizations (CSOs) have engaged in learning and education with disaster-affected communities. CSOs have also collaborated with local governments for timely the resumption of education services by providing the necessary resources, especially for people in remote areas where external support is inadequate. In the case of Pakistan after the 2010 floods, it is suggested that local departments of education should lead the recovery process in collaboration with local government, CSOs, and other organizations.

Lessons gained from past recovery mistakes can be utilized in order to inform future DRR efforts in the education sector. However, since recovery from recent mega-disasters like the 2004 Indian Ocean Earthquake and Tsunami and the 2011 East Japan Earthquake and Tsunami is still on going, it might still be too early to list all the recovery lessons from large-scale events that can be useful for the education sector. Research needs to be continued in the next few years in order to benefit from the recovery experiences. It is expected that there will be numerous issues needed for both short and long-term educational recovery. The shifting role of schools as center in the community recovery process might need to be investigated further. A future possibility in educational recovery is the multi-functional school model, where schools are positioned at the center of community recovery and combined with other public facilities, as what is being attempted in the case of Kamaishi City in Japan.

21.3.2 Common Issues

The discussion group agreed that education recovery has been given a high priority after a disaster anywhere in the world. Standards for education continuity, such as those prepared by the Inter-Agency Network for Education in Emergencies (INEE), are already prepared, but are difficult to follow due to lack of resources and local know-how. In some cases, there are shortages of qualified construction manpower. As a result resumption of classes can be grossly behind target as school reconstruction takes time.

Sustainability is also a common issue among the case studies. After external assistance stops flowing in, disaster recovery slows down or even stops. One bottleneck identified in the recovery process for schools is the loss of community ties due to displaced people. Affected families are forced to evacuate to other places for weeks, and in worst cases, families cannot return to their original communities but need to relocate to other places, keeping students from the respective schools. If the community ties continue to be weakened, recovery of school as well as communities becomes very challenging. In addition, interventions conducted by external agencies without community participation leave the communities disempowered and this also affects the sustainability of recovery efforts.

It is recognized that limitations and constraints in the post-disaster situation should not become excuses for the provision of education that is below acceptable standards. To address the issue of lack of resources and sustainability, one possible solution would be to engage the local NGOs and businesses in school and community recovery. Overdependence on external assistance should be avoided for sustainable recovery and development. Instead, reliance on local resources and initiatives should be encouraged as observed in the success story of community *danran*, which provided welfare services in the post-Hanshin Awaji Earthquake, and the *bousai fukushi (bokomi)* communities, which are self-organized community associations for disaster reduction and social welfare centered around elementary schools. Both danran and bokomi are examples from Kobe City. If the local community has ownership of the recovery process, they will be more dedicated to continue exerting support for recovery initiatives for a longer period of time, if not indefinitely.

21.3.3 Cases of Used and Misused Opportunities

Used opportunities in education disaster recovery include training of teachers as role models for the students and community residents in the recovery process. It was found that many teachers in Japan with their high sense of responsibility sometimes function beyond their given responsibility as educators to being protectors of their students during emergencies. Teachers also connect the school with the outside world (e.g., a vice principal of an elementary school in Kessenuma has been disseminating lessons learned from EJET). Practitioners on the ground have a chance of directly observing the recovery process and sharing them to others who need this kind of information, like other places prone to the same natural hazards. However, because teachers usually are overwhelmed with many responsibilities and expectations from the community, a support system, such as in examining workload of teachers to avoid having them over-fatigued by additional duties in the recovery as well as pre-disaster period should be put in place.

Making schools to become hubs, which interlink resources that are needed to revitalize communities, can be another *used development opportunity*. As seen in the Kamaishi example, public facilities such as children's daycare center, public library, community center and DRR facilities can be centered on the school to serve and facilitate community ties. This consolidated approach can also address other chronic issues, such as aging population and lack of safe highlands, to realize more efficient and effective use of limited resources available to the community.

Mobile Knowledge Resource Centres is another *used development opportunity* in education disaster recovery. These mobile centres have been able to overcome physical and location barriers to benefit communities that are often at very high risks from natural disasters. Support based on innovative ideas from external agencies (e.g., civil society, community based organizations, NGO/NPOs) can strengthen community capacity, facilitate disaster recovery and contribute to sustainable development.

Misused opportunities in education disaster recovery, on the other hand, include not giving priority to infrastructure rebuilding, which resulted to high dropout rates, as in the case of Pakistan after the 2010 floods. There is also the case of not taking a multi-hazard approach in school reconstruction. Because of this earthquake-retrofitted schools that are not resistant to hydrometeorological and other geological hazards are built in new locations such as coastal areas where they are exposed to storm surges and tsunami. In this case, in trying to address one hazard, the school has become exposed to other hazards.

Lastly, the over-standardization of disaster prevention education may become the trigger for misused opportunities in school recovery. There should be room for customization to better take measures to unexpected disasters and incorporating more of the local context in planning school recovery. Rigidity in the form or content of disaster prevention education should not come in the way of teaching the students how to make their own decision and protect themselves and others when faced with disasters and emergencies.

21.4 Technology and Recovery

21.4.1 Overview

Disaster recovery has the potential to be used to reduce the risk of future disasters by building resilience in people and by reducing their exposure to hazard risks. Increasingly, there is evidence of good examples of how this can be achieved through the use of technology and innovation. A modern technology such as Earth Observation (EO) products and services offer innovative and comprehensive solutions that can address critical information needs for mapping and monitoring risks. Innovative technologies are critical and essential tools for disaster risk reduction and management (including recovery), which are more widely available and accessible than they were even a few years ago. They have become powerful and accessible tools that can lead to the protection of lives, assets and contribute in achieving development needs. However, disaster recovery plans must not neglect on capitalizing the existing traditional wisdom, since it has been tested over generations and may be better suited to the local environment and culture. Technology and innovations should be introduced where necessary, but in minimalistic ways, so as to add value to the traditional systems and make them more resilient in the face of new threats of disasters. Therefore, the use of modern technology and indigenous innovations has to be balanced to build holistic community resilience for sustainable recovery and development.

The discussions from the Technology and Innovations Issues of Recovery group ranged from the review and lessons of post-disaster housing reconstruction from Aceh, Yogyakarta, West-Java and West-Sumatera Earthquakes in Indonesia; challenges and innovations of coastal zone management in Tamil Nadu, India, post-Aila community recovery innovations and ecological planning in Bangladesh, the use of technological and innovative measures to improve flood disaster recovery following 2005 Mega-flood Mumbai in India to incorporating traditional knowledge to cope with climate change in post disaster recovery in Rajasthan and Leh, India.

An ex-post review of the past experiences and challenges in post-disaster housing reconstruction after earthquakes in Aceh (2004), Yogyakarta (2006), West-Java

(2009), and West-Sumatera (2009) was presented. The study reveals three main conclusions such as firstly, in terms of institutional approach; the governance issues depend on the scale of the disaster. Secondly, a strategic technical input and approach in implementing safer housing reconstruction are needed for achieving "build back better". Thirdly, a community driven approach in reconstruction is important, but it has to suit and tailor to the local customs.

Another discussion was on the concept of Integrated Coastal Zone Management (ICZM) in Tamil Nadu, India. The ICZM concept has blended the use of remote sensing and spatial analysis tool (GIS) technology as a decision support system with incorporating stakeholder's perception and field data. This concept has helped considerably the community in effective coastal management in Tamil Nadu, India.

The next discussion was on new innovation named the Disaster Resilient Habitat (DRH) in Bangladesh after the Aila Super Cyclone. The DRH aims to strengthen the houses and infrastructure of an entire village as a response to reduce the loss of assets to poor communities as well as to ensure the sustainability of the measure. The DRH ensured the conversion of each structure into a cyclone shelter and minimize the destruction caused, which would allow the community to recover from the disaster. The DRH is a bottom-up approach as it involves the community's indigenous knowledge with technical aid and ultimately makes the community self-sustaining in the longer run.

A technological and innovative measure to improve flood disaster recovery based on post-2005 Mumbai Flood was discussed next. A flood early warning tool is developed, where the rain gauges have been programmed to relay the rainfall intensity in real time through LAN to the city disaster emergency control room. This information is periodically uploaded to the prominent website, which could easily be accessed by the public.

Finally, the experience in tapping the indigenous knowledge of community in tropical desert of Rajasthan and mountain desert of Leh, India to cope with climate change was shared. The traditional wisdom based on local practices, such as weather forecasting, water harvesting, snow capture, as well innovative practices such as artificial glaciers and local climate knowledge management are noteworthy to preserve. These show that post-disaster recovery programs must capitalize on existing traditional wisdom, which may contain innovative ideas since it has been tested over generations and is best suited to the local context.

21.4.2 Common Issues

Although the discussion group elaborated upon the different cases from their study areas, there were related issues that form common goals and strategies, which can be employed for recovery.

- Appropriate technology interventions: These interventions must be derived from appropriate choice of technology and its application should be at small scale, labor intensive, energy efficient, environmentally sound and can be controlled locally.

- Improvement in community resilience: An increase in community's resilience would have a positive effect on the recovery process of the community. Therefore measures to increase the resilience of the community should be proposed and adopted.
- Combined knowledge of professionals and community: The indigenous knowledge is not always the cost-effective or sustainable solution. The same is true for the professional solution, which may not be locally contextualized. Therefore, the indigenous wisdom from the community should be amalgamated with professional knowledge to devise just and applicable solutions for disaster recovery.
- Informing education: The appropriate solutions for disaster recovery and the related information should be made available to all the stakeholders. This can also be useful for other disaster-affected communities to learn and apply the solutions. This can be achieved through "informing education" through various platforms, such as education sector, mass media and social media.

The four issues are culminated into a concept of sustainable recovery through the just use of technology and innovations. There is need to apply threefold strategy including all the above said issues to achieve sustainable recovery which will ensure that not only the disaster recovery is quick, but also that the next disaster is dealt with increased resilience.

Mainstreaming and risk communication of technology and innovations in disaster recovery are the main issues raised during the discussion. So far, the examples of technology and innovations applied in the discussion, such as the Disaster Resilient Habitat (DRH) of post-Aila Super Cyclone in Bangladesh and the incorporation of traditional local climate knowledge management in Rajasthan and Leh, India are part of the recovery process, yet have not been mainstreamed in future policies. In the future, these innovations have to be mainstreamed in all sectors and at all levels. Current situation denote towards the imperatives of creating and animating the disaster risk management concepts and processes in the local environment. Responding to above issue, technology and innovations in disaster recovery as part of disaster management is strongly suggested to be consciously and deliberately fused or mainstreamed in the local development planning processes. Reviewing the case of the concept of DRH, it provides the opportunity in adopting in the national development planning of Bangladesh. Since the country is prone to storm surges and cyclones, a disaster resilient habitat could be manifested in the national development process and budget, and translated into the real actions and resilience activities at the local level (cities and villages). The DRH may transfer this concept to the building regulation, policy, and plan by strengthening and conversing the houses and other important infrastructure (schools) into family and community cyclone shelters development. These elevated constructions allow the community safety as well as secured livelihoods, where they could still protect themselves and the space underneath to bring in cattle(s) if a cyclone warning is issued. Mainstreaming this concept into development planning and budget will considerably help other coastal communities, who face the same post-cyclone disaster recovery issue.

In terms of risk communication, the conveying process of the use of modern ways of technology can adopt a user-friendly communication interface. This shows in the case of Mumbai Flood 2005. The Municipal of Corporation of Greater Mumbai (MCGM) had already initiated the procurement and completed the installation of 30 automatic weather stations by June 2006. The weather station included tipping bucket rain gauges that is capable of giving rainfall data every minute and an audible alarm at preset rainfall intensity values (in this case when the rainfall exceeded 40 mm/h). The weather stations are linked to the central control room via internet, whereas the MCGM officials monitor the updated rainfall every 15 min from each location and issue alerts from the central control room. This has enabled the MCGM to issue warnings to the public through mass media and display the appropriate information on the internet during the four monsoon months on prominent websites as means of wider public outreach. Through the aforementioned example, a user-friendly modern technology innovation could address the risk communication issue of sophisticated ones.

21.4.3 Cases of Used and Misused Opportunities

Some of the used opportunities are as follow:

- *Resilient housing and habitat*: In the cases of cyclone in coastal Bangladesh and Barmer-Leh flash floods, the housing provision has been utilized as an opportunity to provide the resilient habitat as well. This ensures communities to be resilient and at the same time the recovery can be accelerated to provide livelihood and shelter.
- *Institutionalizing and decision support system*: In the cases of Aceh post disaster housing reconstruction with regards to coastal zone management in Chennai and urban floods in Mumbai, the opportunity to institutionalize the technology and innovations has been utilized and converted into a decision support system for better use by the community and the government. This allows input for community driven reconstruction and judicious deployment of resources.

The misused opportunities can be summarized as follow:

- *Imposed Appropriate Technology*: The pre-fabricated housing was imposed on the victims of Leh floods in 2010 in India, and more importantly an opportunity to improve upon and innovate the vernacular architecture techniques was lost.
- *Top-down Non-contextualized approach*: The housing reconstruction after disasters fail to include the local contextual issues—social, cultural, environmental, and non-user friendly contextualized. This was evident in Aceh reconstruction phase, where many international organizations tried to implement the housing solutions from their earlier experiences to a new location with different context. For example, construction of permanent accommodation for the victims, fails to consider issue of environmental sanitation.

21.5 Cross Cutting Issues and Recovery

21.5.1 Overview

The spectrum of the discussions from the "Cross-cutting Issues of Recovery" group extended from mangrove management in Kutchh, India, oil spills in Guimaras, Philippines, environmental management and urban recovery from the post disaster experiences of Kobe (Japan) and Banda Aceh (Indonesia), integrated healthcare system in Japan to a comparative analysis of role of community radio in disaster recovery in Indonesia and Japan.

A case study from the western most state of Gujarat in India that is prominently facing cyclones since 1999 was presented and found that degradation of mangrove forests is increasing the disaster risks. Following the 2001 earthquake in Gujarat, a great amount of industrial and infrastructural investment has been taken place that promoted development along the coasts. The local government lately realized the environmental cost of mangroves and promoted regeneration of mangroves, however, the regeneration areas are very small compared to what has been lost during the coastal infrastructural development. Since the community dependence on mangrove has reduced significantly, they are no longer interested in protecting the mangroves; hence the future of the coastal mangroves remains at stake. At this juncture, the need of a comprehensive mangrove management plan and consolidated local level actions was envisaged.

Secondly, a study on the adaptive and transformative capacities of communities after a manmade disaster in form of oil spills in Guimaras, Philippines in 2006, which damaged the marine environment and severely affected the livelihood of the local communities, was discussed. The oil spill demanded a massive cleanup drive; however, inadequate and insufficient technology posed as the major hindrance leading to stagnation of oil in coastal water for months. The study mainly focused on the assessment of the adaptive capacity of the local households through an adaptive capacity index, which indicates that the households, aside from the loss of livelihood, are fairly adaptable to the oil spill event. The external support played an important role following the oil spill; however, it was informed that the transformative capacities of the local communities did not improve due to lack of attention to the livelihood issues, lack of good governance and poor functioning of the social networks.

Two interesting case studies on urban recovery in Kobe (Japan) and Banda Aceh (Indonesia) was put forward in this workshop that highlighted the fact that even though there are substantial investments during the response phase, the long term recovery is often overlooked. In some cases, recovery has regenerated vulnerability observed in the past, and therefore, reducing vulnerabilities from the pre-disaster to post disaster recovery phase was emphasized. In case of urban recovery, community and local government partnership plays an important role which was a case from the Kobe experience where such kind of self-help groups applied themselves to impart training, organize mock drills and town watching programs together with the local government. On the other hand, housing recovery initiatives in Banda Aceh was proved to be very challenging because of the absence of a comprehensive recovery plan and recovery approaches were found to be unconsolidated with minimal community acceptance due to lack of good partnerships. On a concluding note, mainstreaming environmental recovery proved to be an integral part of urban recovery.

A paper on integrated healthcare system in Japan was explained with a brief background of how psychological isolation of certain groups of the aging community is impacting emergency services. From the present experience it is observed that even in non-emergency cases, mental isolation of aged people is affecting the emergency services. To address this problem, a new approach of integrated healthcare system, which aims to build community network by combining conventional healthcare system with non-conventional health care system with the concept of self-help and mutual health care system amongst the community was recommended. It was concluded that good social network coupled with an environment that offers to build good mental health may strengthen community's coping capacity to disasters. Some good examples of integrated health care system are currently being documented from the recovery experience of EJET.

A unique experience in operating the community radio for disaster response and recovery process was shared in this discussion. In the post disaster period, it helps minimize the challenges in providing accurate and essential information to the victims and in recovery phase, provides opportunities for victims to participate in radio activities and share their experiences. Apart from that, community radio establishes two-way communication between the aid providers and victims and raises awareness of the community regarding recovery plans of the national government and local NGOs. In case of EJET, 20 temporary emergency radio stations were set up in Tohoku, however, despite of being managed by the community, some radio stations were subjected to censorship of the local government. Others have faced challenges in continuing their activities because of the non-renewal of broadcasting license. Experience of community radio in Indonesia was also reported. It was mentioned that almost 800 such radio stations are in operation, but majority of these functions as "hobby radios" and do not necessarily fulfill the requirements of the essential functioning of community radio.

21.5.2 Common Issues

Even though the entire group discussion revolved around various different topics, there were some common issues, which might be inferred when viewed over a broader perspective. At first, the issue of unavailability of baseline information for the vulnerable areas before disasters was highlighted. In fact, it was prominent in case of mangroves degradation in Kutchh, urban recovery case in Banda Aceh and also in the case of oil spill in the Philippines. Absence of clear road map for recovery process provided the major hindrance in these particular cases. Another commonality was the effectiveness of community participation in disaster recovery process, which proved to be significant in determining the success and failure of recovery. In all the cases, especially in the case of community based mangrove management and poor social networking leading to poor transformative capacities of communities in oil spill events, a need for better community participation was envisaged. Absence or inadequacy of linkage among different sectors such as government, industry and community has been another common hindrance addressed by the majority of the case studies. On the other hand, urban recovery experience from Kobe highlighted the effectiveness of community self-help groups in the recovery process.

On a different note, since the majority of the discussions addressed coastal issues and hazards, the common feature of poor and weak implementation of laws, by laws and acts in land use planning and coastal zone regulations was highlighted, particularly in the developing countries such as India, Indonesia and the Philippines. In majority of cases, despite of existing legislative arrangement, coastal zone regulation are not being followed properly. In this regard, striking similarities between the mangrove degradation in Gujarat and the oil spill event in the Philippines was observed. Oil-spill in Philippines, which was a human induced disaster, raises concerns of similar threats that might occur in the Indian coasts due to rapid industrialization. Poor implementation of coastal zone regulation is not only destroying the mangroves, but also mounting risk of oil spills and water pollution in the coastal areas of Gujarat. Lastly, common linkage on livelihood in issues regarding urban growth, coastal resources management and oil spill events was identified. Finally, it was concluded that there exists the need of mainstreaming environmental recovery as a component of disaster recovery to reduce vulnerability from future disasters.

While addressing the differences amongst the discussed studies, it was observed that there exist differences or diversity in local context which calls for different disaster recovery approaches. Particularly, the political process influences disaster management and post disaster recovery. For example, In Aceh, during 2005–2009, Rehabilitation and Reconstruction were greatly affected by the decentralization and such conditions differ from one area to others. On the other hand, different local government has different priority settings. It was also highlighted that the response mechanism amongst countries and various level of government (local, provincial and federal) varies significantly according to their long term and short-term priority.

In the first case of Mangrove Management and cyclone risk reduction in Kutchh, Gujarat, the key issue was to identify the priorities of a developing nation that is whether the local government should promote development or conserve vulnerable coastal ecosystems. It was reported that such priority settings is different and varies significantly between the provincial government and the federal government, the key issue remains with this is how to minimize these gaps and also formulate the methodology to find the balance between economic development and natural resource conservation. Secondly, how the local wisdom can be effectively involved for sensitive ecosystem management in case of Gujarat. Finally, it was suggested to the author that amidst of the rapid development and urbanization in coastal areas all over the world, some good examples of coastal urbanization in sensitive areas may be documents and cited as guidelines.

In the Environmental Management and Urban Recovery case study, the key issue was to find out the minimum threshold time to start recovery process for which the local context varies significantly. An enquiry was made whether the stabilization and scattering of population in the post disaster period in Kobe necessarily relates to recovery process or as a natural phenomena observed in similar kind of urban areas. On the other hand, it was mentioned in the discussion that there was much delay in mobilizing the recovery process in case of Aceh. In this regard, a point came out about the timing of the disaster which coincides with the decentralization process in Indonesia and also Aceh is a place where there was 20 years of civil war which might have affected the recovery process. In this regard the role of Kechamathan Development Program (KDP) was discussed, a nationwide village level institution formed by the World Bank, which was instrumental for decentralized disaster recovery in Banda Aceh.

Related to the Integrated Health Management in Japan, it was highlighted that fire service which also provides the emergency medical support is very much unique; however, it was enquired whether strong social network under the proposed Integrated Health care system can really be an alternative to the trained professionals which the fire department offers. It was opinioned that such initiatives require training of manpower at community level and training may be provided by the Resident Associations through trained officials with proper certification. So, in that case, each community will have some trained people who shall be available during an ordinary medical emergency.

The last paper on the Role of community radio in post disaster recovery largely revolved around the gaps between the community members and the local government in operating the community radio. It was suggested that the effective measures to reduce the gap between the Local government and community need to be identified. Also, there were some specific comments that the social media such as Facebook can be an effective platform for information interchange amongst the community members and other stakeholders.

21.5.3 Used and Misused Opportunities

In case of mangrove management in Kutchh, it was observed that since the earthquake in 2001 and consequent cyclones in the Gulf of Kutchh, both the government and industries realized the importance of mangroves which the author cited as *used opportunity*, on the contrary, unplanned industrial and coastal infrastructural development after the earthquake in 2001 have not only led to depletion of mangroves but have also caused concerns on resources like water and marine pollution—which appeared, in some places, as *misused development opportunity*.

The oil spill event in Philippines in 2006 was instrumental for legislating an important policy in form of "The Oil Pollution Compensation Act of 2007" which actually addresses the economic concerns of the victims. This was referred as the *used opportunities* from the Philippines. On the other hand, even though the external funding was available, the central key issue of sustainable livelihood was not addressed during the recovery from the oil spill which was cited as *misused development opportunity*.

In case of Environmental Management and Urban Recovery, some general cases were highlighted where disasters have impacted positively in better preparedness and formulation of Disaster Management Authority. For example, In India, after 1999 cyclone and earthquake in 2001—DRR Institutionalization picked up very well including DRM Act, DDMA and Guidelines on specific hazards etc. These were investment for better recovery and can be referred as *used opportunities*. On the other hand, it also pointed out that external funding in case of Indonesia damaged the social capital which may be inferred *as misused opportunity*.

In case of the role of community radio in Disaster Recovery it was observed that following the East Japan Earthquake and Tsunami, decision making for recovery was done by all stakeholders regardless of gender, generations, employment and degree of affected people, which typically appears as *used opportunity*. However, community's dissatisfaction of the over censorship by local government over broadcasted contents revealed that the local government, in some cases, did not reflect community needs in the reconstruction process, which can be considered *as misused opportunity*.

Finally, in case of Integrated Health care approach, it was inferred that the existing ambulance services in fire department is a *used opportunity* to diversify the fire department services which simultaneously takes care of the medical emergency. However, it requires further fine-tuning and judicial use of emergency resources.

21.6 Recovery as a Development Opportunity

To make the recovery process as a successful development opportunity, it is required to make a balanced approach of governance, education and technology. These issues also need to be linked to the cross cutting issues. The following part describes the key learning from the cases presented in the book. The key words are mentioned in Fig. 21.1.

- Integration of budget into the governance, as proper financial system is the key element to run the disaster recovery process effectively and more efficiently.
- Inclusion of vulnerable and special needs groups, gender, children, elderly and disable etc. in governance mechanisms for disaster recovery. Special groups are truly vulnerable at the time of disaster recovery process. Hence, special groups should be kept in mind in governance mechanisms for disaster recovery.
- Involvement of community people is crucial in governance mechanism as community people take the lead first into the disaster recovery process. Furthermore, without participation of community people it is bit difficult to achieve the goal of good governance in disaster recovery.

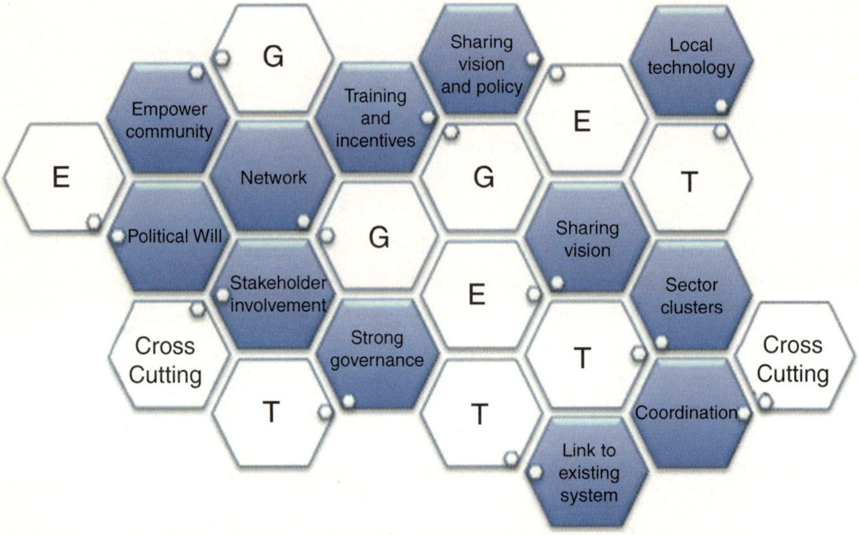

Fig. 21.1 Inter-relationship of different key words related to recovery

- It is also necessary to incorporate pre-disaster governance conditions, which can be effective in time of disaster.
- There is a need to analyze bad experiences in disaster recovery in the education sector, not just good practices. Investigating bad experiences will provide a list of things to avoid in the future as well as facilitate establishing benchmarks for actions necessary for recovery in various disaster situations.
- Recovery of the education sector should not just be about schools but should also take into consideration other aspects of the community, like building community ties, looking after the psychological health of the disaster victims and addressing other chronic community issues like gender balance in school enrolment. These examples point to the need for a more holistic approach that can have long-term impacts not just to the school, but also to the entire community. Effectiveness of school-based community recovery requires additional evidence-based studies as this new concept is now being implemented.
- Community participation in education recovery process has been recognized as important, but there is still no clear guidance on how members of the community, like the PTA (Parents-Teachers Association), local universities, local businesses, NGOs, religious organizations, the different age groups, etc. can take part in the process. Documentation and reporting of the roles and responsibilities of community members as well as the arrangements or protocols governing their participation in successful recovery experiences need to be conducted.
- Monitoring and evaluation of education was not raised in any of the case studies presented in the group (most likely because recovery is still on-going). As this exercise can provide important inputs that can be used in the future and in other

places, monitoring and evaluation should be given consideration in subsequent research.

- Recovery lessons for the education sector need to be revisited after several years or when new lessons are picked from new disasters to keep the recovery lessons updated and relevant.
- Mainstreaming the above technologies and innovations in all sectors and at all levels, as well as institutionalizing them.
- Community knowledge is not fully utilized to the potential and should be upgraded with the help of technology and innovation
- Low cost technology should be made available to the public for wider impact
- Multi-stakeholder approach would enhance the effectiveness of the technological innovation through focused strategies.
- Government and NGOs in formulating disaster recovery plan and translating them to local level actions.
- Need for judicial arrangements of emergency resources and external aids with respect to economic and social well being of the affected community in disaster recovery
- Disaster Recovery should not be regenerating vulnerable conditions in the past, but should be an opportunity for improving coping mechanisms to future disaster risks. Also, environmental management should be given priority in the recovery processes.
- Proper emphasis should be given to carry out disaster recovery and welfare simultaneously and the momentum should be maintained for effective development of the society.
- Local-level wisdom for intrinsic ecosystem management should be harnessed.

Very unfortunately, post disaster recovery process see an unhealthy competition of different agencies, even within the same umbrella organization. This is especially seen in the international agencies, many of which see the disaster situations as the opportunity to raise funds in their respective organizations. It is quite difficult to control this situation, however, the entry of external organizations to the disaster-hit areas can be controlled by a strong local governance system. It is not right to say that all the external agencies have their own specific intention. However, a strong governance system can define the external aid to sever the most to the local people. It should also be recognized that external aid does not only bring money, but a significant amount of knowledge resources and technological interventions. Unless these resources are properly channelized to suit the local needs, it often goes under utilized.

Acknowledgements The author acknowledges JSPS (Japan Society for Promotion of Science), CWS (Church World Services) Asia Pacific, and MERCY Malaysia for financial support.

About the Editor

Rajib Shaw is a Professor in the Graduate School of Global Environmental Studies of Kyoto University, Japan. He worked closely with the local communities, NGOs, governments and international organization, including United Nations, especially in the Asian countries. He is currently the Chair of the United Nations Asia Regional Task Force for Urban Risk Reduction, and the President of Asian University Network of Environment and Disaster Management (AUEDM). His research interests are: community based disaster risk management, climate change adaptation, urban risk management, and disaster and environmental education. He has published several books in the field of disaster and environmental management. He is also the Chief Editor of Asian Journal of Environment and Disaster Management.

Printed by Printforce, the Netherlands